Mechanics							
Quantity	Symbol	Dimensions	To obtain the value in mks units:	Multiply the value in cgs units:	By:	Multiply the value in English units:	By:
Mass	m	M	kg	g	10^{-3}	lb	0.4536
						slugs	14.60
Length	Various	L	m	cm	10^{-2}	in.	0.02540
						ft	0.3048
Time	t	T	sec	sec	1	sec	1
Velocity	$\mathbf{u}, \mathbf{U}$	LT^{-1}	m/sec	cm/sec	10^{-2}	ft/sec	0.3048
						miles/hr	0.6214
Force	$\mathbf{f}, \mathbf{F}$	MLT^{-2}	newtons	dynes	10^{-5}	lb	4.448
				g	9.80×10^{-3}	poundals	0.1383
Pressure	p, P	$ML^{-1}T^{-2}$	newtons/m²	dynes/cm²	10^{-1}	lb/in.²	0.6894×10^{4}
				g/cm²	98.0		
Density	ρ_v	ML^{-3}	kg/m³	g/cm³	10^{3}	lb/in.³	2.768×10^{4}
						lb/ft³	16.02
						slugs/ft³	515.4
Work or energy	Various	$ML^{2}T^{-2}$	joules	ergs	10^{-7}	ft-lb	1.356
				g-cm	9.80×10^{-5}	hp-hr	2.686×10^{6}
Power	Various	$ML^{2}T^{-3}$	watts	ergs/sec	10^{-7}	ft-lb/sec	1.356
				g-cm/sec	9.80×10^{-5}	hp	745.7
Power flux	Various	MT^{-3}	watts/m²	ergs/sec-cm²	10^{-3}	ft-lb/sec-ft²	1.954
Compressibility	K	$M^{-1}LT^{2}$	m-sec²/kg	cm-sec²/kg	10	in.-sec²/slug	1.450×10^{-4}
						unit vol/atm	9.869×10^{-6}
Modulus of elasticity	Y_0, Y_B, μ	$ML^{-1}T^{-2}$	newtons/m²	dynes/cm²	10^{-1}	lb/in.²	6.894×10^{3}

Heat							
Quantity	Symbol	Dimensions	To obtain the value in mks mechanical units:	Multiply the value in metric heat units:	By:	Multiply the value in English heat units:	By:
Temperature	τ, T	τ		°C		°F − 32	$\frac{5}{9}$
Quantity of heat	q, Q	$ML^{2}T^{-2}$	joules	kg-cal	4,185	Btu	1,055
				g-cal	4.185		
Heat flux	$\mathbf{q}, \mathbf{Q}$	MT^{-3}	watts/m²	$\frac{\text{g-cal}}{\text{sec-cm}^2}$	4.185×10^{4}	Btu/hr-ft²	3.130
Thermal conductivity	k	$MLT^{-3}\tau^{-1}$	watts/m-°C	$\frac{\text{g-cal}}{\text{cm-°C-sec}}$	418.5	$\frac{\text{Btu}}{\text{ft}^2\text{-sec-°F/in}}$	519.0
						$\frac{\text{Btu}}{\text{ft}^2\text{-hr-°F/ft}}$	1.730
Specific heat per unit mass	S	$L^{2}T^{-2}\tau^{-1}$	joules/kg-°C	$\frac{\text{g-cal}}{\text{g-°C}}$	4,185	Btu/lb-°F	4,185
Thermal diffusivity	D_t	$L^{2}T^{-1}$	m²/sec	cm²/sec	10^{-4}	ft²/hr	2.581×10^{-5}
Surface emissivity	h	$MT^{-3}\tau^{-1}$	watts/m²-°C	$\frac{\text{g-cal}}{\text{cm}^2\text{-sec-°C}}$	4.185×10^{4}	$\frac{\text{Btu}}{\text{hr-ft}^2\text{-°F}}$	3.130

Traveling-wave Engineering

McGRAW-HILL ELECTRICAL AND ELECTRONIC ENGINEERING SERIES

FREDERICK EMMONS TERMAN, *Consulting Editor*
W. W. HARMAN AND J. G. TRUXAL,
Associate Consulting Editors

AHRENDT AND SAVANT · Servomechanism Practice
ANGELO · Electronic Circuits
ASELTINE · Transform Method in Linear System Analysis
ATWATER · Introduction to Microwave Theory
BAILEY AND GAULT · Alternating-current Machinery
BERANEK · Acoustics
BRACEWELL · Transform Methods in Linear Systems
BRENNER AND JAVID · Analysis of Electric Circuits
BROWN · Analysis of Linear Time-invariant Systems
BRUNS AND SAUNDERS · Analysis of Feedback Control Systems
CAGE · Theory and Application of Industrial Electronics
CAUER · Synthesis of Linear Communication Networks
CHEN · The Analysis of Linear Systems
CHEN · Linear Network Design and Synthesis
CHIRLIAN · Analysis and Design of Electronic Circuits
CHIRLIAN AND ZEMANIAN · Electronics
CLEMENT AND JOHNSON · Electrical Engineering Science
COTE AND OAKES · Linear Vacuum-tube and Transistor Circuits
CUCCIA · Harmonics, Sidebands, and Transients in Communication Engineering
CUNNINGHAM · Introduction to Nonlinear Analysis
EASTMAN · Fundamentals of Vacuum Tubes
EVANS · Control-system Dynamics
FEINSTEIN · Foundations of Information Theory
FITZGERALD AND HIGGINBOTHAM · Basic Electrical Engineering
FITZGERALD AND KINGSLEY · Electric Machinery
FRANK · Electrical Measurement Analysis
FRIEDLAND, WING, AND ASH · Principles of Linear Networks
GEPPERT · Basic Electron Tubes
GHAUSI · Principles and Design of Linear Active Circuits
GHOSE · Microwave Circuit Theory and Analysis
GREINER · Semiconductor Devices and Applications
HAMMOND · Electrical Engineering
HANCOCK · An Introduction to the Principles of Communication Theory
HAPPELL AND HESSELBERTH · Engineering Electronics
HARMAN · Fundamentals of Electronic Motion
HARMAN · Principles of the Statistical Theory of Communication
HARMAN AND LYTLE · Electrical and Mechanical Networks
HARRINGTON · Introduction to Electromagnetic Engineering
HARRINGTON · Time-harmonic Electromagnetic Fields
HAYASHI · Nonlinear Oscillations in Physical Systems
HAYT · Engineering Electromagnetics
HAYT AND KEMMERLY · Engineering Circuit Analysis
HILL · Electronics in Engineering
JAVID AND BRENNER · Analysis, Transmission, and Filtering of Signals
JAVID AND BROWN · Field Analysis and Electromagnetics
JOHNSON · Transmission Lines and Networks
KOENIG AND BLACKWELL · Electromechanical System Theory
KRAUS · Antennas

KRAUS · Electromagnetics
KUH AND PEDERSON · Principles of Circuit Synthesis
LEDLEY · Digital Computer and Control Engineering
LEPAGE · Analysis of Alternating-current Circuits
LEPAGE AND SEELY · General Network Analysis
LEY, LUTZ, AND REHBERG · Linear Circuit Analysis
LINVILL AND GIBBONS · Transistors and Active Circuits
LITTAUER · Pulse Electronics
LYNCH AND TRUXAL · Introductory System Analysis
LYNCH AND TRUXAL · Principles of Electronic Instrumentation
LYNCH AND TRUXAL · Signals and Systems in Electrical Engineering
MILLMAN · Vacuum-tube and Semiconductor Electronics
MILLMAN AND SEELY · Electronics
MILLMAN AND TAUB · Pulse and Digital Circuits
MISHKIN AND BRAUN · Adaptive Control Systems
MOORE · Traveling-wave Engineering
NANAVATI · An Introduction to Semiconductor Electronics
PETTIT · Electronic Switching, Timing, and Pulse Circuits
PETTIT AND MCWHORTER · Electronic Amplifier Circuits
PFEIFFER · Linear Systems Analysis
REZA · An Introduction to Information Theory
REZA AND SEELY · Modern Network Analysis
ROGERS · Introduction to Electric Fields
RYDER · Engineering Electronics
SCHWARTZ · Information Transmission, Modulation, and Noise
SEELEY · Electromechanical Energy Conversion
SEELY · Electron-tube Circuits
SEELY · Electronic Engineering
SEELY · Introduction to Electromagnetic Fields
SEELY · Radio Electronics
SEIFERT AND STEEG · Control Systems Engineering
SISKIND · Direct-current Machinery
SKILLING · Electric Transmission Lines
SKILLING · Transient Electric Currents
SPANGENBERG · Fundamentals of Electron Devices
SPANGENBERG · Vacuum Tubes
STEVENSON · Elements of Power System Analysis
STEWART · Fundamentals of Signal Theory
STORER · Passive Network Synthesis
STRAUSS · Wave Generation and Shaping
TERMAN · Electronic and Radio Engineering
TERMAN AND PETTIT · Electronic Measurements
THALER · Elements of Servomechanism Theory
THALER AND BROWN · Analysis and Design of Feedback Control Systems
THALER AND PASTEL · Analysis and Design of Nonlinear Feedback Control Systems
THOMPSON · Alternating-current and Transient Circuit Analysis
TOU · Digital and Sampled-data Control Systems
TOU · Modern Control Theory
TRUXAL · Automatic Feedback Control System Synthesis
VALDES · The Physical Theory of Transistors
VAN BLADEL · Electromagnetic Fields
WEINBERG · Network Analysis and Synthesis
WILLIAMS AND YOUNG · Electrical Engineering Problems

TRAVELING-WAVE ENGINEERING

Richard K. Moore
Professor and Chairman
Department of Electrical Engineering
University of New Mexico

McGRAW-HILL BOOK COMPANY, INC. 1960
NEW YORK TORONTO LONDON

TRAVELING-WAVE ENGINEERING

 Library of Congress Catalog Card Number 60-9852

THE MAPLE PRESS COMPANY, YORK, PA.

V

42980

Preface

"Traveling-wave Engineering" deals with distributed parameter systems, where the time required for some disturbance to travel from one place to another is significant. It is intended to replace treatments of transmission lines alone with a more general treatment that includes transmission lines along with such "wave" phenomena as vibrating strings and acoustics and such "diffusion" phenomena as heat conduction and molecular and charge-carrier diffusion. Thus, it is, in essence, a book in engineering science. For the practicing engineer, it provides a convenient source of analogies and new approaches.

The transmission line is treated in most detail, since it involves a general combination of the effects found in "lossless" waves, at one extreme, and "diffusion," at the other. Methods widely used for transmission lines are applied to electromagnetic waves, vibrating strings and membranes, acoustic waves, longitudinal and transverse waves in solids, the wave function of wave mechanics, and the various types of diffusion. Plane waves are assumed, where applicable, except in Chapter 10. Although the analogies are of general applicability, only linear systems are discussed, and different methods of analysis are usually required for nonlinear waves.

Basic wave concepts for the transmission line are developed in Chapter 2, and comparable equations for the other waves are derived in Chapter 3. Thereafter, although most new ideas are introduced in terms of transmission lines, examples are given to show their application to other types of waves. The various analogous applications of each effect are introduced as the effect itself is studied to prevent the reader from becoming too firmly committed to thinking of only one type of wave; this approach appears to work better than building a complete knowledge of, say, transmission lines and trying to apply analogous techniques later.

Use of "Traveling-wave Engineering" permits broadening traditional electrical engineering courses in transmission lines to cover other types of waves, in accordance with modern trends. It makes a logical second course in a fields-and-waves sequence. It can also be used effectively in physics courses in waves, and it leads the way to a general course in distributed systems, of value to all engineers.

The text of this book has been used in note form at the University of New Mexico in a three-semester-credit junior-level course for electrical engineering students, which replaced a two-unit transmission-lines course. The students entering have had two semesters of electric circuits and a one-semester junior-level electromagnetic-fields course. For most parts of the book one semester of circuits is sufficient preparation. The electromagnetic-fields prerequisite may be omitted, if desired, since material on this subject is treated in parallel with that on transmission lines (except in Chapter 10).

At the University of New Mexico two one-hour lectures are given per week. A three-hour session each week is devoted to quizzes, problem solving, and laboratory work. Six experiments are conducted during the semester. Experiments involve artificial transmission lines, steady-state alternating heat flow, transient acoustic pulses in a steel bar, and centimeter-wavelength slotted lines.

For a shorter course, Chapters 7, 8, and/or 10 may be omitted. It is also possible to use the book for a short course in transmission lines alone, by omitting Chapter 3 and all the following nonelectrical examples, as well as Chapter 10. Chapter 8, on telephone and power systems, is included because many electrical engineering students have no other opportunity to become acquainted with these parts of their profession. It also serves as a "seventh-inning stretch," since it is mostly qualitative. It may easily be omitted in physics and engineering science courses.

Practicing engineers and scientists will find here many analogies of value. On the one hand, this book should enable those familiar with transmission lines to apply known techniques to a wide range of other problems; on the other, it makes available to those familiar with heat conduction, diffusion, or acoustics the powerful analytical techniques used in transmission-line work. The analogies suggest many modeling techniques of value in obtaining data where analysis is difficult.

The book is organized so that examples illustrating each new principle follow soon after introduction of the principle itself. The examples are clearly designated as such. Problems are included after each chapter, except the first. Some of these are intended to illustrate principles, and others to acquaint the student with practical magnitudes in the various fields.

Rationalized meter-kilogram-second units are used throughout. If one is to comprehend easily the analogies between various phenomena he must not be faced with the additional complication of large numbers of conversion factors. Thus, all forces are in newtons; all energy is expressed in joules and all power in watts; and all densities are in kilograms per cubic meter. In other treatments of the various kinds of waves, forces are expressed in dynes, pounds weight, and poundals; energies in ergs, calories, Btu, and foot-pounds; and densities in grams per cubic centimeter, moles per cubic foot or centimeter, and pounds per cubic foot. The result is that most material constants presented here have been converted from different units to the mks units.

Since this is an *introduction* to waves, the interaction between charge streams and waves is not covered, and the discussions of waveguides and wave mechanics are abbreviated. Likewise, the complex problem of oblique reflections in solids is not treated in detail. References are given, however, to more complete discussions of these advanced topics.

For the same reason, the discussion here is restricted to plane waves reflected at normal incidence, except in Chapter 10. All the pertinent principles can be learned with plane waves, and introduction of cylindrical and spherical geometries would only tend to complicate the mathematics without adding insight into the physical processes.

Two limiting cases of excitation are treated in detail—steady-state alternating sources and simple transient sources. Although a transform method for handling generalized transient excitation is outlined, details are not discussed because of their complexity and the lack of generality of the resulting solutions.

Vector analysis is used where it is needed. Since each principle is illustrated (prior to Chapter 10) with a one-space-dimensional wave, a knowledge of vector analysis, although helpful, is not required.

In Chapters 5 and 6, steady-state a-c solutions are discussed from the standpoint of phasor diagrams. Both examples and problems emphasize this approach. The Smith impedance chart is intentionally delayed until Chapter 9. It is believed that the student should be thoroughly familiar with the *principles* of standing waves before he starts using the laborsaving but meaning-obscuring Smith chart.

Derivations of the wave equations for various phenomena are in terms of simple intensity relations. It would have been possible to make a more unified set of derivations using Lagrange's equations, but it appears more desirable at this level to sacrifice the mathematical generality possible with such an approach to the physical pictures possible with the derivations used here. This is especially true because the field formulations of energy storage and kinetic energy, which would be required, are not well known to most readers.

From these comments it can be seen that "Traveling-wave Engineering" provides a unified introduction to the subject of waves and distributed parameter systems. Such a treatment is efficient in its use of the student's time and is in line with recent trends toward engineering science. Since it brings together in one place a wider range of analogies between wave phenomena than can be found elsewhere, it should also be particularly valuable to practicing engineers, no matter what their specialty.

A book such as this could not have been written without the cooperation of many persons. I have had stimulating discussions on the various types of waves with members of all the engineering departments at the University of New Mexico. In particular, the comments of Profs. R. E. Dove and V. J. Skoglund of the mechanical engineering department and Prof. P. E. Bocquet of chemical engineering have been helpful. I am particularly indebted to Dale Sparks and Allen Edison, who worked closely with me in their handling of multiple laboratory sections and who contributed criticisms of the text.

The patience and constructive criticisms of students of the classes of 1959 to 1961, who had to struggle through a course with dittoed text that frequently arrived late, are much appreciated, as are the efforts of the departmental secretarial staff in running off large numbers of copies on short notice.

This would have been a hopeless task without the valuable assistance of my wife, who spent many hours transcribing dictated text (including equations!) and typing ditto masters and manuscript.

Richard K. Moore

Contents

Nomenclature List

B_C	Capacitive susceptance
B_m	Matching susceptance
C	Capacitance per unit length
C	Torsional rigidity
C	Phasor concentration
C_t	Thermal capacity
D	Diffusion coefficient
D_t	Thermal diffusivity
$\mathbf{E}$	Electric field vector
$\mathbf{E}^+$	Incident electric field
$\mathbf{E}^-$	Reflected electric field
$\mathbf{E}^t$	Transmitted electric field
F_w	Phasor force in w direction
G	Conductance per unit length
$\mathbf{H}$	Magnetic field vector
$\mathbf{H}^+$	Incident magnetic field
$\mathbf{H}^-$	Reflected magnetic field
$\mathbf{H}^t$	Transmitted magnetic field
I	Phasor electric current
I_R	Phasor receiving-end current
I_s	Phasor sending-end current
I^+	Phasor incident current at load
I^-	Phasor reflected current at load
$\mathbf{J}_i$	Phasor current density of ith constituent
$J_0(x)$	Bessel function of first kind, zero order
$J_0'(x)$	$\frac{d}{dx} J_0(x)$
K	Magnitude of reflection coefficient
K	Compressibility
L	Inductance per unit length
L	Line length

M_{vx}	x component of phasor mass flow rate per unit area
N	Number of sections in lumped-parameter model
$\mathbf{N}$	Wave normal unit vector
N_h	Phasor concentration of holes
N_i	Phasor concentration of ith constituent
$\mathbf{P}$	Poynting vector
P	Phasor pressure
P_x	Phasor x component of stress (longitudinal)
P_y	Phasor y component of stress (shear)
$\mathbf{Q}$	Phasor heat-flow density
Q	Quality factor of a resonator
R	Series resistance per unit length
$\mathbf{R}$	Vector from origin of coordinate system
R_f	Resistance to fluid flow
R_g	Generator internal resistance
S	Specific heat per unit of mass
T	Period of a sine wave
T	Tension in a string
T	Phasor temperature
T	Time of travel
$\mathbf{U}$	Amplitude of sinusoidal velocity
V	Phasor voltage on transmission line
V_g	Phasor generator emf
V_R	Phasor receiving-end voltage
V_s	Phasor sending-end voltage
V^+	Phasor incident voltage at load
V^-	Phasor reflected voltage at load
X	Displacement from rest position in x direction
X_s	Sending end reactance
Y	Admittance on transmission line
Y_B	Bulk modulus of elasticity
Y_0	Modulus of elasticity for thin rod
Y_0	Characteristic admittance of transmission line
Z_g	Generator internal impedance
Z_0	Characteristic impedance of transmission line
Z_{0T}	Characteristic impedance (iterative) of T section
Z_R	Receiving-end, or load, impedance
Z_s	Input, or sending-end, impedance
Z_y, Z_z	Directional impedances in y and z directions
a	Cross-sectional area of a tube
a	Height of a waveguide
b	Reservoir area per unit length
b	Width of a waveguide

b_m	Matching susceptance in units of Y_0
c	Instantaneous concentration (per unit volume)
d	Distance from load (or from point of reflection)
e	Base of natural logarithms
f	Frequency
$\mathbf{f}$	Instantaneous force
f_0	Resonant frequency
g	Gravitational acceleration
h	Surface heat-transfer coefficient
i	Instantaneous current
i_R	Instantaneous current in load
i_s	Instantaneous sending-end current
i^+	Instantaneous incident current
i^-	Instantaneous reflected current
j	$\sqrt{-1}$
k	Thermal conductivity
k	Boltzmann's constant
k	Angle of reflection coefficient
l	Length of transmission line
$\mathbf{m}_v$	Instantaneous mass-flow rate per unit area
n_i	Instantaneous number of particles of ith constituent per unit volume
n_{i0}	Equilibrium value of n_i
p	Instantaneous deviation of pressure from ambient value
p_a	Ambient pressure
$\mathbf{p}$	Instantaneous stress
$\mathbf{q}$	Instantaneous heat flux
q_i	Charge on ith constituent
r	Resistance in units of Z_0
s	Complex decrement in Laplace transform
t	Time
$\mathbf{u}$	Instantaneous particle velocity
v	Instantaneous voltage
v_g	Instantaneous generator emf
v_g	Group velocity
v_p	Velocity of propagation (phase velocity)
v_R	Instantaneous voltage at receiving end of line
v_s	Instantaneous voltage at sending end of line
v^+	Instantaneous incident voltage
v^-	Instantaneous reflected voltage
$v_{px,y,z}$	Directional propagation velocities
w	Instantaneous transverse displacement of a point on a string
x	Distance on transmission line from sending end

x	Distance along string
x	Rectangular coordinate (longitudinal)
x	Reactance in units of Z_0
y	Transverse rectangular coordinate
y	Admittance in units of Y_0
z	Rectangular coordinate
z	Impedance in units of Z_0
$\mathcal{I}$	Sound intensity
$\mathcal{I}^+$	Incident alternating current in phasor form
$\mathcal{I}^-$	Reflected alternating current in phasor form
$\mathcal{J}_{ix}$	Instantaneous x component of diffusion-current density of ith component
$\mathcal{V}^+$	Incident alternating voltage in phasor form
$\mathcal{V}^-$	Reflected alternating voltage in phasor form
Γ_R	Reflection coefficient for voltage and its analogs
Γ_s	Reflection coefficient, from generator, for voltage and its analogs
Γ_t	Transmission coefficient for voltage and its analogs
$\Gamma_{R,\text{eff}}$	Effective value of reflection coefficient, as modified by losses
Γ_{Rx}	Reflection coefficient for components associated with wave travel in the x direction
Γ_{tx}	Transmission coefficient for components associated with wave travel in the x direction
Δ	Increment, always used with another letter
α	Attenuation coefficient
β	Phase-shift coefficient
β_x	Phase-shift coefficient in x direction
γ	Propagation constant
γ_g	Ratio of specific heats
γ_T	Propagation constant for T section (iterative)
γ_x	Propagation constant in x direction
δ	Skin depth
ϵ	Permittivity
η	Intrinsic impedance
θ	Angular displacement
θ	Angle of coordinate rotation
θ_i	Angle of incidence
θ_t	Angle of transmission, or refraction
θ_{rt}	Angle of reflection, transverse wave
θ_{tt}	Angle of transmission, transverse wave
λ	Wavelength
λ_0	Wavelength at resonance
λ_0	Cutoff wavelength for waveguide
λ_g	Guide wavelength

λ_x	Directional wavelength in x direction
μ	Permeability
μ	Viscosity
μ_i	Mobility of the ith constituent
ν_i	Deviation of charge-carrier density from ambient
ρ_L	Mass per unit length
ρ_A	Mass per unit area
ρ_v	Mass per unit volume
σ	Electrical conductivity
σ	Poisson ratio
τ	Instantaneous temperature
τ_0	Ambient temperature
ψ	Matter wave potential
ω	Angular frequency
ω	Angular velocity
ω_0	Resonant angular frequency
∇	Vector differential operator
erf	Error function
Im	Imaginary part of
Re	Real part of
$\mathbf{1}_x, \mathbf{1}_y, \mathbf{1}_z$	Unit vectors in the x, y, and z directions

1. Introduction

Traveling-wave engineering deals with situations in which it is necessary to consider the finite time required for some change to travel from one place to another. In ordinary electric-circuit theory, it is assumed that closing a switch at one point of a circuit causes currents to flow and voltages to appear at all points of the circuit at once. In elementary mechanics (rigid-body dynamics), all parts of a body are assumed to feel the effects of an applied force or blow simultaneously. Actually, the effects of any change, such as the closing of a switch or the striking of a solid object with a hammer, take a finite time to reach a different part of the circuit or object.

With the traveling-wave phenomena of heat and chemical diffusion, we never think in terms of simultaneous occurrence at all points in a body. For example, we know that, when we place a pot on a stove, the bottom of the pot becomes hot on the outside before it does on the inside. Similarly, we know that it takes time for a sugar solution to diffuse throughout a cup of coffee.

1-1. The Traveling-wave Idea

Even excluding the special cases of heat and chemical diffusion, the idea of traveling waves is familiar in modern life. Everyone recognizes, for example, that there is a time delay between a sound and its echo, because of the finite velocity of the sound in traveling to some reflecting surface and back. When a stone is dropped into a pool of water, waves are set up in the water, and appreciable time passes before the first wave hits the edge of the pool. Considerable time is required between the passage of a boat in the center of a stream and the arrival of its wake at the shore.

The term *light-year,* a unit of distance in astronomy, implies the finite velocity with which light (an electromagnetic disturbance) travels.

Light travels so much faster than sound that, to our senses, light anywhere within range of sound transmission appears to travel to us in zero time. Most of us, knowing that sound travels 1 mile in about 5 sec, have measured the time between a lightning flash and thunderclap to determine the distance to the lightning—when the time is too short, the distance is too small for comfort, and we jump!

In astronomy, the traveltime of the light wave may be many years, because the distances are so much greater. In fact, we know that heavenly events we see today actually took place hundreds, thousands, or even millions of years ago. Stars exploding today are sending out bursts of light that will not be seen on earth for a long time.

When we measure the distance to a stroke of lightning by timing the arrival of the thunder, we frequently observe delays of tens of seconds. Also, when sound waves travel through the water to locate submarines (sonar), there are frequently delays of many seconds between transmission of the pulse and reception of an echo. It takes minutes and frequently hours for heat to penetrate to any distance. In roasting a large turkey there may be a delay of as much as 10 hr between applying heat to the skin and raising the internal temperature enough for the center of the turkey to become fully cooked.

Traveling waves involving delays of seconds and minutes, and sometimes longer, occur in traffic. It is exasperating to be lined up in front of a traffic light and have the "car-starting wave" fail to reach you before the light has changed back to red!

With most electrical phenomena (and many mechanical ones, too) the delay is much shorter than the times mentioned above. In telephone transmission and intercontinental radio, the delays may exceed a *milli*second (10^{-3} sec). Radio transmission to the moon takes seconds, but solely terrestrial transmissions do not. The traveltime for the shock wave due to the striking of a metallic object by a hammer (or a bullet) is also measured in milliseconds. Many times, however, we shall be thinking of delays measured in *micro*seconds (10^{-6} sec) or less. Sometimes we shall even consider delays of less than a *millimicro*second (10^{-9} sec)!

To determine whether traveling-wave concepts must be used, rather than ordinary circuit theory or rigid-body dynamics, we have to consider the magnitude of the delay time which can be tolerated. When we talk about sine waves, it is usually necessary to use some kind of traveling-wave technique if the delay represents a significant part of the *period* of oscillation. The meaning of "significant" in this statement depends on the application. With some highly precise electric devices, such as narrow-beam antennas and traveling-wave bridges, phase shifts of less than a degree (time delay less than 0.3 per cent of a period) may be significant. In other situations, almost a quarter-cycle delay may be permitted before traveling-wave techniques are needed.

With pulse excitation, the distinction between traveling-wave and elementary techniques frequently depends on whether the traveltime is a large or a small portion of the pulse length. The term "large" is also subject to interpretation here.

Since traveling-wave techniques must be used when, say, a delay of a tenth of a period is involved, the minimum distances at which these methods are required vary widely, depending on the frequency used. At 60 cps, a tenth period is 1.67 msec. For waves traveling at the free-space velocity of light (velocity on many transmission lines is close to this), the corresponding distance is $3 \times 10^8 \times 1.67 \times 10^{-3}$ m—about 500 km. In radar work, frequencies of 10^{10} cps are common. At this frequency, a tenth period is 10^{-11} sec, and the corresponding distance is only 3 mm. When we consider that a hundredth period may often be significant, we see that the distances at which traveling-wave techniques must be used at 10^{10} cps are very short indeed!

On the other hand, power *distribution* lines at 60 cps are always short enough so that traveling waves need not be considered. Table 1-1 shows this.

TABLE 1-1. MAXIMUM DISTANCES FOR ORDINARY CIRCUIT TECHNIQUES

Application	Frequency, cps	Distance, 1/10 period significance	Distance, 1/100 period significance
Power transmission	60	500 km	50 km
Voice transmission	3,000	10 km	1 km
Broadcast band	1.5×10^6	20 m	2 m
TV broadcast (VHF)	1.5×10^9	20 cm	2 cm
Air-borne radar	10^{10}	3 mm	0.3 mm

Similar tables might be drawn up for other types of traveling wave, but instead of preparing a separate table for each, a number of examples have been combined in Table 1-2. It can be seen that, normally, traveling-wave techniques must be applied for much shorter distances with acoustic and thermal than with electromagnetic waves. Furthermore, the lower frequencies involved in these other types of wave are evident. The use of traveling-wave techniques may frequently be important even with *annual* temperature variations.

The above discussion has dealt with sine waves, but since Fourier analysis permits us to find sinusoidal components of any wave, these criteria may also be used with other types of wave.

TABLE 1-2. DISTANCES REQUIRING TRAVELING-WAVE TECHNIQUES

Phenomenon	Frequency	Distance, 1/10 period significance	Distance, 1/100 period significance
Sound wave in air	60 cps	58 cm	5.8 cm
	3,000 cps	1.16 cm	0.116 cm
Sound wave in water	60 cps	2.5 m	25 cm
	3,000 cps	5 cm	0.5 cm
Sound wave in glass	60 cps	10 m	1.0 m
	3,000 cps	20 cm	2.0 cm
Heat in copper	60 cps	0.49 mm	0.049 mm
	3,000 cps	0.07 mm	0.007 mm
	1 cps	3.78 mm	0.378 mm
	1 cycle/hr	22.6 cm	2.26 cm
Heat in a concrete dam	60 cps	0.047 mm	0.0047 mm
	1 cps	0.365 mm	0.0365 mm
	1 cycle/hr	2.19 cm	2.19 mm
	1 cycle/year	2.05 m	20.5 cm

Frequently, in electrical and, even more often, in nonelectrical traveling-wave problems, it is desirable to use time-delay criteria directly rather than to Fourier-analyze the transients. For example, "ghost" images appear on television screens because of multiple transmission paths having different lengths. The criterion for annoyance in this case could be determined by Fourier analysis. However, it is much easier to base it directly on the time and, hence, on space resolution in the horizontal direction on the television cathode-ray tube. In most problems involving heat or diffusion, the excitation is nowhere near sinusoidal, and resolution into sinusoidal components has little meaning to workers in the field. Hence, for these problems, criteria based on time are more meaningful than those based on frequency.

1-2. Historical Background

Undoubtedly, the earliest traveling-wave phenomena to be recognized were surface waves on water, but complete mathematical treatment of water waves with relatively large amplitude and with wind excitation is difficult. Hence, the earliest traveling-wave phenomena to receive extensive attention by mathematicians were those associated with vibrating strings and, to some extent, vibrating drumheads. D'Alembert solved the partial differential equation associated with the vibrating string in 1750.

By this time, experiments had already been carried out to determine the velocity of sound transmission through air, and also through water and through solids. An interesting account of these investigations is given in "Theory of Sound."[1]

Fourier developed the general expression for conduction of heat early in the nineteenth century (and then developed the well-known Fourier series so that he could solve the equation he had developed). Since Fourier's time, the heat-conduction equation has been solved for a wide variety of geometric arrangements, both of the conductor and of the heat sources. Only heat conduction is discussed in this text, but it should be realized that the engineering science of heat transfer involves transfer by radiation and convection. Neither of these is treated here by the methods of traveling-wave engineering, although thermal radiation involves traveling electromagnetic waves, and wave processes are present in convection.

The study of electric traveling waves has a shorter history than does the study of mechanical and heat waves, but it is an interesting one. The first electric circuits of any significant size were telegraph lines. Early telegraph lines were so short that time-delay effects were negligible. Transmission to great distances was accomplished by relaying through other stations. Sometimes relaying was done by automatic devices; but, even then, the mechanical delays in the relaying devices were sufficient to mask the electrical delays.

When the first transoceanic cables were laid in the 1850s, it was found that very significant delays occurred. Not only were time delays of seconds noticeable to the operators, but the difference in velocity of propagation of different frequencies was such that the high-frequency components of the telegraph signals arrived before the lower-frequency ones. This resulted in distortion and hence restricted the signaling speed of the early cables to about 60 characters per second.

When long-distance telephony was instituted just before the turn of the century, this difference in velocity between the frequencies made it necessary to limit the circuits to a few tens of miles. Prof. Michael Pupin, at Columbia University, obtained patents covering the addition of inductors ("loading coils") in series with a line at regular intervals to equalize the velocity for the various frequencies. Thus, adding resistance and inductance, which would seem to make things worse, actually made possible long-distance telephony. Pupin's work was done in 1899; by 1901, it was already in use, and long-distance telephony was on its way. Similar suggestions made earlier by Heaviside had not been put to practical use.

Loading coils also reduce the maximum frequency of the signal on a line; so modern systems using carrier frequencies high above the voice

[1] Lord Rayleigh, "Theory of Sound," Dover Publications, New York, 1945.

frequencies must compensate for this difference in velocity in other ways. This "equalizing" problem has led to the development of much of modern circuit theory.

In 1864, Maxwell published the famous paper in which he combined the earlier experimental work of Ampere, Faraday, and Gauss into a unified theory of electromagnetic waves. Maxwell predicted that such waves would travel through the atmosphere with finite velocity, but it was not until about 20 years later that Heinrich Hertz performed his well-known experiments demonstrating the correctness of Maxwell's prediction. Hertz caused an arc discharge to take place in a gap in a loop of metal, actually a traveling-wave resonant circuit. A similar loop on the other side of the room, with a very narrow gap, was seen to spark "simultaneously."

Later, Marconi and others working on radio communication made practical application of this wave traveling through space. Not only is the wave through space a traveling wave, but the traveling-wave concept and its companion, the standing-wave concept, are the basis for explaining the behavior of the antennas necessary to launch and capture radio waves. Whereas Hertz worked with waves having a frequency of about 500 Mc, Marconi and the others who worked with radio for many years used lower frequencies, such that times of travel were significant only for the antenna system and the actual transmission through space.

Radio amateurs, forced to operate on frequencies of several megacycles per second, found that these supposedly worthless frequencies were ideal for long-distance communication. Consequently, megacycle frequencies became widely used, and there was also greater use of transmission lines whose length caused time delays from a significant portion of one cycle to many cycles—mostly in antenna feeder systems. In the early 1930s the demand for more frequencies and the advantages of more compact directional antennas caused many investigations of the very-high-frequency (VHF) and ultrahigh-frequency (UHF) regions. Here it was found that the delay times in the wires connecting different circuit elements were such that transmission-line (traveling-wave) techniques had to be used within the oscillator, amplifier, and detector circuits themselves. Concentric- and parallel-wire transmission lines replaced lumped inductance and capacitance as tuning elements. Transmission lines also came to be used for the interconnection of units, since the ordinary arrangements suitable at lower frequencies would not work. Actually, at these frequencies the "ordinary" connections within a circuit must be treated as short transmission lines.

In the late 1930s, research progressed to higher frequencies, where it became feasible to "box in" the parallel-wire line and to eliminate the center conductor of the coaxial line. This new method of transmission by *waveguides* sent electromagnetic energy through hollow pipes. It is interesting to note that some of the early workers in radio used similar structures, but they were lost to view for many years while interest centered at the lower frequencies.

At frequencies above 3,000 Mc, waveguides are used almost exclusively, not only for transmission from one place to another but as tuned circuits (cavity resonators). When there are two easily defined conductors, circuit techniques may ordinarily be used; but field techniques are required for waveguides. Many concepts developed in the circuit application, however, may be used both for waveguides and for waves in space. In fact, the two-conductor line itself guides waves, but the name *waveguide* is usually reserved for structures not having two separate conductors.

The techniques used in traveling-wave tuned circuits are the same as those which have been used in acoustics for centuries, except for the differences due to the vector nature of the electromagnetic wave and the scalar nature of the acoustic wave. Techniques used in tuning microwave devices are much the same as techniques used in tuning wind instruments and organs. The resonance of a violin string is mathematically the same as the resonance of a short-circuited half-wavelength transmission line.

More recently, wave-guiding structures have been developed to slow the velocity of a traveling wave until it is comparable with velocities which may be achieved in electron streams. The result is a whole family of microwave tubes which amplify by interaction between traveling waves and electron streams. Similar techniques may someday be used to extract power from controlled fusion reactions.

Little has been said about power lines because the greatest variety of electrical applications of the traveling-wave concept is to communication lines and to other high-frequency devices. Nevertheless, some power lines, like the famous one from Hoover Dam, on the Colorado River, to Los Angeles, are long enough so that the time delay in a disturbance traveling from one end to the other is a significant portion of a period at 60 cps. The application of traveling-wave concepts to these lines explains some of the voltage variations and power losses in the lines. A line with wave delay approaching a quarter of a period can have some astonishingly high voltages on an unloaded receiving end—with disastrous results to expensive equipment—unless the situation is remedied by lowering the sending-end voltage or changing the load impedance.

It should be noted that the propagation of transient disturbances on power lines involves traveling-wave notions to a much greater extent than do the steady-state phenomena. The time required for a circuit breaker to "learn" of a short circuit may be quite significant. For example, a really heavy surge may trip more than one circuit breaker, because the traveltime to the backup breaker is short compared with the time required for the first breaker reached by the disturbance to open.

1-3. Some Problems of Traveling-wave Devices

Both advantageous and undesirable effects may be obtained with traveling-wave devices, simply because of the finite time of travel. For example, it is possible to delay a signal a known amount of time by passing it through a known length of transmission line, and this may be very desirable in computers and other types of pulse devices. On the other hand, it is frequently required that coincident events occurring at different places be observed simultaneously at some one point. Unless the signals from each event pass through the same length of line, the simultaneity will not be preserved at the observation point.

Acoustic devices for determining the position of a submarine or the depth of water under a ship depend on the time of travel of an acoustic wave and would not be possible if sound traveled with an infinite velocity. On the other hand, it is annoying, in a large auditorium or a large outdoor gathering, to have the speaker's words arrive at our ear while we can see his lips saying later words.

We keep our houses at a relatively even temperature day and night with the assistance of the time delay involved when the thermal wave travels through the walls. Furthermore, if we bury our water pipes deep enough in the ground, the thermal wave due to low winter temperatures does not reach them and they will not freeze. In cooking, on the other hand, it would be very convenient if the heat wave were to travel very fast, so that the inside and outside of a roast, say, could be at the same temperature. There would be a similar advantage in the setting of concrete.

A phenomenon which can cause a great deal of trouble but which also forms the basis for many traveling-wave devices is reflection. A wave arriving at either end of a line or at any point where line characteristics change may be reflected and travel back to its source. This reflection may interact with the sending device and cause it to malfunction. Strong reflections frequently render microwave oscillators inoperative. Strong reflections from a television transmitting antenna may require considerable reductions in the power applied to the transmitter; in addition, they may cause "ghosts" to appear on the screen of all receivers tuned to the station. Reflections on a telephone line carrying signals both ways can result in "echoes" or, worse, in oscillation ("singing").

Reflection, of course, is the basis of the acoustic devices mentioned above. Some reflection of sound waves from the walls of a room is desirable, but if the walls reflect too much, we say that the room acoustics are bad.

Usually, reflection is not a significant effect in thermal and diffusion problems.

The operation of traveling-wave resonant circuits, including antennas and the tuning circuits used in transmitters and receivers, is dependent on strong reflections. The resonators work because signals bounce back and forth between one end and the other, with energy storage alternating between magnetic and electric fields, just as in a resonant lumped-constant circuit.

Traveling-wave resonant circuits are also widely used in acoustics. All wind instruments, from the piccolo to the organ, obtain their tones from traveling-wave resonant circuits. Furthermore, the traveling-wave resonance on a string is the basis of stringed instruments, from the violin to the piano and harp.

Traveling-wave resonance in mechanical systems can also lead to trouble. The famous crash of the Tacoma Narrows bridge is a case in point. Designers of airplane wings must also take this into account, as must designers of crankshafts for automobile engines. An interesting example of traveling-wave resonance (which is controlled so as not to cause destruction) can be seen in the rotor of a helicopter.

The necessity of equalizing the velocity of propagation at the various frequencies has been mentioned in connection with transoceanic cables and telephone cables. It is particularly important in television, where a small amount of distortion is easy to recognize in the picture. A great deal of money and effort has been expended in the telephone plant to compensate for unequal propagation velocities.

Efficiency is important in transmitting large amounts of power, and efficiency calculations are made not only for power lines but also for the transmission lines associated with high-power radio and radar transmitters. Reducing reflections on either kind of line increases efficiency.

1-4. Applications of Traveling-wave Techniques

A number of the applications of traveling-wave techniques have been suggested above. Others are outlined below, but it should be apparent that the following list is far from complete:

A. Applications of transmission lines and waveguides
- 1. Power distribution
  - *a*. Transmission of power
  - *b*. Interconnection of units in radio, radar, and television
  - *c*. Connecting transmitters and receivers to antennas

2. Information transmission
    a. Transmission of manual and automatic telegraph signals
    b. Transmission of telephone signals
    c. Transmission of television signals
    d. Transmission of telemetering signals
3. Circuit applications
    a. Tuning antennas
    b. Tuning other transmission lines
    c. Impedance matching
    d. Tuning circuits in receivers and transmitters
    e. Filter circuits at VHF and above
    f. Bridges at VHF and above
    g. Time-delay circuits
    h. Microwave tubes

B. Application of traveling-wave techniques in the study of electromagnetic propagation through space
1. Plane-wave transmission and reflection
2. Spherical and cylindrical waves as applied to transmission from one antenna to another
3. Antennas
4. Extraction of power from traveling ion streams in tubes, thermonuclear reactors, and radiation from sunspots

C. Applications of traveling-wave techniques in acoustics
1. Resonant pipes—organs and wind instruments
2. Geophysical prospecting (mostly for oil)
3. Location of submarines and the sea bottom
4. Room acoustics
5. Operation of the human ear
6. Travel of stress waves in structures

D. Traveling-wave techniques for transverse vibrations (acoustic waves are longitudinal vibrations)
1. Vibrations of strings in musical instruments
2. Structural vibrations in such things as airplane wings and bridges
3. Vibration of drumheads
4. Loudspeaker-diaphragm operation
5. Ocean waves

E. Typical applications of traveling heat waves
1. Finding temperature beneath the ground
2. Finding temperature within buildings
3. Cooling of massive concrete structures
4. Temperature determination in an internal-combustion engine
5. Heat treatment of metals

6. Operation of heat exchangers in boilers, refrigeration units, and power plants
7. Cooling of electric machinery

F. Applications of traveling-wave techniques in chemical-diffusion problems

1. Mixture of gases in reacting vessels
2. Passage of liquids through porous substances
3. Intermixture of solids
4. Case-hardening of steel
5. Operation of transistors and diodes

It should also be noted that neutron diffusion in nuclear reactors and bombs obeys the same law as chemical diffusion.

The Schrödinger equation of quantum mechanics is another version of a traveling-wave equation, and "matter waves" are discussed in modern physics. Solutions of this equation where reflection is involved explain many phenomena associated with the structure of matter.

It can be seen that the traveling-wave concept has extremely wide application.

1-5. Organization of the Book

The complete transmission-line equation is the most general traveling-wave equation we shall discuss. For this reason, most new ideas will be presented in terms of transmission lines. Since the transmission-line equation itself is a one-dimensional equation and since many traveling-wave phenomena involve three space dimensions, the three-dimensional electromagnetic equation is frequently given in parallel with the transmission-line equation. A brief appendix on the vector relations for electromagnetics is included for students who are unfamiliar with this subject. It is important to remember, where transmission lines and field equations are shown side by side, that the field equation is simply the three-dimensional equivalent of the one-dimensional transmission-line equation and that the vector space-differential operators are the three-dimensional equivalents of their companion one-dimensional operators.

The basic idea of traveling waves is developed in Chap. 2 for transmission-line and space electromagnetic waves. Chapter 3 gives parallel developments for nonelectric waves.

The general problem of the transient traveling wave is not treated. However, the special cases of transient traveling waves in which losses are negligible and of such phenomena as heat waves or traveling electromagnetic waves in a conductor are treated in Chap. 4.

Although reflection is discussed in Chap. 4, it is considered more completely (for the steady-state case) in Chap. 5. To simplify the discussion, Chap. 5 deals only with waves in which losses are negligible, and Chap. 6 treats the modifications necessary where losses must be taken into account.

It is frequently convenient to solve practical traveling-wave problems with electric circuits or comparable lumped-constant models, and vice versa. These are discussed in Chap. 7.

Chapter 8 deals with applications of transmission lines in United States communication and power systems, and Chap. 9 covers special techniques used with radio-frequency transmission lines, waveguides, and acoustic resonators.

Prior to Chap. 10, plane-wave reflections are considered only from plane surfaces normal to the transmission direction. In Chap. 10, the situation is generalized, and a brief introduction is given to spherical waves, which are important when the source of a wave may be considered as a point.

1-6. Units

For all electromagnetic problems the rationalized meter-kilogram-second (mks) system is used. In this system the units of charge, potential, and current are the coulomb, volt, and ampere, respectively. The unit of magnetic flux is the weber. This system is in common usage and has been officially adopted by several international congresses.

The same system is used for the nonelectrical problems, although it is not in common usage in the United States by specialists in such problems. A few of the homework problems use the English system of units, and some use the centimeter-gram-second (cgs) system.

1-7. Summary

When the time of travel of an electrical, or other, disturbance from one point to another is significant, the techniques of traveling waves must be used. With sine waves, traveling-wave techniques must be used if the time of travel is a significant portion of a period of the oscillation. The traveling waves considered in electromagnetics are similar to those encountered when a pebble is dropped into a pond, when sound travels through the air, or when heat penetrates a wall.

Historically, the first uses of electric traveling waves were in transmission lines carrying information. Today this still remains one of the most important applications, but transfer of power and tuning are also important applications. Since acoustic, thermal, vibrational, diffusion, and other waves are similar to electromagnetic waves, many of the techniques developed in this book are equally applicable to these waves.

2. Traveling Electromagnetic Waves

Traveling waves carry energy from one place to another, even though their primary purpose may be information transfer. Standing waves, which are considered in Chap. 5, contain stored energy. In this respect, standing waves are comparable with resonant *LC* circuits and spring-mass vibrating systems. Traveling and standing waves are combined in many applications. In fact, standing waves result from a linear superposition of traveling waves going in opposite directions.

2-1. Basic Principles

A simple example will illustrate the traveling-wave idea. Consider the electric circuit of Fig. 2-1. A battery is connected to a two-wire

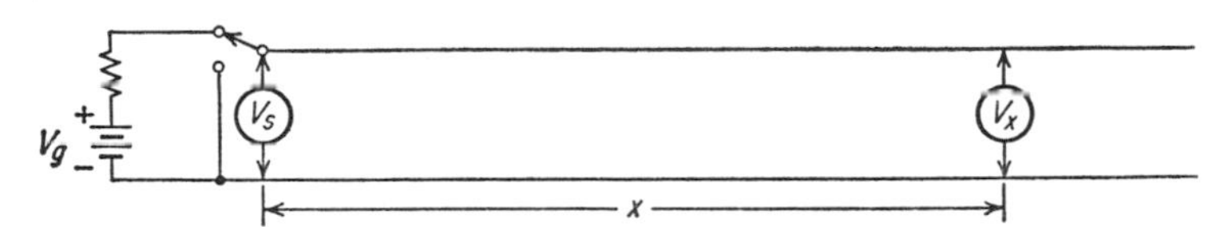

Fig. 2-1. Transmission-line circuit.

transmission line through a two-position switch. The other position of the switch connects a short circuit across the line in place of the battery.

Before studying the example, note the representation of the traveling-wave problem by a parallel-wire transmission line. Because of its convenience, this form is used often in this book, even though the actual physical configuration may be quite different. Figure 2-2 shows some of the common transmission-line forms. Even the parallel-wire transmission line does not ordinarily have the length-to-width ratio shown in Fig. 2-1. In practical cases, the length is much greater than the spacing, but our pictures seldom show the length more than a few times the spacing. The normal situation is illustrated in Fig. 2-2*a*. Another very common practical form of transmission line consists of concentric cylinders, as shown in Fig. 2-2*b*. Figure 2-2*g* shows a rectangular waveguide (a pipe through which the wave travels). This also is a common type of transmission line. Although the actual forms indicated in Fig. 2-2 should be kept in mind, the schematic representation of Fig. 2-1 will be most useful for our purposes here.

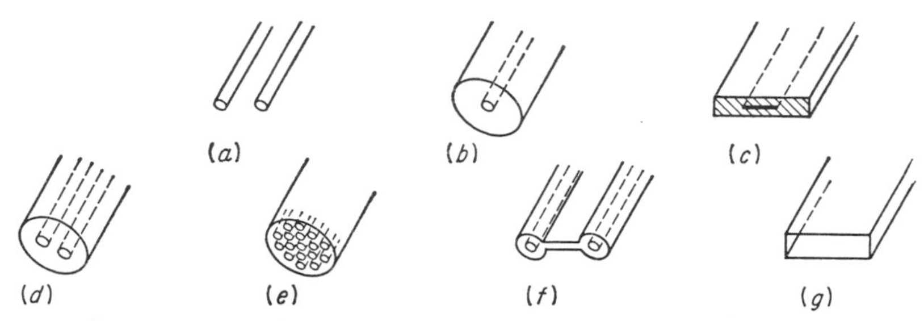

FIG. 2-2. Common transmission lines: (*a*) parallel wire; (*b*) coaxial; (*c*) strip line; (*d*) shielded parallel wire; (*e*) telephone cable; (*f*) twin-lead; (*g*) rectangular waveguide.

In Fig. 2-1 voltmeters are indicated at the "sending end" of the line (left-hand end) and at a point a distance x down the line from this end. The meter V_s immediately registers a voltage when the switch is thrown from the short circuit to the battery (this voltage is not equal to the battery emf, as is shown later). At the time the switch is thrown, and for some time after, the reading of meter V_x is zero, for it takes a finite time for this meter to "learn of the closing of the switch." At time t_x this meter begins to register the same voltage as the other meter. The amount of time taken for the second meter to learn of the closing of the switch is determined by the *velocity of propagation*. This velocity is the ratio of distance traveled to time required and is given by

$$v_p = \frac{x}{t_x}$$

The questions which immediately arise are, Why does the signal require a finite time to travel down the line? Why does it travel with finite velocity? Why did not both meters begin to register at the same time? In addition to answering in terms of our intuitive feeling that nothing can travel with infinite velocity and our well-established ideas about the velocity of light, we may explain the effect in terms of a circuit model of the transmission line involving lumped inductance and capacitance. This explanation is very rough and should not be construed as an accurate picture of wave travel on a line; nevertheless, it gives a physical explanation of a difficult concept.

A certain amount of inductance is associated with each short length of wire in a transmission line—Ampere's law tells us that magnetic fields are set up around the wires, and Faraday's law states that changes in this field will induce a back emf in the wire. Furthermore, any two conductors, such as short lengths on opposite sides of the line at the same distance from the end, have some capacitance; that is, if equal and opposite charges are placed on them, there is a potential difference between them. Hence, each section of line, no matter how short, has some series inductance and some parallel capacitance.

Figure 2-3*a* attempts to show this, although one could, of course, subdivide the line into smaller elements than could possibly be sketched. Here the inductance is indicated in the usual manner on each side of the line, and a capacitance is shown for each turn of the inductance. Resistance is not shown, since it would complicate the analysis.

It is difficult to analyze a circuit such as that shown in Fig. 2-3*a*, so the circuit has been redrawn for analysis in the form of Fig. 2-3*b*. This circuit is an approximation, with each inductor representing the inductance of a short section of line (length Δx) and each capacitor corresponding to the capacitance of the same length of line. As the distance Δx is decreased, the number of elements necessary to approximate a given length of line is increased, and the circuit model depicts the true situation more exactly.

Consider what happens when the previously short-circuited line is suddenly connected to the battery. Initially, C_1 is uncharged (assured by the short circuit). Since the charge on a capacitor cannot be changed

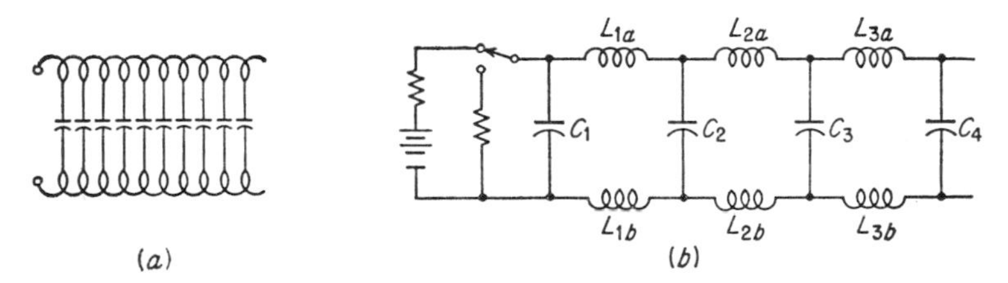

FIG. 2-3. Circuit equivalent of a transmission line.

instantaneously, there is no voltage across C_1 to drive current through L_{1a} and L_{1b}. When a voltage is established across the capacitor, current tries to start in the inductors; but the current in an inductance cannot be changed instantaneously, so that it takes a while for the current to become established. When it does start, it flows into C_2, for this is uncharged and therefore appears as a short circuit. When this capacitor finally acquires a charge, its voltage tries to start a current in the inductors L_2—but this takes time. When it finally starts, it flows into C_3, which, being uncharged, appears to the current starting in L_2 as a short circuit. This process continues on down the line, taking time to happen, just as it takes time to tell! It is evident that C_3 cannot begin to charge at once, for it must wait until the preceding capacitors have started to charge and until a current can be established through the inductive path.

When the switch disconnects the battery and short-circuits C_1, through the resistance, C_1 begins to discharge. This causes a reversal of the current in L_{1a} and L_{1b}, but the reversal cannot take place instantaneously. When this reversal has, in fact, been accomplished, it tends to discharge C_2. When this happens, the current in the L_2 reverses, but not instantaneously. This process continues, so that eventually C_4 "discovers that the battery has been removed." Since the velocity at which the wave travels in discharging the capacitors is the same as that of the wave charging the capacitors, both continue down the line. Hence, the discharging wave never overtakes the charging wave, and a *pulse* travels down the line.

Some may find this sort of explanation easier to understand when it is applied to a mechanical wave. Consider a sound impulse traveling through a pipe. Consider the air in each short length of pipe Δx as a separate cell having mass and compressibility, just as lumped inductance and capacitance simulate the actual transmission-line parameters.

Suddenly the air at one end of the pipe is set in motion. The first effect is to compress the air in cell 1. As it is compressed, it starts to move, but inertia requires this to take some time. When the air in cell 1 moves, it pushes against the side of cell 2. First the air in cell 2 is compressed. When the effect of inertia is overcome, it too starts to move and exerts pressure against the side of cell 3. In time, the air of cell 3 starts moving, pushing against cell 4, and the wave progresses down the pipe. The air in cell 3 does not start to move as soon as that in cell 1, for it cannot move until the air in cell 2 begins to push on it.

Let us now consider the picture of the *transmission-line* pulse as it travels down the actual line (not the *lumped-constant* model). Figure 2-4 shows a series of pictures of the voltage on the line at different fixed times, and Fig. 2-5 shows the voltage as a function of time at fixed points on the line. Note that the voltage of Fig. 2-4 cannot be read by a single voltmeter, for it must be read simultaneously at many points. To make this picture experimentally would require that readings be made simultaneously on a large number of voltmeters located at different points on the line and that a curve be drawn through the points so obtained. It is interesting to speculate how this might actually be done—for simultaneous photography of all the meters from a given point would require light of infinite velocity!

At time zero, the switch is closed and the battery is connected to the line. Figure 2-4*a* shows the voltage on the line immediately after this.

By time t_1, the leading edge of the wave has reached x_1, and the voltage is the same for all points to the left of x_1. At time t_2, the switch is opened and the input short-circuited. Figure 2-4*c* shows the voltage pulse on the line immediately after this. Still later, at time t_3, the pulse has progressed to the position shown in Fig. 2-4*d*. Because both ends of the pulse travel with the same velocity, both have traveled a distance

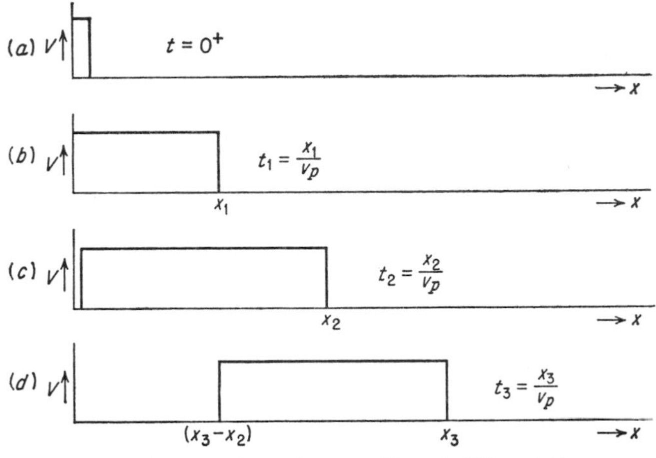

FIG. 2-4. Picture of a pulse on a line at different times.

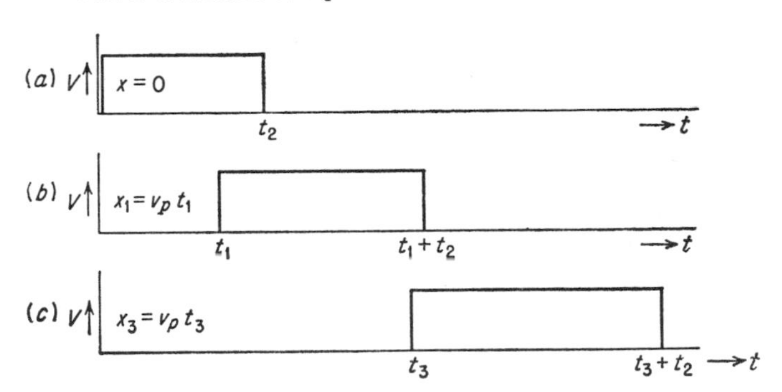

FIG. 2-5. Picture of a pulse at different points on a line.

$x_3 - x_2$ in the time $t_3 - t_2$ since the switch was opened. The leading edge is now at x_3, and the trailing edge is at $x_3 - x_2$. The distance occupied on the line by the pulse is still x_2. Subsequent pictures would show it unchanged in shape and size but farther to the right.

It is characteristic of traveling waves that both the time and the space pictures must be shown for complete understanding. For the rectangular pulse shown, the waveshape is the same on both plots. When a nonsymmetric pulse is used, it can be seen that the space and time plots are mirror images of each other. Sometimes only one picture is shown, but the reader can and should sketch the other if it is needed for understanding.

The example shown is typical of pulses where damping is negligible. This occurs with low-loss transmission lines, with plane waves in space, and with many vibrating-membrane and acoustic problems. Where damping (R and G for electric circuits) must be considered, the velocity of sinusoidal components at different frequencies is rarely the same. Thus, the various Fourier components of a square pulse travel at different velocities (and are reduced in size by different amounts), so their superposition at the *receiving end* of the line is not the same as at the *sending end*. This is a form of *distortion*. Thus, on a lossy line, the pulse shown in Fig. 2-5*c* would have a different shape from that in Fig. 2-5*a*. This situation also prevails for plane waves in conducting media, for all heat waves, for diffusion processes, and in other traveling-wave applications; it is discussed further in Chaps. 6 and 8.

2-2. Derivation of General Transmission-line Equations

The general partial differential equation describing travel of voltage waves down a transmission line is developed in this section. Parallels are drawn where appropriate to the comparable relations governing traveling electromagnetic waves in space. Similar derivations for non-electromagnetic waves are given in Chap. 3.

There are two principal reasons for concentrating first on the equations of the transmission line. First, the transmission-line equations involve only one space dimension (distance along the line) and time—they are therefore two-dimensional. Space wave equations of all types involve three space dimensions and time—they are four-dimensional equations. Second, the transmission-line wave equation is in some ways more general than the other types of wave equation.

As usual, we shall use the two-wire line for illustrative purposes. Consider such a line subdivided into small pieces of length Δx. Each such section has a certain amount of series resistance and inductance associated with each of the wires. In addition, each has a certain shunt capacitance and conductance. Since, in a practical line, the conductors must be made of practical metals, the resistance cannot be zero, as assumed in the previous example. In a practical line, some mechanical means must be found to separate the conductors, and anything used for this purpose must have some conductance.

Various representations of the transmission-line parameters for a single short section are shown in Fig. 2-6. Figure 2-6*a* shows the series resistance and inductance on both sides of the transmission line as they are in practice. The shunt elements are shown at the right-hand side of the section. In Fig. 2-6*b* the series elements have been combined on one side; this is permissible when no external connections cause external currents to flow in response to series voltage drops in the line. Another common model, the T section, is represented in Fig. 2-6*c*. Here the series elements are split, and the shunt elements are in the middle. Figure 2-6*d* shows another common representation—the π section. Here the shunt element is split. If Δx is sufficiently large, the differences between these four circuit models (and others which might be used) can be significant. For purposes of deriving the transmission-line equations, however, the form used makes no difference; the results are all the same when Δx approaches zero. In each of the models shown, the elements have been given in terms of four parameters of the line:

R = series resistance per unit length
L = series inductance per unit length
G = shunt conductance per unit length
C = shunt capacitance per unit length

These basic parameters are determined by the physical configuration of the line. Their evaluation is a static-field problem which is not treated here. It is important to note, however, that $R\,\Delta x$, $L\,\Delta x$, $C\,\Delta x$, and $G\,\Delta x$ are the *exact* values for resistance, inductance, capacitance, and conductance of the section of line.

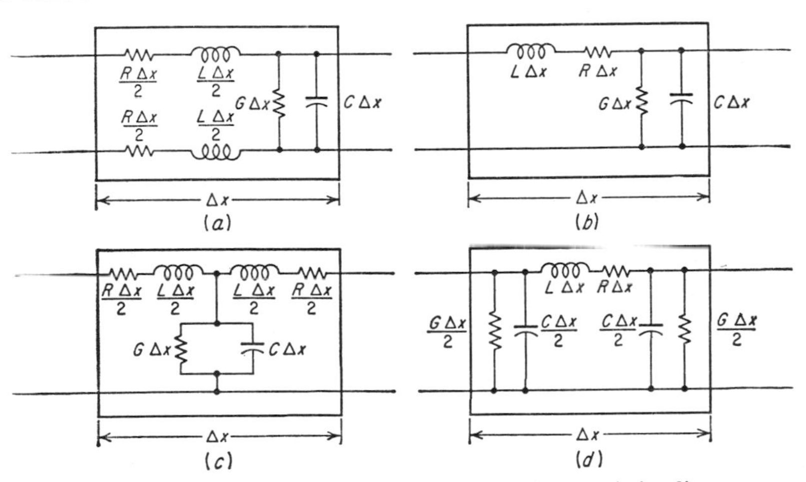

Fig. 2-6. Circuit models of a short length of transmission line.

The "telegrapher's equations" which govern wave propagation on the line are derived using the circuit model of Fig. 2-6*b*, since it is the simplest shown. The shunt elements could be to the left of the series elements without changing the final result. The derivation would be slightly different if they were to the left, however.

Consider Fig. 2-7. Here the section of Fig. 2-6*b* is shown in its relation to the rest of the transmission line. The voltages appearing across the line at the entry and exit of the section are shown, along with the entering and leaving currents. It is assumed that the section of line of length Δx is located a distance x from the sending end. The equations to be derived relate the two voltages shown and the two currents shown.

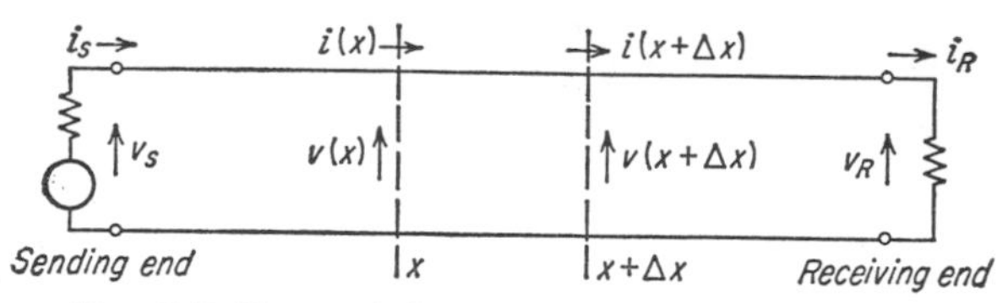

Fig. 2-7. Transmission-line voltages and currents.

The input voltage and current to the short section are $v(x)$ and $i(x)$. The output voltage and current are $v(x + \Delta x)$ and $i(x + \Delta x)$. Because of the drop in the series elements,

$$v(x + \Delta x) = v(x) - R\,\Delta x\, i(x) - L\,\Delta x\,\frac{\partial i(x)}{\partial t}$$

Thus the ratio of the difference in output and input voltage to the length of the section is

$$\frac{v(x + \Delta x) - v(x)}{\Delta x} = -Ri(x) - L\,\frac{\partial i(x)}{\partial t}$$

Passing to the limit as the length of the section Δx approaches zero, this becomes

Field equivalent

(2-1) $$\frac{\partial v}{\partial x} = -\left(R + L\,\frac{\partial}{\partial t}\right) i \qquad\qquad \nabla \times \mathbf{E} = -\mu\,\frac{\partial \mathbf{H}}{\partial t}$$

The field equivalent of the transmission-line equation is shown at the right. It is the vector differential form of Faraday's law[1] of electromagnetic induction. Here

$\nabla \times \mathbf{E}$ = curl of $\mathbf{E}$
$\mathbf{E}$ = electric field strength, volts/m
μ = permeability, henrys/m
$\mathbf{H}$ = magnetic field strength, amp/m

Since the curl is a first-order space derivative, the units of the field equation are the same as those of the transmission-line equation, except for the "per meter" in each quantity, and the equations are of the same basic form. This type of parallelism is always present between trans-

[1] William H. Hayt, Jr., "Engineering Electromagnetics," p. 240, McGraw-Hill Book Company, Inc., New York, 1958.

mission-line equations and the corresponding equations for waves in space, though it is not pointed out explicitly where the two are shown side by side in this book.

In general, given the basic experimental relations, it is possible to derive the three-dimensional field equations by a process analogous to that used in deriving the line equations. Details of these derivations are not presented here. Rather, the field equations are presented without proof at various points of interest. A thorough understanding of the vector operations of curl, gradient, and divergence should make the parallels obvious. To assist in understanding, however, it may be desirable to consider the case where **E** has only a y component and **H** only a z component. In this case, the field equation becomes

$$\frac{\partial E_y}{\partial x} = -\mu \frac{\partial H_z}{\partial t}$$

which is obviously analogous to the line equation with $R = 0$. A comparable process, not presented in this book, may assist in understanding later field-line parallels.

To obtain the current equation corresponding to (2-1), observe that the current leaving the section is less than that entering by just the amount that "leaks off" through the shunt elements. Thus,

$$i(x + \Delta x) = i(x) - G\,\Delta x\, v(x + \Delta x) - C\,\Delta x \frac{\partial v(x + \Delta x)}{\partial t}$$

Here we are using the model of Fig. 2-6*b*, so the voltage across the shunt elements is $v(x + \Delta x)$. Writing $v(x + \Delta x)$ as a Taylor series expansion about the point x,

$$v(x + \Delta x) = v(x) + \frac{\partial v(x)}{\partial x} \Delta x + \cdots$$

Substituting this into the equation for the current, we have

$$i(x + \Delta x) = i(x) - G\,\Delta x\, v(x) - C\,\Delta x \frac{\partial v(x)}{\partial t} - G(\Delta x)^2 \frac{\partial v(x)}{\partial x} - C(\Delta x)^2 \frac{\partial^2 v(x)}{\partial x\, \partial t} + \cdots$$

Rearranging to show the ratio of the difference to the length,

$$\frac{i(x + \Delta x) - i(x)}{\Delta x} = -Gv(x) - C\frac{\partial v(x)}{\partial t} - \Delta x \left[G \frac{\partial v(x)}{\partial x} + C \frac{\partial^2 v(x)}{\partial x\, \partial t}\right] + \cdots$$

In the limit, as $\Delta x \to 0$,

Field equivalent

(2-2) $$\frac{\partial i}{\partial x} = -\left(G + C\frac{\partial}{\partial t}\right) v \qquad\qquad \nabla \times \mathbf{H} = \left(\sigma + \epsilon \frac{\partial}{\partial t}\right) \mathbf{E}$$

Note that the term to the right in the preceding equation dropped out in the limiting process. This term was due to assuming that the shunt elements came after rather than before the series elements, as shown in the model of Fig. 2-6*b*. Since this term is negligible anyway, it can be seen that any of the models of Fig. 2-6 could have been used for the short section.

It can be seen that the corresponding field equation is Ampere's law,[1] with σ the conductivity in mhos per meter and ϵ the permittivity in farads per meter.

Equations (2-1) and (2-2) are known as the *telegrapher's equations*. They are first-order equations involving both voltage across the line and current in the line. Comparable equations are developed in Chap. 3 for most of the other types of waves studied here. Because they are so important, these telegrapher's equations are repeated together here:

Field equivalent

(2-1) $$\frac{\partial v}{\partial x} = -\left(R + L\frac{\partial}{\partial t}\right)i \qquad \nabla \times \mathbf{E} = -\mu\frac{\partial \mathbf{H}}{\partial t}$$

(2-2) $$\frac{\partial i}{\partial x} = -\left(G + C\frac{\partial}{\partial t}\right)v \qquad \nabla \times \mathbf{H} = \left(\sigma + \epsilon\frac{\partial}{\partial t}\right)\mathbf{E}$$

Equations (2-1) and (2-2) may be combined into a single equation in terms of either voltage or current. To obtain the voltage equation, differentiate both sides of (2-1) with respect to distance, obtaining

$$\frac{\partial^2 v}{\partial x^2} = -\left(R + L\frac{\partial}{\partial t}\right)\frac{\partial i}{\partial x}$$

Note that $\partial i/\partial x$ is given by (2-2), so the voltage equation is

Field equivalent

$$\frac{\partial^2 v}{\partial x^2} = \left(R + L\frac{\partial}{\partial t}\right)\left(G + C\frac{\partial}{\partial t}\right)v \qquad \nabla^2\mathbf{E} = \mu\frac{\partial}{\partial t}\left(\sigma + \epsilon\frac{\partial}{\partial t}\right)\mathbf{E}$$

or

(2-3) $$\frac{\partial^2 v}{\partial x^2} = RGv + (RC + LG)\frac{\partial v}{\partial t} + LC\frac{\partial^2 v}{\partial t^2} \qquad \nabla^2\mathbf{E} = \mu\sigma\frac{\partial \mathbf{E}}{\partial t} + \mu\epsilon\frac{\partial^2 \mathbf{E}}{\partial t^2}$$

The comparable field equation is obtained by a similar process, the space differentiation being in the form of the Laplacian operator. It is correct only when $\nabla \cdot \mathbf{E} = 0$.

It is important to examine this "wave equation" carefully, for it is of very general utility. Many types of waves may be expressed by similar equations, with one or more of the terms missing. For example, note that the field equation does not contain a term like the first term

[1] *Ibid.*, p. 246.

on the right-hand side of the line equation. The reason is that there is nothing which compares for fields in space with the series resistance on a transmission line.

2-3. Solution of the General Transmission-line Equations—Steady State

This section deals with a steady-state solution to the wave equation (2-3). The general solution to this equation is discussed in Chap. 4. The steady-state solution is one for which both voltage and current on the transmission line are sinusoidal, with angular frequency ω.

Let the instantaneous values of voltage and current be given by[1]

$$v = \text{Re}\, Ve^{j\omega t} = V \cos \omega t \qquad i = \text{Re}\, Ie^{j\omega t} = I \cos \omega t \tag{2-4}$$

As is customary in steady-state a-c problems, the complex forms will be used rather than instantaneous values. Thus, the voltage of (2-4) may be substituted in (2-3). The symbol Re means "real part of." Since the operation of taking the real part of the complex number is common to all terms, since the term exp $(j\omega t)$ is common to all terms, and since (2-3) is a linear equation, they need not be shown. Carrying out the indicated time differentiation and dropping these common factors, (2-3) becomes

$$\frac{d^2V}{dx^2} = RGV + (LG + RC)j\omega V + (j\omega)^2 LCV \tag{2-3a}$$

This simplifies to

$$\frac{d^2V}{dx^2} = (R + j\omega L)(G + j\omega C)V \qquad \textit{Field equivalent} \quad \nabla^2\mathbf{E} = j\omega\mu(\sigma + j\omega\epsilon)\mathbf{E} \tag{2-5a}$$

This is frequently written as

$$\frac{d^2V}{dx^2} = \gamma^2 V \qquad \textit{Field equivalent} \quad \nabla^2\mathbf{E} = \gamma^2\mathbf{E} \tag{2-5b}$$

The complex number γ is called the *propagation constant.* It is defined as

$$\gamma = \sqrt{(R + j\omega L)(G + j\omega C)} \qquad \textit{Field equivalent} \quad \gamma = \sqrt{j\omega\mu(\sigma + j\omega\epsilon)} \tag{2-6}$$

[1] Capital letters for voltage, current, and analogous nonelectrical quantities refer to the *phase vectors,* or *phasors,* associated with sinusoidal time variation. Lower-case letters refer to instantaneous values. This distinction is not made for the field vectors, since capital letters are commonly used for both instantaneous values of the field and associated phasors.

Equation (2-5*b*) can be recognized as the harmonic equation. Its well-known solution is of the form

(2-7) $$V = Ae^{-\gamma x} + Be^{\gamma x}$$

where A and B are constants as yet undetermined.

A wave equation comparable to (2-3) could be developed for current and solved in the same manner. It is just as easy, however, to obtain the current from the voltage by application of (2-1). Of course, a similar technique may be used to determine the magnetic field from the electric field. Thus,

$$(R + j\omega L)I = -\frac{dV}{dx}$$

$$I = \frac{-1}{R + j\omega L}\frac{dV}{dx} = +\frac{1}{R + j\omega L}(A\gamma e^{-\gamma x} - B\gamma e^{\gamma x})$$

(2-8) $$I = \sqrt{\frac{G + j\omega C}{R + j\omega L}}\,(Ae^{-\gamma x} - Be^{\gamma x})$$

Note that the multiplying factor has units of admittance, so we may define as an impedance

Field equivalent

(2-9) $$Z_0 = \sqrt{\frac{R + j\omega L}{G + j\omega C}} \qquad \eta = \sqrt{\frac{j\omega\mu}{\sigma + j\omega\epsilon}}$$

On the transmission line the quantity Z_0 is known as the *characteristic impedance.* The comparable quantity for waves in space is the *intrinsic impedance* η.

Using (2-9) and (2-8) and rewriting (2-7), we have for the voltage and current on the line

(2-7) $$V = Ae^{-\gamma x} + Be^{\gamma x}$$

(2-10) $$I = \frac{1}{Z_0}(Ae^{-\gamma x} - Be^{\gamma x})$$

The fact that the sign of the second term is plus for the voltage and minus for the current is quite important. This will become clear when reflections are discussed in Chap. 4.

Electromagnetic waves in space, acoustic waves, thermal waves, and others discussed here often take the idealized form of *plane waves.* These are waves for which there is variation in intensity in only one direction (the x direction in this case). That is,

$$\frac{\partial}{\partial y} = 0 \qquad \frac{\partial}{\partial z} = 0$$

where the operators may operate on any quantity associated with the wave. This means that there is neither phase shift nor amplitude change in the y and z directions. That is, planes parallel to the YZ plane are surfaces in which everything about the wave is constant.

For a plane wave with electric field in the y direction, the solution of the wave equation (2-5) is

$$E_y = Ae^{-\gamma x} + Be^{\gamma x} \tag{2-7a}$$

Applying (2-1) to this solution, subject to the plane-wave conditions, the magnetic field is

$$H_z = \frac{1}{\eta}(Ae^{-\gamma x} - Be^{\gamma x}) \tag{2-10a}$$

Alternative forms for the solution to (2-3) are in terms of hyperbolic functions. Since hyperbolic functions are defined as linear combinations of the exponentials, the solution in hyperbolic functions merely amounts to a rearrangement of the constants in Eqs. (2-7) and (2-10). Rather than show the multipliers for the hyperbolic functions in terms of the constants for the exponential solution, new constants are used in

		Field equivalent
(2-11)	$V = C \cosh \gamma x + D \sinh \gamma x$	$E_y = C \cosh \gamma x + D \sinh \gamma x$
(2-12)	$I = C' \cosh \gamma x + D' \sinh \gamma x$	$H_z = C' \cosh \gamma x + D' \sinh \gamma x$

Many writers on transmission lines make extensive use of the hyperbolic form of the solution. The present author feels, however, that the hyperbolic functions tend to destroy the meaning of the solution, so they are rarely mentioned in this book.

2-4. Wave Parameters and the Infinite Line

Initially, it is convenient to study the above solutions for an infinitely long transmission line, that is, a line from a generator at $x = 0$ extending without bound in the direction of increasing x. The propagation constant γ [defined in (2-6)] is always complex. If both R and G were equal to zero, it would be pure imaginary, but this cannot be in any practical case, even though we sometimes talk about "lossless lines." A lossless line is really a mathematical simplification for a line in which losses are small; practically, perfect conductors and insulators cannot be constructed. Even air is not a perfect insulator, so γ associated with waves in air has a slight real part.

If γ has a real part, it is positive, so the B terms of (2-7) and (2-10) become infinite as x becomes infinite, *unless* B is zero. Since it is not plausible to expect a finite generator to set up an infinite voltage at an infinite distance away, B must be zero for the infinite line. Thus, on an infinite line we have

		Field equivalent
(2-13)	$V = Ae^{-\gamma x}$	$E_y = Ae^{-\gamma x}$
(2-14)	$I = \frac{A}{Z_0} e^{-\gamma x}$	$H_z = \frac{A}{\eta} e^{-\gamma x}$

The ratio of voltage to current on the infinite line is therefore given by

		Field equivalent
(2-15)	$\frac{V}{I} = Z_0$	$\frac{E_y}{H_z} = \eta$

Thus the characteristic impedance of a transmission line is simply the ratio of voltage to current at a point on an infinite line, and the intrinsic impedance is the ratio of electric to magnetic fields in an infinite plane wave.

Obviously, an infinite line is impossible to achieve in practice, but the infinite-line concept has value in the discussion of waves traveling in only one direction on any line, such as transients starting down the line, and in its use with "properly terminated" lines. Consider the transmission line shown in Fig. 2-8. In part *a* this is shown as a segment

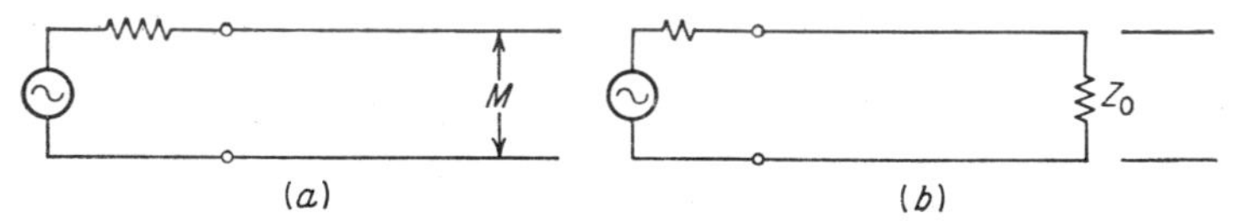

FIG. 2-8. Infinite and properly terminated lines.

of an infinite line. In part *b* it is shown cut at point M, with an impedance equal to the characteristic impedance inserted as a termination. Since the ratio of voltage to current in the load, or terminating, impedance is the same as in the infinite line, there is no difference in the current which flows in the line with termination from that which flows in the endless line. Hence, at a point to the left of M, an observer has no way of telling whether the line is infinite or terminated in its characteristic impedance. A line so terminated is said to be *properly terminated.*

As is shown later, properly terminated lines are desired for many purposes. They are more efficient; they do not have echoes, which may cause trouble in television and telephone; and the maximum voltage on the line for a given power carried is less with a properly terminated line.

Recall that the propagation constant γ is given by

		Field equivalent
(2-6)	$\gamma = \sqrt{(R + j\omega L)(G + j\omega C)}$	$\gamma = \sqrt{j\omega\mu(\sigma + j\omega\epsilon)}$

Since it is usually complex, it is convenient to describe the propagation constant by

$$\gamma = \sqrt{(R + j\omega L)(G + j\omega C)} = \alpha + j\beta \tag{2-16}$$

where α is the *attenuation constant* and β is the *phase constant*. It is possible to obtain separate algebraic expressions for these constants. However, it is usually easier to calculate them directly from (2-16), so the algebraic expressions are not given.

Using these constants, we can separate the expressions for voltage and current on the infinite line into magnitude and phase factors:

Field equivalent

$$V = Ae^{-\alpha x}e^{-j\beta x} \qquad E_y = Ae^{-\alpha x}e^{-j\beta x}$$

$$I = \frac{A}{Z_0} e^{-\alpha x}e^{-j\beta x} \qquad H_z = \frac{A}{\eta} e^{-\alpha x}e^{-j\beta x}$$

In ordinary circuit analysis, we frequently neglect the time factor, as has been done here. However, in traveling-wave engineering it is often desirable to consider both the time and the space factors, for either represents a phase shift when the other is held constant. Hence, it is more appropriate to write

Field equivalent

$$Ve^{j\omega t} = Ae^{-\alpha x}e^{j(\omega t - \beta x)} \qquad E_y e^{j\omega t} = Ae^{-\alpha x}e^{j(\omega t - \beta x)} \tag{2-17}$$

$$Ie^{j\omega t} = \frac{A}{Z_0} e^{-\alpha x}e^{j(\omega t - \beta x)} \qquad H_z e^{j\omega t} = \frac{A}{\eta} e^{-\alpha x}e^{j(\omega t - \beta x)} \tag{2-18}$$

Of course, we know that these expressions merely describe the size and angle of phasors and that instantaneous values of voltage and current are

$$v = \text{Re } Ae^{-\alpha x}e^{j(\omega t - \beta x)} = Ae^{-\alpha x} \cos (\omega t - \beta x)$$

$$i = \text{Re } \frac{A}{Z_0} e^{-\alpha x}e^{j(\omega t - \beta x)} = \frac{A}{|Z_0|} e^{-\alpha x} \cos (\omega t - \beta x - \theta)$$

where $Z_0 = |Z_0|e^{j\theta}$. To see the meaning of the time-distance phase factor, let us consider a transmission line with no loss, so that α is negligible. On such a line Z_0 is a resistance, so $\theta = 0$. Figure 2-9*a* shows a plot of the voltage on such a line as a function of distance for three different times, and Fig. 2-9*b* shows comparable plots as a function of time for three different positions on the line.

The plot represented in part *a* of the figure would result if a curve were drawn through simultaneous readings obtained by a large number of instantaneous-response voltmeters attached to the line. The three curves are drawn for instants of time a quarter-cycle apart, so the values for $x = 0$ represent three points on the ordinary circuit plot of voltage vs.

time. Since one of the curves of part b is just this plot, the voltages at the origin in part a represent points on the $x = 0$ curve of b.

Examination of the voltage-vs.-distance plots of Fig. 2-9a shows that the $\omega t = \pi/2$ curve is identical with the $\omega t = 0$ curve except that it is displaced a distance $\beta x = \pi/2$ to the right. The $\omega t = \pi$ curve is displaced an additional $\pi/2$ on the βx scale. Thus, if one could move a voltmeter along the line at the right speed, so that he would just keep up with the movement of these curves, his meter would always read the same. That is, what appears to a fixed observer as a sinusoidal alternating voltage would appear to an observer moving at the right speed as a

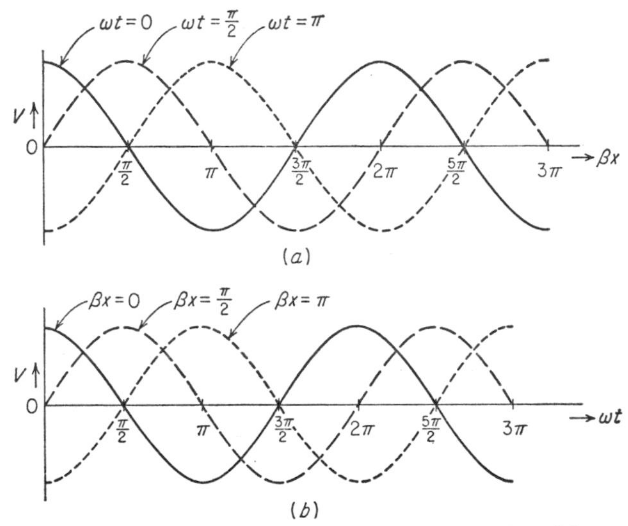

FIG. 2-9. Plots of voltage on a lossless line as a function of position and time: (a) voltage vs. distance; (b) voltage vs. time.

direct unchanging voltage. For example, if the moving observer started at $t = 0$ at $x = 0$ and reached $x = \pi/2\beta$ at $t = \pi/2\omega$, his voltmeter would continue to show a potential difference between the conductors of exactly A volts. Hence, the wave pictured in Fig. 2-9 is a *traveling wave*.

For many transmission lines and for waves in space, the moving observer described in the preceding paragraph would have to travel close to the speed of light (or at it), so this is an experiment limited to science fiction. On the other hand, the experiment is much easier to perform with sound waves, for many modern aircraft actually travel much faster than the speed of sound.

In most traveling-wave problems, as stated previously, it is wise to sketch both the voltage-vs.-distance curves for a set of times and the voltage-vs.-time curves for a set of space positions, as has been done in Fig. 2-9. Note that in both these pictures it is the instantaneous voltage that is plotted, and this is as much a function of position as of time. The plots of Fig. 2-9 indicate that space and time pictures are the same, but this is a result of the use of an even function, the cosine. Actually, the space and time diagrams are mirror images, as is shown in dealing with transients in Chap. 4.

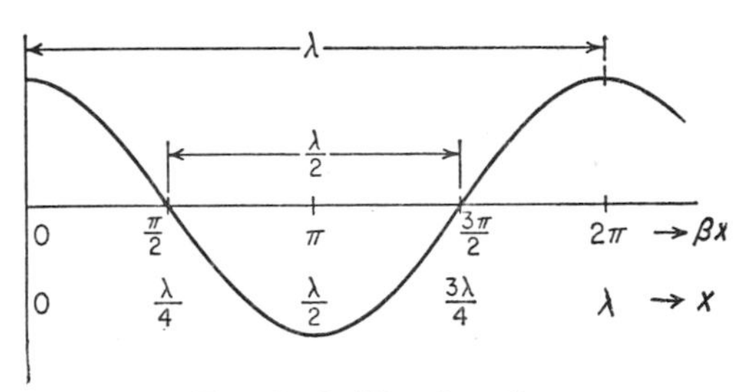

FIG. 2-10. Wavelength.

Just as we sometimes consider the period in a sinusoidally varying time function, the space period must also be considered. The period in space of a traveling wave is called the *wavelength*. A space plot for $t = 0$ is given in Fig. 2-10 to illustrate this. Analogous space and time relations are

$$\beta\lambda = 2\pi \qquad \omega T = 2\pi$$
$$\lambda = \frac{2\pi}{\beta} \qquad T = \frac{2\pi}{\omega} \tag{2-19}$$
$$\beta x = \frac{2\pi}{\lambda} x \qquad \omega t = \frac{2\pi}{T} t$$

with T the period for sinusoidal time variation.

To obtain the phase velocity, that is, the velocity with which the moving observer must travel to observe a stationary voltage, consider that

$$\phi = \omega t - \beta x$$

must be a constant if the voltage

$$v = A \cos \phi$$

is to be constant. Thus, if

$$x = \frac{1}{\beta} (\omega t - \phi)$$

the velocity for the observer, and therefore for the wave, is given by

$$v_p = \frac{dx}{dt} = \frac{\omega}{\beta} \tag{2-20}$$

For electromagnetic waves in space, and on many transmission lines, the phase velocity is the velocity of light, 3×10^8 m/sec. For sound waves in air, the phase velocity is about 350 m/sec.

Use of the relations developed above results in a number of common forms for writing the phase of a traveling wave. Some of these are listed here:

$$V = Ae^{j(\omega t - \beta x)} = Ae^{j(\omega t - 2\pi x/\lambda)} = Ae^{j\omega(t - x/v_p)} = Ae^{j\beta(v_p t - x)} \tag{2-21}$$

An opportunity to calculate velocity and wavelength under various conditions is given in the problems.

No line is truly lossless, and many lines have a very considerable loss. It is therefore necessary to have a unit to measure the effect of these losses on voltage, current, and power. Consider the magnitude in (2-17) and (2-18):

$$|V| = |A|e^{-\alpha x} \tag{2-22}$$

$$|I| = \left|\frac{A}{Z_0}\right| e^{-\alpha x} \tag{2-23}$$

The ratio of voltages or currents for two points at a given spacing is the same as the ratio for two other points on any other part of the line, but with the same spacing. This may be seen by examining two pairs of points adjacent to each other and with equal spacings, as shown in Fig. 2-11. Here we assume an infinite line, with sending-end voltage V_s.

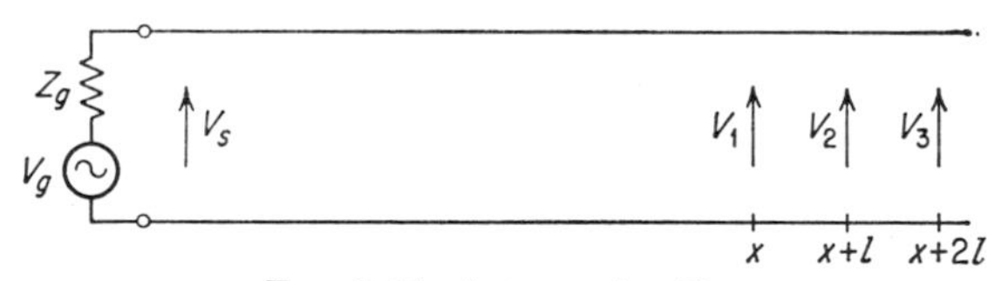

FIG. 2-11. Attenuating line.

At a distance x from the origin we observe a voltage V_1. V_2 is observed at $(x + l)$ and V_3 at $(x + 2l)$. Thus, the distance between V_2 and V_1 is the same as between V_3 and V_2, namely, l. The magnitude of V_1 is

$$|V_1| = |V_s|e^{-\alpha x}$$

that at V_2 and that at V_3 are, respectively,

$$|V_2| = |V_s|e^{-\alpha(x+l)}$$
$$|V_3| = |V_s|e^{-\alpha(x+2l)}$$

Hence the ratio of voltages for the two equal intervals is the same:

$$\left|\frac{V_2}{V_1}\right| = \left|\frac{V_3}{V_2}\right| = e^{-\alpha l} \tag{2-24}$$

even though the centers of the two intervals have different locations.

Although the voltage ratios themselves might be used as a measure of decrease in voltage, and hence of "signal strength," they frequently get quite large, and it is more convenient to use a logarithmic measure. A suitable available measure is the exponent in the exponential describing the voltage ratio. Thus, we may describe this ratio in terms of

$$\alpha l = \log_e \frac{V_1}{V_2} \qquad \text{nepers} \tag{2-25}$$

The unit *neper* is derived from *Naperian*, the name given to logarithms to the base e (also called *natural logarithms*). This unit is a measure of *attenuation;* that is, it measures the amount of decrease of voltage or current in traveling a given distance down a line. Note that the attenuation is positive when the exponent of the voltage ratio is negative—when the voltage farther down the line is less than that nearer the sending end. If the line contains amplifiers, the voltage may be smaller near the sending end, in which case the attenuation is negative. We say that this section of line (and amplifier) has a *gain* of the appropriate number of nepers. Hence

$$\text{Attenuation in nepers} = -\text{ (gain in nepers)} \tag{2-26}$$

Table 2-1 lists the voltage ratios corresponding to different numbers of nepers.

TABLE 2-1. ATTENUATION IN NEPERS

Number of nepers	V_2/V_1	V_1/V_2
1	0.37	2.72
2	0.15	7.39
10	0.000045	22,026

The neper is a rather large unit, as can be seen from the table. A more convenient unit is the decibel. It is important to note that the decibel is defined as a *power* ratio, not as a voltage or current ratio. This means that it is directly related to a voltage ratio only if the two voltages are measured across the same impedance. In transmission-line work this is often the case, but in amplifier studies the input and output impedance may be different by several orders of magnitude.

The decibel (abbreviated db) is a modification of a larger (inconveniently so) unit, the bel—named for Alexander Graham Bell—which was first used in describing the performance of telephone systems. It is defined by the relation

$$\text{Attenuation in db} = 10 \log_{10} \frac{P_1}{P_2} \tag{2-27}$$

where P_1 is the larger power and P_2 the smaller. If both powers are measured at the *same impedance level,*

$$\text{Attenuation in db} = 20 \log_{10} \left| \frac{V_1}{V_2} \right| \tag{2-28}$$

For the examples above, we have

$$\text{Attenuation in db} = 20 \log_{10} e^{\alpha l} = 20\alpha l \log_{10} e$$

Hence

$$\text{Attenuation in db} = 8.686\alpha l = 8.686 \text{ (attenuation in nepers)} \tag{2-29}$$

It can be seen that the decibel is a more convenient unit than the neper, for it corresponds to a voltage ratio of only 1.12, as compared with the 2.72 ratio for 1 neper. Thus, 1 db represents a 12 per cent voltage change, but 1 neper represents a 172 per cent change.

Some idea of the number of decibels corresponding to certain power and voltage ratios is given in Table 2-2.

TABLE 2-2. ATTENUATION IN DECIBELS

Decibels	Voltage ratio	Power ratio
1	1.12	1.26
3	1.41	2.0
6	2.0	4.0
10	3.18	10.0
16	6.32	40.0
20	10.0	100.0
100	10^5	10^{10}

Example 2-1. An open-wire telephone line is made up of 2 of the 40 wires on a pole line. The AWG 10 wires (0.264 mm diameter) are spaced 20 cm on the crossarm. Resistance, inductance, and capacitance are calculated to have the values shown, and the conductance was measured in dry weather. The basic line constants are as follows:

$R = 6.30$ ohms/km
$L = 2.11 \times 10^{-3}$ henry/km
$G = 6.8 \times 10^{-8}$ mho/km
$C = 0.00565$ μf/km

Find α, β, Z_0, v_p, and λ for a frequency of 1,000 cps.

$$\begin{aligned}
\gamma &= \sqrt{(R + j\omega L)(G + j\omega C)} \\
&= \sqrt{(6.30 + j2\pi \times 10^3 \times 2.11 \times 10^{-3})} \\
&\qquad \times \sqrt{(6.8 \times 10^{-8} + j2\pi \times 10^3 \times 5.65 \times 10^{-9})} \\
&= \sqrt{(6.30 + j13.25)(6.8 \times 10^{-8} + j3.55 \times 10^{-5})}
\end{aligned}$$

$$= \sqrt{(14.65\underline{/64.6°})(3.55 \times 10^{-5}\underline{/89.9°})}$$
$$= 2.29 \times 10^{-2}\underline{/77.3°} = \alpha + j\beta$$
$$\alpha = 5.05 \times 10^{-3} \text{ neper/km} = 8.686 \times 5.05 \times 10^{-3} = 0.044 \text{ db/km}$$
$$\beta = 2.23 \times 10^{-2} \text{ radian/km}$$
$$v_p = \frac{\omega}{\beta} = \frac{2\pi \times 10^3}{2.23 \times 10^{-2}} = 2.81 \times 10^5 \text{ km/sec}$$
$$\lambda = \frac{2\pi}{\beta} = 281 \text{ km}$$
$$Z_0 = \sqrt{\frac{R + j\omega L}{G + j\omega C}} = \sqrt{\frac{14.65\underline{/64.6°}}{3.55 \times 10^{-5}\underline{/89.9°}}} = 641\underline{/-12.7°}$$
$$= 629 - j141 \text{ ohms}$$

It can be seen that the effect of the conductance is negligible. This is characteristic of open-wire transmission lines, for both telephone and power. The attenuation is quite small—in 100 km, attenuation is only 0.505 neper and the voltage ratio is exp (0.505) = 1.65. The velocity of propagation is less than the velocity of light by about 6 per cent, with the result that the wavelength is correspondingly less than that for plane waves in air.

Example 2-2. A telephone toll cable has greatly increased capacitance compared with the open-wire line. Smaller conductors are used, so the resistance is higher. For example, consider an AWG 19 pair in a cable, for which the following parameters apply at 1,000 cps:

R = 52.2 ohms/km
L = 0.69 mh/km
G = 0.623 μmho/km
C = 0.038 μf/km

The attenuation, phase shift, impedance, wavelength, and velocity may be calculated as follows:

$$\gamma = \sqrt{(52.2 + j2\pi \times 10^3 \times 6.9 \times 10^{-4})} \times \sqrt{(0.623 \times 10^{-6} + j2\pi \times 10^3 \times 3.8 \times 10^{-8})}$$
$$= \sqrt{(52.2 + j4.33)(0.623 \times 10^{-6} + j2.39 \times 10^{-4})}$$
$$= 0.112\underline{/47.5°} = 0.076 + j0.083$$
$$\alpha = 0.076 \text{ neper/km} \qquad \beta = 0.083 \text{ radian/km}$$
$$v_p = \frac{\omega}{\beta} = \frac{2\pi \times 10^3}{0.083} = 7.57 \times 10^4 \text{ km/sec}$$
$$\lambda = 75.7 \text{ km}$$

It can be seen that the attenuation for the cable is much greater than for the open-wire line. In fact, the attenuation and phase constants are almost equal for the cable. In the cable, the wavelength and velocity are much less than the velocity of light—only about 25 per cent of the 3×10^5 km/sec. Series resistance and shunt capacitance are the principal contributors to the propagation constant.

The attenuation for a 100-km cable is 66 db, or 7.6 nepers. Thus, if the sending-end voltage is 10, the output voltage is only $10e^{-7.6} = 10 \times 0.5 \times 10^{-3}$ volt $= 5$ mv.

The impedance in this case comes out to $345 - j319$ ohms.

Example 2-3. At high radio frequencies, coaxial transmission lines are customary. At these frequencies, the attenuation constant is usually small compared with the phase constant, and the characteristic impedance is almost the same as that for a lossless line:

$$Z_0 = \sqrt{\frac{L}{C}}$$

For example, consider a common flexible coaxial cable, designated RG8U, at a frequency of 100 Mc. At this frequency,

$R = 0.79$ ohm/m
$L = 0.262$ μh/m
$G = 1.93 \times 10^{-6}$ mho/m
$C = 96.8$ $\mu\mu$f/m

$$\begin{aligned}
\gamma &= \sqrt{(0.79 + j2\pi \times 10^8 \times 0.262 \times 10^{-6})} \\
&\qquad \times \sqrt{(1.93 \times 10^{-6} + j2\pi \times 10^8 \times 0.968 \times 10^{-10})} \\
&= \sqrt{(0.79 + j164)(1.93 \times 10^{-6} + j6.08 \times 10^{-2})} \\
&= j\sqrt{(1.64 \times 6.08)\left(1 - j\frac{0.79}{1.64} \times 10^{-2}\right)\left(1 - j\frac{1.93}{6.08} \times 10^{-4}\right)} \\
&= j3.15\sqrt{(1 - j4.8 \times 10^{-3})(1 - j3.18 \times 10^{-5})} \\
&\approx j3.15(1 - j4.8 \times 10^{-3})^{\frac{1}{2}}
\end{aligned}$$

The binomial theorem states

$$(1 + x)^{\frac{1}{2}} = 1 + \tfrac{1}{2}x + \cdots$$

so

$$\begin{aligned}
\gamma &\approx j3.15(1 - j2.4 \times 10^{-3}) \\
\alpha &= 7.6 \times 10^{-3} \text{ neper/m} = 8.686 \times 7.6 \times 10^{-3} \\
&\qquad = 6.62 \times 10^{-2} \text{ db/m} \\
\beta &= 3.15 \text{ radians/m} \\
\lambda &= \frac{2\pi}{\beta} = 1.99 \text{ m} \\
v_p &= 1.99 \times 10^8 \text{ m/sec} \\
Z_0 &= \sqrt{\frac{0.79 + j164}{1.93 \times 10^{-6} + j6.08 \times 10^{-2}}} \approx \sqrt{\frac{164}{6.08 \times 10^{-2}}} = 52 \text{ ohms}
\end{aligned}$$

Although α is much smaller than β, it is larger than the values for Examples 2-1 and 2-2. If this cable were used for 100 km, like the cable of Example 2-2, the attenuation would be 6,620 db. For the 10-volt input, the output would be 10^{-330} volts.

The velocity in RG8U is about two-thirds that of a wave in space, but it is almost the same as that for a plane wave in a dielectric having the same permittivity as the insulation of the cable.

It can be seen that Z_0 is very closely approximated by $\sqrt{L/C}$.

Example 2-3 was for a "low-loss" line, even though the attenuation in 100 km was phenomenal, for a low-loss line is defined as one for which the attenuation in a *wavelength* is small. This is the same as saying that the attenuation in nepers is much less than the phase shift in radians. When this is true, an approximate formula may be used for the attenuation constant α.

Suppose that the resistance is very much less than the reactance and that the conductance is very much less than the susceptance. Thus,

$$R \ll \omega L \qquad G \ll \omega C \tag{2-30}$$

Equation (2-6) may be written

$$\gamma = \left[(j\omega L)(j\omega C) \left(1 + \frac{R}{j\omega L}\right)\left(1 + \frac{G}{j\omega C}\right)\right]^{\frac{1}{2}}$$
$$= j\omega \sqrt{LC} \left[1 - j\left(\frac{R}{\omega L} + \frac{G}{\omega C}\right) - \frac{RG}{\omega^2 LC}\right]^{\frac{1}{2}}$$

Applying the binomial theorem and dropping higher-order terms, we have

$$\gamma \approx j\omega \sqrt{LC}\left[1 - j\frac{1}{2\omega}\left(\frac{R}{L} + \frac{G}{C}\right)\right]$$

so that

Field equivalent

$$\alpha = \frac{1}{2}\left(R\sqrt{\frac{C}{L}} + G\sqrt{\frac{L}{C}}\right) \qquad \alpha = \frac{1}{2}\left(\sigma\sqrt{\frac{\mu}{\epsilon}}\right) \tag{2-31}$$
$$\beta = \omega\sqrt{LC} \qquad \beta = \omega\sqrt{\mu\epsilon}$$

But, when (2-30) holds, the characteristic impedance is approximately

Field equivalent

$$Z_0 = \sqrt{\frac{L}{C}} \qquad \eta = \sqrt{\frac{\mu}{\epsilon}} \tag{2-32}$$

so (2-31) becomes

Field equivalent

$$\alpha = \frac{1}{2}\left(\frac{R}{Z_0} + GZ_0\right) \qquad \alpha = \tfrac{1}{2}\sigma\eta \tag{2-33}$$

The wavelength and velocity for these low-loss lines are given by simple approximate formulas. Thus,

Field equivalent

$$\lambda = \frac{2\pi}{\beta} = \frac{1}{f\sqrt{LC}} \qquad v_p = \frac{\omega}{\beta} = \frac{1}{\sqrt{LC}} \qquad v_p = \frac{1}{\sqrt{\mu\epsilon}} \tag{2-34}$$

Here f is the frequency in cycles per second ($\omega = 2\pi f$). Examples of the application of these relations are given in the problems.

2-5. Summary

Waves travel down a transmission line by successively charging capacitance, establishing current in inductance, charging more capacitance, establishing more current, and so on to the end of the line. By setting up circuit models for the line (Fig. 2-6), we derived the *telegrapher's equations:*

Field equivalent

$$\frac{\partial v}{\partial x} = -\left(R + L\frac{\partial}{\partial t}\right)i \qquad \nabla \times \mathbf{E} = -\mu\frac{\partial \mathbf{H}}{\partial t} \tag{2-1}$$

$$\frac{\partial i}{\partial x} = -\left(G + C\frac{\partial}{\partial t}\right)v \qquad \nabla \times \mathbf{H} = \left(\sigma + \epsilon\frac{\partial}{\partial t}\right)\mathbf{E} \tag{2-2}$$

These were combined to obtain the *wave equation:*

Field equivalent

$$\frac{\partial^2 v}{\partial x^2} = RGv + (RC + LG)\frac{\partial v}{\partial t} + LC\frac{\partial^2 v}{\partial t^2} \qquad \nabla^2\mathbf{E} = \mu\sigma\frac{\partial \mathbf{E}}{\partial t} + \mu\epsilon\frac{\partial^2 \mathbf{E}}{\partial t^2} \tag{2-3}$$

This is a very general equation. The parallel development for electromagnetic waves in space results in, at most, two of the four terms on the right-hand side of the transmission-line equation. Most acoustic-wave equations neglect all but the last term, and thermal and molecular-diffusion equations retain the middle term only. These equations are, however, more general, in the sense that they use the Laplacian operator (three-dimensional) in place of the second-order space partial derivative on the left-hand side.

The steady-state sinusoidal solution of the wave equation gives

Field equivalent (plane wave)

$$Ve^{j\omega t} = Ae^{j\omega t-\gamma x} + Be^{j\omega t+\gamma x} \qquad E_z e^{j\omega t} = Ae^{j\omega t-\gamma x} + Be^{j\omega t+\gamma x} \tag{2-35}$$

$$Ie^{j\omega t} = \frac{A}{Z_0}e^{j\omega t-\gamma x} - \frac{B}{Z_0}e^{j\omega t+\gamma x} \qquad H_y e^{j\omega t} = \frac{A}{\eta}e^{j\omega t-\gamma x} - \frac{B}{\eta}e^{j\omega t+\gamma x} \tag{2-36}$$

These are modified forms of (2-7) and (2-10). In these equations

Field equivalent

(2-6) $$\gamma = \sqrt{(R + j\omega L)(G + j\omega C)} \qquad \gamma = \sqrt{j\omega\mu(\sigma - j\omega\epsilon)}$$

(2-9) $$Z_0 = \sqrt{\frac{R + j\omega L}{G + j\omega C}} \qquad \eta = \sqrt{\frac{j\omega\mu}{\sigma + j\omega\epsilon}}$$

For an infinite line, only the first terms of (2-35) and (2-36) remain, for the second terms would be infinite at infinite distance even for a finite source. The same effects as for an infinite line are obtained by terminating a finite line in its characteristic impedance Z_0, for in this situation the voltage-to-current ratio is the same as for the infinite line at every point where the line exists. Such a line is said to be properly terminated.

Study of

(2-16) $$\gamma = \alpha + j\beta$$

yields interesting information. Using this split, we may separate phase and amplitude for voltage and current:

Field equivalent (plane wave)

(2-17) $$Ve^{j\omega t} = Ae^{-\alpha x}e^{j(\omega t - \beta x)} \qquad E_y e^{j\omega t} = Ae^{-\alpha x}e^{j(\omega t - \beta x)}$$

(2-18) $$Ie^{j\omega t} = \frac{A}{Z_0}e^{-\alpha x}e^{j(\omega t - \beta x)} \qquad H_z e^{j\omega t} = \frac{Ae^{-\alpha x}}{\eta}e^{j(\omega t - \beta x)}$$

If we are to keep track of a given phase (constant value of the imaginary exponent), x must increase as t does, resulting in a phase velocity

(2-20) $$v_p = \frac{\omega}{\beta}$$

This is the velocity with which an observer would have to travel down the line if he wished to see a constant, direct voltage.

The distance for a phase shift of 2π radians is defined as the wavelength λ, so that

(2-19) $$\lambda = \frac{2\pi}{\beta}$$

Combining (2-19) and (2-20) results in a very important relation to remember:

(2-37) $$f\lambda = v_p$$

This relation is true for all types of wave. It is used more often than (2-19) to determine λ, for v_p is frequently known when β is not.

Attenuation is the logarithmic measure of reduction in voltage, current, and power due to losses in the line. The attenuation in nepers on an infinite line is the quantity αx of (2-17) and (2-18). It is the natural logarithm of the ratio of the sending-end voltage to the voltage at a point on the line. It is frequently more convenient to use a smaller unit based on logarithms to the base 10. This unit is called the decibel and is defined as

$$\text{Attenuation in db} = 10 \log_{10} \frac{P_1}{P_2} \tag{2-27}$$

where P_1 is the power sent and P_2 is the power received.

Approximate values for α, β, λ, v_p, and Z_0 for a line with $\alpha \ll \beta$ (low-loss line) are given by

Field equivalent

$$\text{(2-31)} \quad \alpha = \frac{1}{2}\left(R\sqrt{\frac{C}{L}} + G\sqrt{\frac{L}{C}}\right) \qquad \alpha = \frac{1}{2}\sigma\sqrt{\frac{\mu}{\epsilon}},\ \beta = \omega\sqrt{\mu\epsilon}$$

$$\beta = \omega\sqrt{LC}$$

$$\text{(2-34)} \quad \lambda = \frac{1}{f\sqrt{LC}} \qquad v_p = \frac{1}{\sqrt{LC}} \qquad \lambda = \frac{1}{f\sqrt{\mu\epsilon}},\ v_p = \frac{1}{\sqrt{\mu\epsilon}}$$

$$\text{(2-32)} \quad Z_0 = \sqrt{\frac{L}{C}} \qquad \eta = \sqrt{\frac{\mu}{\epsilon}}$$

Since many transmission-line, space-wave, and acoustic problems satisfy the criterion for them to hold, these approximate formulas are very useful.

In dealing with any traveling-wave problem, the quantities analogous to R, G, L, and C must be determined so that the derived quantities of the preceding paragraph may be calculated. Chapter 3 deals with the derivation of the telegrapher's and wave equations for various nonelectric waves and with the analogies between these transmission-line parameters and parameters of the nonelectric waves. When these analogies have been mastered, the transmission-line technique is very powerful for solution of a wide range of problems.

PROBLEMS

2-1. A properly terminated low-loss radio-frequency transmission line has a velocity of propagation of 2.5×10^8 m/sec. A sinusoidal wave with a frequency of 10^8 cps is impressed on the end of the line. The voltage on the end of the line is maximum when $t = 0$. Sketch the voltage on the line for a distance of 6 m at $t = 0$, 0.125×10^{-8}, 0.25×10^{-8}, 0.375×10^{-8}, and 0.5×10^{-8} sec.

2-2. For the line of Prob. 2-1, sketch for 1.5 cycles the voltage as a function of time at the following distances from the sending end: 0.3125, 0.625, 1.250, and 1.875 m.

2-3. A wave is traveling in the negative x direction on a transmission line, so that the current is given by

$$i = I\cos(\omega t + \beta x)$$

Sketch the current on the line for 1.5 wavelengths at $t = 0$, $\pi/4\omega$, $\pi/2\omega$, and π/ω.

2-4. For the line of Prob. 2-3, sketch for 1.5 cycles the current as a function of time at $x = 0, \pi/4\beta, \pi/2\beta$, and π/β.

2-5. The voltage at the end of the line of Prob. 2-1 is zero prior to $t = 0$. At that time it rises to 1 volt, at which level it remains for 0.25×10^{-8} sec, after which it returns to zero. Sketch the voltages appearing on the line at the times specified in Prob. 2-1.

2-6. For the voltage pulse of Prob. 2-5, sketch the voltage-vs.-time curves at the places specified in Prob. 2-2.

2-7. A transmission line may be considered to have no attenuation. The line may also be considered to be infinitely long, in both directions. A lightning stroke

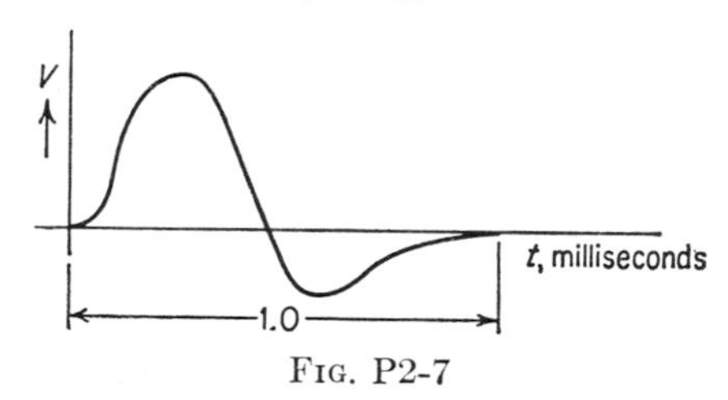

FIG. P2-7

induces the voltage pulse shown at $x = 0$. The velocity of propagation on the line may be considered to be 2×10^8 m/sec. Sketch the voltage on the line at the following times: 0.5, 1.0, 2.0, and 3.0 msec.

2-8. For the lightning waves of Prob. 2-7, sketch the voltage vs. time at $x = 100$ km and $x = -50$ km.

2-9. When a capacitor is discharged into an infinite lossless transmission line, the current at the sending end of the line is given by

$$i = \frac{V_c}{Z_0} e^{-t/Z_0 C}$$

where V_c is the initial voltage across the capacitor and C is its capacitance. For a line whose characteristic impedance is 50 ohms and whose velocity of propagation is 3×10^8 m/sec, sketch the voltage wave due to a 10^{-7}-farad capacitor initially charged to 500 volts. The sketches should show the current vs. time at the end of the line and at a point 1 km away and should also show the current vs. distance at times of 1, 2, and 10 μsec.

2-10. At high frequencies, the inductance, capacitance, and resistance per unit length of a coaxial transmission line are given by[1]

$$L = \frac{\mu}{2\pi} \log_e \frac{b}{a} \qquad \text{henrys/m}$$

$$C = \frac{2\pi\epsilon}{\log_e (b/a)} \qquad \text{farads/m}$$

$$R = \frac{1}{2} \sqrt{\frac{f\mu}{\pi\sigma}} \left(\frac{1}{a} + \frac{1}{b}\right) \qquad \text{ohms/m}$$

Here b is the inner radius of the outer conductor and a is the outer radius of the inner conductor. Assume $\mu = \mu_0 = 4\pi \times 10^{-7}$ henry/m, $\epsilon = \epsilon_0 = 1/36\pi \times 10^{-9}$ farad/m. For copper, $\sigma = 5.8 \times 10^7$ mhos/m. Consider a coaxial line having $b = 2.7$ mm and $a = 1.0$ mm, used at a frequency of 10^{10} cps. Determine α, β, λ, v_p, and Z_0.

[1] Walter C. Johnson, "Transmission Lines and Networks," pp. 87–88, McGraw-Hill Book Company, Inc., New York, 1950.

2-11. The inductance, capacitance, and resistance per unit length on parallel-wire transmission lines may be shown to be[1]

$$L = \frac{\mu}{4\pi}\left(1 + 4 \log_e \frac{D}{r}\right) \qquad \text{henrys/m}$$

$$C = \frac{\pi\epsilon}{\log_e [(D - r)/r]} \qquad \text{farads/m}$$

$$R = 83.2 \times 10^{-9} \frac{\sqrt{f}}{r} \qquad \text{ohms/m for copper}$$

D is the center-to-center spacing of the parallel wires of radius r. For a frequency of 10^6 cps, where $D = 20.2$ cm and $r = 1$ mm, find α, β, v_p, λ, and Z_0. Assume $\mu = \mu_0(4\pi \times 10^{-7})$ and $\epsilon = \epsilon_0[(1/36\pi) \times 10^{-9}]$.

2-12. A standard telephone line, which uses 8-in. spacing for wire with 0.165 in. diameter, has the following characteristics at 1,000 cps. Determine α, β, λ, v_p, and Z_0.

$R = 2.56$ ohms/km
$L = 1.94$ mh/km
$C = 0.00622$ μf/km
$G = 0.068$ μmho/km

2-13. Assuming that the resistance of the line of Prob. 2-12 is independent of frequency, compute and plot the variation of α over the voice-frequency range 300 to 4,000 cps.

2-14. For the same line and frequency range, calculate and plot v_p. If a difference in time delay between the low and high frequencies must not exceed 0.1 period at the highest frequency, what is the maximum transmission distance on this line?

2-15. A coaxial telephone cable of the type used for intercity transmission has the following characteristics at 100 kc. Determine α, β, λ, v_p, and Z_0 for this cable.

$R = 32$ ohms/mile
$L = 0.47$ mh/mile
$C = 0.0773$ μf/mile
$G = 46$ μmhos/mile

2-16. A pair of wires in a telephone cable have the following characteristics. Determine α, β, λ, v_p, and Z_0 for this cable at 1,000 cps.

$R = 6.50$ ohms/km
$L = 0.435$ mh/km
$C = 0.0622$ μf/km
$G = 1.5$ μmhos/km

2-17. Do the computation of Prob. 2-13 for the cable of Prob. 2-16 instead of the line of Prob. 2-12.

2-18. Do the computation of Prob. 2-14 for the cable of Prob. 2-16 instead of the line of Prob. 2-12.

2-19. RG17A/U is a low-attenuation radio-frequency coaxial cable. A table lists the following data on this cable:

Nominal characteristic impedance = 50.0 ohms
Nominal capacitance per foot = 29.5 $\mu\mu$f
Attenuation at 100 mc = 1.0 db/100 ft

Determine its inductance and resistance per unit length, assuming that the low-loss-cable formulas may be used and that G is negligible. Determine its velocity of propagation.

[1] J. J. Karakash, "Transmission Lines and Filter Networks," p. 21, The Macmillan Company, New York, 1950.

2-20. A power line is made up of 900,000-circular mil steel-reinforced aluminum wire. The cable has 54 aluminum and 7 steel conductors, with an over-all diameter of 1.162 in. The parameters of such a line are determined with "geometric mean distances."[1] The equivalent spacing between wires for one phase is 20 ft. The following parameters apply:

R = 0.119 ohm per conductor per mile at 60 cps and 50°C
X_L = 0.757 ohm per conductor per mile inductive reactance at 60 cps
B_C = 5.60 μmhos per mile capacitive susceptance at 60 cps
G = negligible

Find α, β, v_p, λ, and Z_0 at 60 cps. Compare these with comparable values for a communication line.

2-21. If the line of Prob. 2-20 is properly terminated and the power delivered to the load is 50×10^6 watts, find the voltage and current at both ends of a 50-mile line. How much power is lost in the line? Compare this with the value obtained by using the load current and the total series resistance.

2-22. Discuss the field-line analogy in Eq. (2-2).

2-23. Derive an equation similar to (2-3) for current instead of voltage.

2-24. Go through in detail the steps involved in developing Eq. (2-3*a*) from Eq. (2-3).

2-25. Derive the expression for voltage when it is assumed that the current on a line is

$$I = Ce^{-\gamma x} + De^{+\gamma x}$$

2-26. Consider the telephone cable of Example 2-2. Determine the voltage received after the following distances, assuming a 1-volt input: 100, 1,000, and 3,000 km. If such lines are used in transcontinental telephony, how do you suppose such a large ratio of input to output signal is compensated for?

2-27. Show that a plane wave traveling in the x direction and having only a z component of **E** has only a y component of **H**.

2-28. Derive the telegrapher's equations using the model of Fig. 2-6*c*.

2-29. Derive the telegrapher's equations using the model of Fig. 2-6*d*.

2-30. Derive the telegrapher's equations using a model like that of Fig. 2-6*b* but assuming that the shunt conductance and capacitance appear at the left-hand side rather than the right-hand side of the element.

BIBLIOGRAPHY

Hayt, William H., Jr.: "Engineering Electromagnetics," McGraw-Hill Book Company, Inc., New York, 1958.

Johnson, Walter C.: "Transmission Lines and Networks," McGraw-Hill Book Company, Inc., New York, 1950.

Ramo, Simon, and John R. Whinnery: "Fields and Waves in Modern Radio," 2d ed., John Wiley & Sons, Inc., New York, 1953.

Rogers, Walter E.: "Introduction to Electric Fields," McGraw-Hill Book Company, Inc., New York, 1954.

Skilling, Hugh H.: "Fundamentals of Electric Waves," 2d ed., John Wiley & Sons, Inc., New York, 1948.

———: "Electric Transmission Lines," McGraw-Hill Book Company, Inc., New York, 1951.

[1] E. W. Kimbark, "Electrical Transmission of Power and Signals," pp. 61–64, John Wiley & Sons, Inc., New York, 1949.

3. Nonelectromagnetic Wave Equations

The derivations of the telegrapher's equations and the wave equations for various nonelectromagnetic waves are carried out in this chapter. As in Chap. 2, the derivations themselves deal with a single space dimension, but the two- and three-dimensional results are also indicated where appropriate. The greatest emphasis is placed on vibrations in strings, on acoustic waves, and on thermal and diffusion processes.

3-1. Introduction

The derivations in this chapter are all carried out in terms of the nonelectromagnetic equations which correspond directly to the electromagnetic equations of Chap. 2. In each case a set of telegrapher's equations is developed which is comparable with

Field equivalent

(2-1) $$\frac{\partial v}{\partial x} = -\left(R + L\frac{\partial}{\partial t}\right)i \qquad \nabla \times \mathbf{E} = -\mu\frac{\partial \mathbf{H}}{\partial t}$$

(2-2) $$\frac{\partial i}{\partial x} = -\left(G + C\frac{\partial}{\partial t}\right)v \qquad \nabla \times \mathbf{H} = \left(\sigma + \epsilon\frac{\partial}{\partial t}\right)\mathbf{E}$$

In each case these are combined to obtain a wave equation which compares with

Field equivalent

(2-3) $$\frac{\partial^2 v}{\partial x^2} = RGv + (RC + LG)\frac{\partial v}{\partial t} + LC\frac{\partial^2 v}{\partial t^2} \qquad \nabla^2\mathbf{E} = \mu\sigma\frac{\partial \mathbf{E}}{\partial t} + \mu\epsilon\frac{\partial^2 \mathbf{E}}{\partial t^2}$$

(2-5b) $$\frac{\partial^2 V}{\partial x^2} = \gamma^2 V \qquad \nabla^2\mathbf{E} = \gamma^2\mathbf{E}$$

Results comparable to

		Field equivalent
(2-6)	$\gamma = \sqrt{(R + j\omega L)(G + j\omega C)}$	$\gamma = \sqrt{j\omega\mu(\sigma + j\omega\epsilon)}$
(2-16)	$\gamma = \sqrt{(R + j\omega L)(G + j\omega C)} = \alpha + j\beta$	

are obtained in each case where appropriate.

The quantities analogous to R, L, G, and C are determined in each case, and the result is used to obtain the characteristic impedance from relations like

		Field equivalent
(2-9)	$Z_0 = \sqrt{\dfrac{R + j\omega L}{G + j\omega C}}$	$\eta = \sqrt{\dfrac{j\omega\mu}{\sigma + j\omega\epsilon}}$
(2-15)	$\dfrac{V}{I} = Z_0$	

The wavelength and velocity are also important and are obtained from

$$\lambda = \frac{2\pi}{\beta} \qquad T = \frac{2\pi}{\omega} \tag{2-19}$$

$$v_p = \frac{dx}{dt} = \frac{\omega}{\beta} \tag{2-20}$$

Acoustic waves are low-loss waves like those discussed at the end of Sec. 2-4, and approximate values for the equations listed above are used for acoustic waves. Since acoustic waves are treated as lossless, R and G are missing in Eqs. (2-1) to (2-3) in this analogy.

3-2. Transverse Waves on Membranes and Strings

Historically, the first wave equation derived was for a vibrating string in which the excursion is small compared to length. Interestingly enough, the wavelength, velocity of propagation, and characteristic impedance are independent of any property of the string itself, other than its mass per unit length, but the tension applied to the string by some external agency is important.

Waves on strings are well known, since we are accustomed to both violin strings and piano strings. Less familiar are the mechanical waves which travel along conductors on transmission lines when they are excited by the wind or by something touching one of the lines. Traveling waves of the type found on strings are also found in suspension bridges as a load moves across the bridge.

The equivalent of the infinite transmission line is seldom found with waves on strings. Frequently, the "terminating impedance" for a string fixes the position of the ends at zero, which corresponds to the short circuit on a transmission line. This section consists essentially of the derivation, with no discussion of the significant practical results due to reflection from "imperfect" terminations.

Consider Fig. 3-1, which shows a length Δx of a string displaced a short distance w from its undisturbed position. The length Δx is located a distance x from the origin of coordinates. There is a tension T in the string, and the mass per unit length is given by ρ_L. The string is considered to have *no stiffness;* that is, it is perfectly "limp."

To obtain the first "telegrapher's equation" let us apply Newton's second law (familiar to many as $f = ma$). At the position x, there is a

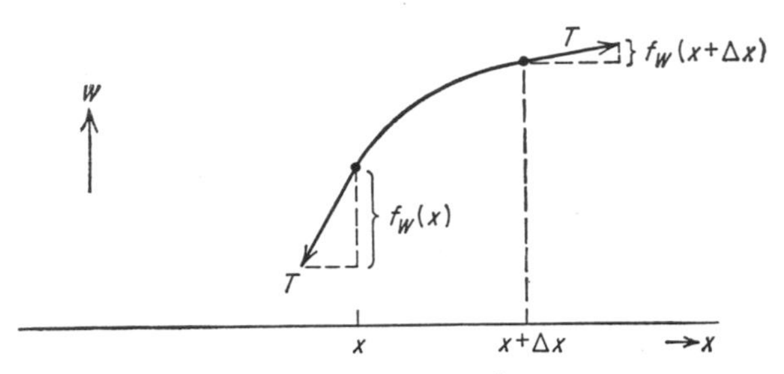

FIG. 3-1. String forces and geometry.

downward force, corresponding to the downward component of the tension. We measure force in the same direction as w; hence, this is a negative force and is given by

$$f_w(x) = -T\frac{\partial w}{\partial x}(x)$$

where w is the displacement and T is the tension. At the other end there is an upward force, which is the vertical component of the tension at that point and is given by

$$f_w(x + \Delta x) = T\frac{\partial w}{\partial x}(x + \Delta x) \approx T\frac{\partial w}{\partial x}(x) + T\frac{\partial}{\partial x}\left(\frac{\partial w}{\partial x}\right)\Delta x$$

This equation uses the first two terms of a Taylor expansion of $\partial w/\partial x$ about x. Note that curvature of the string within the distance Δx makes the slope at one end different from that at the other, and it is this which causes the vertical component of the tension to be different at the two ends of the string. Applying D'Alembert's principle, we have

$$f_w(x + \Delta x) - f_w(x) = \frac{\partial}{\partial t}(mu_w) = \rho_L\,\Delta x\,\frac{\partial u_w}{\partial t}$$

where m is the mass in Δx, ρ_L is the mass per unit length, and u_w is the vertical velocity of Δx. Here we have used the fact that the mass in a length Δx is given by the mass per unit length times that length. We may divide both sides by Δx and pass to the limit as x approaches 0, in which case

Transmission-line analogy[1]

$$\frac{\partial f_w}{\partial x} = \rho_L \frac{\partial u_w}{\partial t} \qquad \frac{\partial i}{\partial x} = -C \frac{\partial v}{\partial t} \tag{3-1}$$

To obtain the other telegrapher's equation, observe that

$$\frac{\partial w}{\partial x} = \frac{f_w}{T}$$

is the equation defining the vertical component of force. Differentiating both sides with respect to time, this becomes

$$\frac{\partial}{\partial t}\left(\frac{\partial w}{\partial x}\right) = \frac{1}{T}\frac{\partial f_w}{\partial t}$$

Interchanging the order of differentiation on the left-hand side and noting that

$$u_w = \frac{\partial w}{\partial t}$$

we may write

Transmission-line analogy

$$\frac{\partial u_w}{\partial x} = \frac{1}{T}\frac{\partial f_w}{\partial t} \qquad \frac{\partial v}{\partial x} = -L\frac{\partial i}{\partial t} \tag{3-2}$$

It can be seen that (3-1) and (3-2) are comparable with the telegrapher's equations.

It is important to realize that the telegrapher's equations for the string are approximations. It was assumed that the motion of each section of string was only transverse, whereas it actually contains both transverse and longitudinal components. It was also assumed that the tension was the same at all points on the string, but for strings under the influence of gravity this is not, in general, true. Elastic effects were neglected, and it was assumed that no energy was lost in the vibrating process; in fact, energy is lost internal to the string, and if the string is in air, energy is lost as acoustic radiation. These approximations are quite satisfactory for small displacements and most applications. With large displacements, however, effects of gravity, longitudinal motion, and at times elastic effects must be considered. For many purposes energy loss is unimportant, but if it did not exist, a piano note would last forever—of course, we should not hear it if no energy were lost to the acoustic wave.

[1] Line analogies are indicated where considered appropriate, without further explanation.

TABLE 3-1. ANALOGIES BETWEEN VIBRATING STRINGS AND TRANSMISSION LINES

Electrical quantity	*Mechanical quantity*
i	f_w
v	u_w
C	$-\rho_L$
L	$-1/T$

Before going further, it is appropriate to list the analogous electrical and mechanical quantities; this is done in Table 3-1. As treated here, the vibrating-string problem corresponds to that of the lossless transmission line; so there are no quantities corresponding to R and G, and the lossless forms of the telegrapher's equations are analogous to the vibrating-string equations. A comparison of (3-1) and (3-2) with (2-1) and (2-2) shows the analogies listed in the table. The telegrapher's equations for the string may be combined to give the wave equation for either force or velocity, as (2-1) and (2-2) were combined to get (2-3). The results are

Transmission-line analogy

$$\frac{\partial^2 f_w}{\partial x^2} = \frac{\rho_L}{T}\frac{\partial^2 f_w}{\partial t^2} \qquad \frac{\partial^2 i}{\partial x^2} = LC\frac{\partial^2 i}{\partial t^2} \tag{3-3}$$

$$\frac{\partial^2 u_w}{\partial x^2} = \frac{\rho_L}{T}\frac{\partial^2 u_w}{\partial t^2} \qquad \frac{\partial^2 v}{\partial x^2} = LC\frac{\partial^2 v}{\partial t^2} \tag{3-4}$$

The solutions of these equations are directly analogous to the solutions of the corresponding electrical equations, so that we may write

$$U_w = Ae^{-\gamma x} + Be^{\gamma x} \tag{3-5}$$

$$F_w = \frac{A}{Z_0}e^{-\gamma x} - \frac{B}{Z_0}e^{\gamma x} \tag{3-6}$$

where we have assumed $u_w = \text{Re}\, U_w e^{j\omega t}$, $f_w = \text{Re}\, F_w e^{j\omega t}$.

Here the propagation constant γ is given by

$$\gamma = j\beta = j\omega\sqrt{\frac{\rho_L}{T}} \tag{3-7}$$

with the result that the velocity of propagation is

$$v_p = \frac{\omega}{\beta} = \sqrt{\frac{T}{\rho_L}} \tag{3-8}$$

and the wavelength for a given frequency f is

$$\lambda = \frac{2\pi}{\beta} = \frac{1}{f}\sqrt{\frac{T}{\rho_L}} \tag{3-9}$$

The characteristic impedance, of course, is the ratio of the velocity to the force when B is equal to zero in (3-5) and (3-6), so it is given by

$$Z_0 = \frac{U_w}{F_w} = \frac{1}{\sqrt{\rho_L T}} \tag{3-10}$$

The quantities enumerated in (3-7) to (3-10) were all obtained by direct substitution of analogous quantities from Table 3-1 into (2-19), (2-20), (2-31), and (2-32).

A word about the units is in order. Velocity is in meters per second, provided that tension is in newtons and mass per unit length is in kilograms per meter. It can be seen that the impedance must have units of velocity divided by force, or meters per newton-second. Of course, cgs units or English units may be used, and, in fact, English units are used in the example below.

One would expect that it would take a larger force to get a given velocity for a larger tension or mass, so the quantities in the impedance expression seem reasonable intuitively. It is, perhaps, less apparent that the velocity relation should take the form that it does, but some reflection will show that it is indeed in accord with our experience with vibrating strings.

In an actual vibrating string some energy is lost because of internal friction, so the lossless equations used here are only approximate.

Example 3-1. This is an example of a vibrating string. The standard of musical pitch is the A above middle C on a piano (440 cps). In our example the tension is 400 lb, and the string has a weight of 4 lb/1,000 ft. We should find the velocity of propagation, the wavelength, and the characteristic impedance.

Use of pounds for force requires the use of slugs for mass. Thus the mass per 1,000 ft is

$$\text{Mass in slugs/1,000 ft} = \frac{4\ \text{lb/1,000 ft}}{32.2\ \text{ft/sec}^2} = 0.124\ \text{slug/1,000 ft}$$

Using (3-8), we find for the velocity of propagation

$$v_p = \sqrt{\frac{400}{0.124 \times 10^{-3}}} = \sqrt{3.2 \times 10^6} = 1{,}790\ \text{ft/sec}$$

The wavelength is given by

$$\lambda = \frac{1}{f}\sqrt{\frac{T}{\rho_L}} \tag{3-9}$$

which may be expressed more conveniently as

$$\lambda = \frac{v_p}{f} = \frac{17.90}{4.4} = 4.08\ \text{ft}$$

The characteristic impedance is given by (3-10) and is

$$Z_0 = \frac{1}{\sqrt{(0.126 \times 10^{-3})(4 \times 10^2)}} = 4.47 \text{ ft/lb-sec}$$

To see the units involved, note that 400 lb corresponds to 1,780 newtons and that 0.126×10^{-3} slug/foot corresponds to 1.81×10^{-3} kg/ft, or 5.9 g/m.

Example 3-2. An aluminum-steel power cable for a heavy transmission line is 1.45 in. in diameter and has a weight of 2.04 lb/ft. On a given line this is under a tension of 2,000 lb, which allows it to sag a total of 20 ft between towers spaced 400 ft apart. Find the velocity and wavelength as well as the characteristic impedance.

$$v_p = \sqrt{\frac{T}{\rho_L}} = \sqrt{\frac{2 \times 10^3}{2.04/32.2}} = 178 \text{ ft/sec}$$

If this cable is oscillating with a period of 10 sec, its wavelength is 1,780 ft. The characteristic impedance is

$$Z_0 = \sqrt{\frac{32.2}{2.04} \times \frac{1}{2 \times 10^3}} = 0.082 \text{ ft/lb-sec}$$

In metric units,

$$\begin{aligned}
\rho_L &= 2.04 \text{ lb/ft} \times 0.454 \text{ kg/lb} \times 3.28 \text{ ft/m} \\
&= 3.04 \text{ kg/m} \\
T &= 2{,}000 \text{ lb} \times 4.45 \text{ newtons/lb} = 8{,}900 \text{ newtons} \\
v_p &= \sqrt{\frac{8{,}900}{3.03}} = 54 \text{ m/sec} \\
Z_0 &= \frac{1}{\sqrt{3.04 \times 8.9 \times 10^3}} = 0.006 \text{ m/newton-sec}
\end{aligned}$$

Vibrations of a membrane, such as a drumhead or the skin of an airplane wing, are governed by an equation in two space dimensions which compares to the one-dimensional vibrating-string equation. This equation is derived by Rayleigh.[1] The wave equation which results is

$$\nabla^2 u_w = \frac{1}{v_p^2} \frac{\partial^2 u_w}{\partial t^2} \tag{3-11}$$

(Note that the Laplacian operator here can only be two-dimensional, since we are talking about vibration of a surface.) This equation can be derived by a method similar to that for the one-space-dimensional equation. The tension in the string is replaced by a tension per unit width

[1] Lord Rayleigh, "Theory of Sound," vol. I, p. 306, Dover Publications, New York, 1945.

on the sheet and the mass per unit length by a mass per unit area. Two-dimensional first-order equations comparable with the telegrapher's equations are derived and combined to give (3-11). The corresponding expressions for velocity of propagation and impedance are given by

$$v_p = \sqrt{\frac{T_1}{\rho_A}} \qquad Z_0 = \frac{1}{\sqrt{\rho_A T_1}} \tag{3-12}$$

where ρ_A = mass per unit area
T_1 = tension per unit width

3-3. Acoustic Waves in Fluids

Acoustic, or sound, waves may travel through gases, liquids, or solids. This section deals with their progress through gases and liquids.

These sound waves are the first nonelectromagnetic *plane waves* discussed in this book. Later, mechanical waves in solids, temperature waves, and diffusion processes are also discussed from the standpoint of plane waves.

Of the two similar derivations for sound waves in a fluid, that for gas is treated first. Because of the relatively high speed and small distance traveled by individual particles in a sound wave, there is insufficient time between the compression and expansion portions of the cycle for heat to diffuse throughout the volume, and the process is essentially adiabatic.[1] For an adiabatic process, pressure and volume are related by

$$p_0 V_0^{\gamma_g} = \text{constant}$$

where p_0 is pressure associated with the volume V_0, V_0 represents a specific number of gas molecules, and γ_g is the ratio of specific heats at constant pressure and constant volume. Differentiating this equation to see the effects of changing pressure on the volume, and vice versa, we find

$$\frac{dp_0}{p_0} = -\gamma_g \frac{dV_0}{V_0} \tag{3-13}$$

Consider now the small rectangular volume V_0 shown in Fig. 3-2. It is oriented so that two of its faces are parallel to the YZ plane (and therefore represent constant amplitude-phase planes of the plane wave). These faces have area ΔA, and the distance between them is Δx. Hence, the volume is

$$V_0 = \Delta x \, \Delta A$$

[1] No heat is transferred to or from a small volume of the gas. See, for example, Erich Hausman and Edgar P. Slack, "Physics," 2d ed., pp. 285–287, D. Van Nostrand Company, Inc., Princeton, N.J., 1939.

This small volume contains a particular number of molecules, and as the gas expands, these molecules take up more space. For a plane wave, this expansion is in only one direction, the x direction. Hence, in a given short time Δt, the expansion in volume is ΔA times the difference in distance traveled by the right-hand and left-hand faces. In terms of the velocity u_x in the x direction, this is

$$\Delta V_0 = [u_x(x + \Delta x)\,\Delta t - u_x(x)\,\Delta t]\,\Delta A$$

But (3-13) states that

$$\Delta V_0 = -\frac{V_0}{\gamma_g}\frac{\Delta p_0}{p_0} = -\frac{\Delta x\,\Delta A\,\Delta p_0}{\gamma_g p_0}$$

so, after some rearrangement, and cancellation of ΔA,

$$\frac{u_x(x + \Delta x) - u_x(x)}{\Delta x} = -\frac{1}{\gamma_g p_0}\frac{\Delta p_0}{\Delta t}$$

Passing to the limit as $\Delta x \to 0$ and $\Delta t \to 0$,

$$\frac{\partial u_x}{\partial x} = -\frac{1}{\gamma_g p_0}\frac{\partial p_0}{\partial t}$$

In dealing with acoustic waves, it is customary to express the *wave* in terms of the difference (time-varying) from the ambient pressure (fixed); that is,

$$\text{Total pressure} = \text{ambient pressure} + p$$

In these terms, the above equation is

Transmission-line analogy

$$\text{(3-14)} \qquad \frac{\partial u_x}{\partial x} = -\frac{1}{\gamma_g p_a}\frac{\partial p}{\partial t} \qquad\qquad \frac{\partial v}{\partial x} = -L\frac{\partial i}{\partial t}$$

where p_a, the ambient pressure, is essentially the same as p_0. This is one of the telegrapher's equations for a sound wave, as indicated by the line analogy.

To get the other telegrapher's equation, we apply D'Alembert's principle to the gas in V_0. If the pressure differential is considered positive in the positive x direction, the forces acting on the body are as shown in Fig. 3-2. The force acting on the left-hand face tends to move the volume to the right. The force acting on the right-hand face is due to the pressure of the adjacent gas and opposes the motion. Because there is no pressure variation in the y and z directions, the forces acting in those directions are equal and opposite, and no motion occurs in those directions.

We may therefore write

$$[p(x) - p(x + \Delta x)]\,\Delta A = \rho_v V_0 \frac{\partial^2 x'}{\partial t^2} = \rho_v\,\Delta A\,\Delta x' \frac{\partial^2 x'}{\partial t^2}$$

Here we use x' to represent the coordinates of the left-hand side of the volume as it moves and x to represent its rest position. The mass per unit volume is given by ρ_v. Dividing both sides by Δx, we obtain

$$\frac{p(x + \Delta x) - p(x)}{\Delta x} = -\rho_v \frac{\partial^2 x'}{\partial t^2}$$

For this purpose, Δx and $\Delta x'$ are essentially the same. The second derivative of the coordinate of the box, however, is just the first derivative of its velocity, which is, in fact, the velocity of the individual

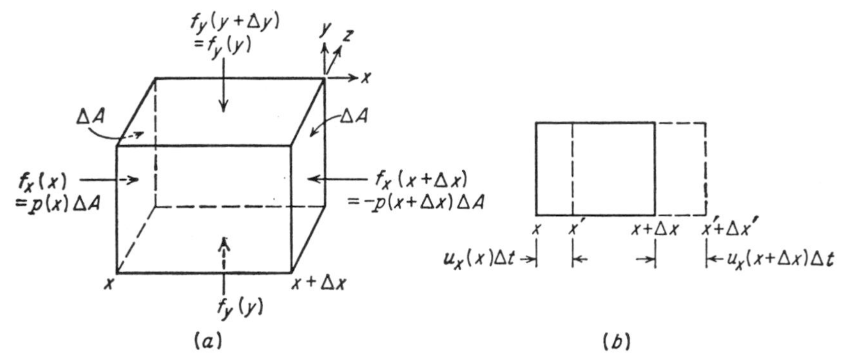

FIG. 3-2. Differential volume in a fluid: (*a*) forces in a cube; (*b*) motion of a cube in a plane wave.

particles in the sound waves. Therefore, we may finally write the equation analogous to the second telegrapher's equation as

Transmission-line analogy

$$\text{(3-15)} \qquad \frac{\partial p}{\partial x} = -\rho_v \frac{\partial u_x}{\partial t} \qquad \frac{\partial i}{\partial x} = -C \frac{\partial v}{\partial t}$$

It can be seen from (3-14) and (3-15) that the analogies between electrical and acoustic quantities are as given in Table 3-2.

TABLE 3-2. ACOUSTICAL-ELECTRICAL ANALOGIES FOR SOUND WAVES IN A GAS

Electrical quantities	*Acoustical quantities*
i	p
v	u_x
C	ρ_v
L	$1/\gamma_g p_a$

We may combine the telegrapher's equations in the usual manner, obtaining the following wave equations:

Transmission-line analogy

$$\text{(3-16)} \qquad \frac{\partial^2 p}{\partial x^2} = \frac{\rho_v}{\gamma_g p_a} \frac{\partial^2 p}{\partial t^2} \qquad \frac{\partial^2 i}{\partial x^2} = LC \frac{\partial^2 i}{\partial t^2}$$

$$\text{(3-17)} \qquad \frac{\partial^2 u_x}{\partial x^2} = \frac{\rho_v}{\gamma_g p_a} \frac{\partial^2 u_x}{\partial t^2} \qquad \frac{\partial^2 v}{\partial x^2} = LC \frac{\partial^2 v}{\partial t^2}$$

The three-dimensional acoustic equation is easily obtained from the one-dimensional equation (3-16) by assigning a unit vector to each side of this equation and developing similar equations for plane waves going in the y and z directions. The three equations so obtained may then be added together to obtain

Electromagnetic field analogy

(3-18) $$\nabla^2 p = \frac{\rho_v}{\gamma_g p_a} \frac{\partial^2 p}{\partial t^2} \qquad \nabla^2 \mathbf{H} = \mu\epsilon \frac{\partial^2 \mathbf{H}}{\partial t^2}$$

The phase velocity for these waves is, of course, given by

(3-19) $$v_p = \sqrt{\frac{\gamma_g p_a}{\rho_v}}$$

The wavelength is

(3-20) $$\lambda = \frac{1}{f}\sqrt{\frac{\gamma_g p_a}{\rho_v}}$$

and the characteristic impedance, the ratio of velocity to pressure, is

(3-21) $$Z_0 = \frac{u}{p} = \frac{1}{\sqrt{\gamma_g p_a \rho_v}} = \frac{U}{P}$$

where $u = \text{Re } Ue^{j\omega t}$, $p = \text{Re } Pe^{j\omega t}$. For lossless waves, Z_0 may be taken as the ratio of either instantaneous or phasor quantities. It was defined in Chap. 2, however, in terms of the phasors.

Frequently these basic quantities are less well known than the velocity of propagation, and it is desirable to obtain the impedance in terms of the velocity of propagation and the density. Combining (3-19) and (3-21)' we obtain

(3-22) $$Z_0 = \frac{1}{\rho_v v_p}$$

Note that the units for the characteristic impedance may be expressed as square meter-seconds per kilogram or as cubic meters per newton-second.

The derivation for acoustic waves in liquid is very similar, except that for the liquid the compressibility is expressed by

(3-23) $$\frac{dV}{V_0} = -K\,dp_0 = -Kp$$

instead of by (3-13). Here the quantity K is known as the compressibility and is expressed in units of meter-(second)^2 per kilogram or in terms of square meters per newton. The value of K for water at ordinary temperatures and pressures is 4.78×10^{-10}.†

† "Reference Data for Radio Engineers," 4th ed., p. 857, International Telephone and Telegraph Corporation, New York, 1956. For a complete discussion of this subject, see Dwight E. Gray (ed.), "American Institute of Physics Handbook," sec. 2n, pp. 2-136–2-164, McGraw-Hill Book Company, Inc., New York, 1957.

Figure 3-2 is applicable to the liquid as well as to the gas derivation. The D'Alembert equation (3-15) is the same for liquid as for gas. Equation (3-14) does not apply to liquid, however, for we must use the expression

$$\Delta V_0 = -KV_0 p$$

whence, after differentiating with respect to time, we find

$$\frac{\partial u_x}{\partial x} = -K \frac{\partial p}{\partial t} \tag{3-24}$$

Hence, the analogies of Table 3-3 apply. As before, we combine

TABLE 3-3. ACOUSTICAL-ELECTRICAL ANALOGIES FOR SOUND WAVES IN LIQUIDS

Electrical quantities	*Acoustical quantities*
i	p
v	u_x
C	ρ_v
L	K

(3-24) and (3-15) to obtain a wave equation for both pressure and velocity. The pressure equation is

$$\frac{\partial^2 p}{\partial x^2} = \rho_v K \frac{\partial^2 p}{\partial t^2} \tag{3-25}$$

For nonplane waves, the resulting equation is

$$\nabla^2 p = \rho_v K \frac{\partial^2 p}{\partial t^2} \tag{3-26}$$

By analogy to the previous discussions, we can get the velocity and characteristic impedance for the acoustic waves in liquid:

$$v_p = \frac{1}{\sqrt{\rho_v K}} \qquad Z_0 = \frac{u}{p} = \sqrt{\frac{K}{\rho_v}} = \frac{U}{P} \tag{3-27}$$

In terms of the velocity, the characteristic impedance is again given by

$$Z_0 = \frac{1}{\rho_v v_p} \tag{3-22}$$

The units of the impedance in this case are, of course, the same as for the sound wave in air, because the impedance in a liquid is also the ratio of the velocity to the pressure.

It should be noted that sound waves are *longitudinal*, whereas vibrating strings and membranes are *transverse*. That is, the particle motion in a plane sound wave is in the direction of wave travel, whereas that on a string is at right angles to the direction of travel. Plane electromagnetic waves are transverse in the sense that **E** and **H** are normal to the direction of travel. Electron motion due to **E** would also be transverse.

Example 3-3. *Sound in Air.* To find the appropriate parameters for a sound wave in air at standard temperature and pressure, we need merely substitute in (3-19) and (3-21). Under standard conditions (at 0°C), the following apply:[1]

$$\gamma_g = 1.40$$
$$p_a = 10^5 \text{ newtons/m}^2 \text{ (14.7 lb/in.}^2\text{)}$$
$$\rho_v = 1.29 \text{ kg/m}^3$$

Using these values, we obtain

$$v_p = \sqrt{\frac{1.40 \times 10^5}{1.29}} = 330 \text{ m/sec}$$

The characteristic impedance is obtained most easily from (3-22) as

$$Z_0 = \frac{1}{\rho_v v_p} = \frac{1}{1.29 \times 330} = 0.00235 \text{ (m/sec)/(newtons/m}^2\text{)}$$
$$= 0.00235 \text{ m}^3\text{/newton-sec}$$

The velocity is a function of temperature, and the value given here for the standard conditions is modified for other temperatures.

Example 3-4. If the rms sound pressure is 1 dyne/cm², find the particle velocity and the wavelength at 1,000 cps.

$$1 \text{ dyne} = 10^{-5} \text{ newton}$$
$$1 \text{ cm}^2 = 10^{-4} \text{ m}^2$$
$$P = \frac{10^{-5}}{10^{-4}} = 0.10 \text{ newton/m}^2$$
$$U = PZ_0 = 0.10 \times 0.00235 = 2.35 \times 10^{-4} \text{ m/sec}$$
$$= 0.0235 \text{ cm/sec}$$
$$\lambda = \frac{v_p}{f} = \frac{330}{10^3} = 0.33 \text{ m} = 33 \text{ cm}$$

The sound pressure of 1 dyne/cm² given in this example is about that corresponding to average conversation. The *threshold of pain* is about 650 dynes/cm², and the particle velocity in this case is 15.5 cm/sec. Calculation of particle displacement is left for a problem.

Example 3-5. *Sound in Water.* The characteristic impedance and phase velocity are calculated from (3-27), using

$$K = 4.78 \times 10^{-10}$$
$$\rho_v = 1 \text{ g/cm}^3 = 10^3 \text{ kg/m}^3$$

[1] From table in Appendix C, which summarizes the physical properties of materials, including acoustic properties of gases.

Hence,

$$v_p = \frac{1}{\sqrt{4.78 \times 10^{-10} \times 10^3}} = 1{,}450 \text{ m/sec}$$

$$Z_0 = \frac{1}{\rho_v v_p} = \frac{1}{10^3 \times 1.45 \times 10^3} = 6.9 \times 10^{-7} \text{ m}^3\text{/newton-sec}$$

For a sound pressure of 1 dyne/cm², as above, the very low impedance due to the near incompressibility of water gives a much lower particle velocity of 6.9×10^{-8} m/sec and a 1-kc wavelength of 1.45 m. Note that the impedance is much less in the liquid than in the air but that the velocity is only greater by a factor of about 4. In ultrasonic cleaning devices, pressures as high as 10^6 dynes/cm² are used at 30 kc. The resulting particle velocity of 0.069 m/sec is of the same order of magnitude as the velocity in air for Example 3-4, but the pressure is a great deal higher.

3-4. Mechanical Waves in Solids

A number of types of mechanical waves which propagate through solids can be likened to transmission-line waves. A wave in which the particles vibrate in the direction of travel (longitudinal wave) is directly analogous to an acoustic wave in a liquid or gas. Transverse, or shear, waves, in which the motion of particles is normal to the direction of wave travel, also occur in solids. A special form of shear waves is of particular interest—the torsional wave associated with the rotation of a solid. In a torsional wave, the direction of particle motion is transverse to the axis of rotation, and the wave travels in the direction of the axis of rotation. A fourth type of mechanical wave in a solid is the bulk transverse vibration of a finite solid, such as the oscillation of a beam under a varying load. This is a more complicated version of the type of wave on a nonrigid vibrating string.

A surface-connected wave is also possible in solids. This is known as the Rayleigh wave.

The longitudinal wave is treated in some detail here. The others are treated in less detail, and the Rayleigh wave not at all.

The following discussion deals with waves in homogeneous, isotropic media, that is, with waves through solids whose properties do not vary with position and whose elastic properties are the same for stresses in all directions. Many relatively pure solids, like metal bars and cylinder walls, satisfy these conditions fairly well. However, mechanical waves are also important in the earth, in connection with seismology (the study of earthquakes and the use of artificial earthquakes for prospecting for minerals and oil). In the earth the elastic constants usually vary with position, since the earth's constitution is not uniform; and they also may be different in different directions at a given point, because of the crystalline structure of some rocks. Therefore, one should realize that the type of treatment given here is strictly an approximation.

It is assumed that the stress-strain relation is *linear*. This is a good approximation for materials such as steel for stresses below the yield point. For some more ductile materials the approximation is valid only over a smaller range of stresses.

The derivation for longitudinal plane waves in solids is much the same as that for acoustic waves in fluids. As with fluids, we consider what happens to a differential volume element with faces oriented normal and parallel to the direction of wave travel. Figure 3-2, therefore, applies to this situation.

Application of D'Alembert's principle is the same as before, so that Eq. (3-15) applies, but since the stresses in a solid may be different in different directions, p is replaced by p_x, a component of the stress vector.

Transmission-line analogy

(3-15*a*) $$\frac{\partial p_x}{\partial x} = -\rho_v \frac{\partial u_x}{\partial t} \qquad\qquad \frac{\partial i}{\partial x} = -C \frac{\partial v}{\partial t}$$

The compressibility of a solid is expressed in terms of Hooke's law, which states that strain (the stretching of a solid) is directly proportional to stress (the stretching pressure—or tension—applied). This law applies reasonably well for most solids for relatively small stresses or changes in stress. However, in every case there is a saturation point beyond which the linear relation expressed by Hooke's law does not apply. Mathematically, we can state Hooke's law for stretching as

$$\frac{\text{Tension}}{\text{Unit area}} = Y_B \frac{\partial x'}{\partial x}$$

The partial derivative listed here is an expression for the strain in terms of the quantities of Fig. 3-2. Y_B is the *bulk modulus of elasticity*, or *bulk Young's modulus*.

Since we normally refer to compression rather than tension in our discussion of acoustic waves, we utilize

$$\frac{\text{Tension}}{\text{Unit area}} = -p_x$$

Hence

(3-28) $$p_x = -Y_B \frac{\partial x'}{\partial x}$$

As usual, we are concerned with velocities rather than displacements, so that we may differentiate both sides of this equation and rearrange it to obtain the second telegrapher's equation for the longitudinal wave in a solid:

Transmission-line analogy

(3-29) $$\frac{\partial u_x}{\partial x} = -\frac{1}{Y_B}\frac{\partial p_x}{\partial t} \qquad \frac{\partial v}{\partial x} = -L\frac{\partial i}{\partial t}$$

It is apparent from (3-15) and (3-29) that the relations of Table 3-4 may be used for determining the electromechanical analogies.

TABLE 3-4. ELECTRICAL-ACOUSTICAL ANALOGS IN SOLIDS

Electrical quantities	*Mechanical quantities*
i	p_x
v	u_x
C	ρ_v
L	$1/Y_B$

As usual, we may combine the two telegrapher's equations to get a wave equation in either p or u. Only the former is written.

(3-30) $$\frac{\partial^2 p_x}{\partial x^2} = \frac{\rho_v}{Y_B}\frac{\partial^2 p_x}{\partial t^2}$$

As shown here, this is a lossless wave equation, although, of course, some losses must be considered in any practical problem.

As usual, we may write for the propagation constant in such a case

$$\gamma = j\beta = j\omega\sqrt{\frac{\rho_v}{Y_B}}$$

From this and other analogies we derive expressions for the velocity and the impedance:

(3-31) $$v_p = \frac{\omega}{\beta} = \sqrt{\frac{Y_B}{\rho_v}} \qquad Z_0 = \frac{u_x}{p_x} = \frac{1}{\sqrt{\rho_v Y_B}} = \frac{U_x}{P_x}$$

where $u_x = \text{Re}\, U_x e^{j\omega t}$ and $p_x = \text{Re}\, P_x e^{j\omega t}$.†

Adding comparable equations for the y and z components, the three-dimensional form for this wave equation is

(3-32) $$\nabla^2 \mathbf{p} = \frac{\rho_v}{Y_B}\frac{\partial^2 \mathbf{p}}{\partial t^2}$$

Of course, three-dimensional forms could have been derived for the telegrapher's equations as well.

The quantity Y_B used in (3-28) is known as the bulk Young's modulus and applies to plane-wave problems. Frequently, however, Young's

† See comment after Eq. (3-21).

modulus is quoted for an equation, like (3-28) or (3-29), that applies for tension or compression of a thin cylinder, rather than for a plane wave. The modulus for such rods is called Y_0, and it differs from the bulk modulus. The reason for this is that stretching a thin cylinder makes it thinner and elongated and compressing a thin cylinder makes it thicker. Because of this flow of matter in a direction other than longitudinal, there is more stretching for a given tension, so Y_0 is always less than Y_B. For most stiff media

$$Y_0 \approx 0.75 Y_B\dagger$$

A similar derivation may be carried out for shear waves. In this case, the stress **p** is tangential to the end surfaces of the block of Fig. 3-2, rather than normal to the ends as for longitudinal waves. For such waves, (3-28) is replaced by

$$p_y = -\mu \frac{\partial y'}{\partial x}$$

where it has been assumed that the motion of the particles and the shear tension is in the y direction, whereas the wave motion continues to be in the x direction. Shear waves therefore exhibit the phenomenon of polarization, as do electromagnetic waves in space. The quantity μ is known as the stiffness constant or shear modulus. The wave equation of (3-30) is correct for the shear wave, if we replace Y_B with μ and consider p_y instead of p_x. The resulting velocity and impedance are given by

$$v_p = \sqrt{\frac{\mu}{\rho_v}} \qquad Z_0 = \frac{1}{\sqrt{\rho_v \mu}} = \frac{u_y}{p_y} = \frac{U_y}{P_y} \tag{3-33}$$

where $u_y = \text{Re } U_y e^{j\omega t}$, $p_y = \text{Re } P_y e^{j\omega t}$. In general, μ is considerably less than Y_B, and the propagation velocity of shear waves is correspondingly smaller.

Wave equations for torque and angular velocity for torsional waves may be derived which are comparable with those for longitudinal waves. The angular velocity equation is given by

$$\frac{\partial^2 \omega}{\partial x^2} = \frac{\rho_v}{C} \frac{\partial^2 \omega}{\partial t^2} \tag{3-34}$$

where ω is the angular velocity and C is a quantity called the torsional rigidity, which is related to Young's modulus, in the case of a thin rod, by

$$C = \frac{Y_0}{2(\sigma + 1)}$$

† Theodore F. Hueter and Richard H. Bolt, "Sonics," pp. 25–27, John Wiley & Sons, Inc., New York, 1955.

Here σ, known as the Poisson ratio, is the ratio of the change in diameter to a given change in length which causes the diameter shrinkage. For hard solids Poisson's ratio is on the order of a quarter to a third, so C varies from $0.4Y_0$ to $3Y_0/8$.†

The transverse vibrations of a rigid solid are governed by a different equation than are those of a "limp" string. In fact, this equation is different from the standard wave equation, because it has a *fourth derivative* with respect to x, though similar solutions are possible in terms of sinusoids.[1] For a cylindrical rod this equation is given by

$$\frac{\partial^4 y}{\partial x^4} = -\frac{4}{r^2}\sqrt{\frac{\rho_v}{Y_B}}\frac{\partial^2 y}{\partial t^2} \tag{3-35}$$

Here r is the radius, y is the displacement of a point at coordinate x on the rod, and the other quantities are as before.

It should be reiterated that a longitudinal wave starting out in an inhomogeneous medium will couple energy into the transverse vibrational wave and also into the shear wave. Furthermore, when a longitudinal wave strikes a boundary at any angle other than normal incidence (a boundary normal to the direction of wave travel), both shear waves and surface waves along the boundary are established, so that some energy is transferred to these. This means, in the case of seismic waves, that one ordinarily receives two, and sometimes three, reflections from a given discontinuity. The first to arrive is the longitudinal wave, which travels at highest velocity. The shear wave travels at a lower velocity, as does the surface wave.

Some typical values for a few of the constants involved in the travel of waves through solids are given in Table 3-5. It should be realized that the values in this table are only representative. The properties of different steels, of copper of different purity, and of aluminum of different purity will vary. The properties of granite vary even more, and granite does not completely obey Hooke's law and the other elastic relations. Of course, there are many kinds of rubber, with many resulting kinds of moduli and velocities.

It is interesting to note that the values of all the constants for the three metals are not far apart. The velocities are somewhat higher in the stiffer metals. The very low shear velocity in rubber is interesting and is related to the very low shear modulus which also exists in rubber. Since the shear modulus is a measure of rigidity, one would expect it to be low in rubber.

† Lord Rayleigh, "Theory of Sound," vol. I, pp. 253–254, Dover Publications, New York, 1945.

[1] *Ibid.*, pp. 255–305.

TABLE 3-5. PROPERTIES OF SOLIDS†

Property‡	Steel	Copper	Aluminum	Rockport granite	Rubber
Y_0	21×10^{10}	12×10^{10}	7×10^{10}	4.5×10^{10}	2.0×10^{6}
Y_B	27.5×10^{10}	18.4×10^{10}	10.8×10^{10}		1.0×10^{10}
ρ_v	7.8×10^{3}	8.9×10^{3}	2.7×10^{3}		0.93×10^{3}
v_p (plane)	5,940	4,560	6,320	5,200	1,040
v_p (shear)	3,220	2,250	3,100	2,700	27

† From Theodore F. Hueter and Richard H. Bolt, "Sonics," p. 346, John Wiley & Sons, Inc., New York, 1955; and L. D. Leet, "Practical Seismology and Seismic Prospecting," pp. 97–99, Appleton-Century-Crofts, Inc., New York, 1938.

‡ Y_0 and Y_B are in newtons per square meter, ρ_v is in kilograms per cubic meter, and v_p is in meters per second.

Example 3-6. A thin steel rod is a part of a machine. It is in contact with a surface in the machine which is vibrating at a frequency of 1,000 cps and with a peak velocity of 1 cm/sec. Assuming that the rod is properly terminated and therefore may be considered as of infinite length, find the internal pressure and the wavelength of the longitudinal pressure wave. The vibration is sinusoidal.

To find the internal pressure, we utilize density and Young's modulus from Table 3-5, obtaining for the characteristic impedance

$$Z_0 = \frac{1}{\sqrt{\rho_v Y_0}} = \frac{1}{\sqrt{2.1 \times 7.8 \times 10^{14}}} = 2.47 \times 10^{-8} \text{ m}^3/\text{newton-sec}$$

From this, the peak pressure is found to be

$$P_x = \frac{U_x}{Z_0} = 4.05 \times 10^5 \text{ newtons/m}^2$$
$$= 9.12 \text{ lb/cm}^2 = 58.9 \text{ lb/in.}^2$$

To determine the wavelength, we first find the velocity of propagation:

$$v_p = \sqrt{\frac{Y_0}{\rho_v}} = \sqrt{\frac{2.1 \times 10^{11}}{7.8 \times 10^3}} = 5{,}190 \text{ m/sec}$$

Note that phase velocity in this rod is less than the velocity for a *plane wave* in steel given in Table 3-5 (5,940 m/sec). The wavelength is obtained by dividing the velocity by the frequency and is therefore 5.19 m.

In any practical application this rod probably would not be terminated in its characteristic impedance, and standing waves would therefore exist on the rod. These are discussed in Chap. 5.

Example 3-7. A steel rod like that of Example 3-6 is 5 m long. Instead of being subject to vibrations, it is struck a sudden blow, having a force of 200 newtons (about 45 lb) distributed over its 4-cm² cross section. For the transient wave so generated, find the particle velocity and the time of travel for both the longitudinal and the shear waves.

The pressure is calculated by dividing the force by the area and is therefore given by

$$p_x = \frac{200}{4 \times 10^{-4}} \text{ newtons/m}^2$$

Hence the particle velocity is given by

$$u_x = p_x Z_0 = \frac{200 \times 2.47 \times 10^{-8}}{4 \times 10^{-4}}$$
$$= 0.0124 \text{ m/sec}$$

The time of travel for the longitudinal wave is given by the length divided by the phase velocity calculated in Example 3-6, so

$$T = \frac{5}{5{,}190} = 0.965 \text{ msec}$$

For the shear wave the velocity is less, and that from Table 3-5 may be used. The time involved with it is

$$T = \frac{5}{3{,}220} = 1.55 \text{ msec}$$

Example 3-8. This example has to do with a highly idealized seismic wave. In geophysical prospecting, explosions are frequently set off near the surface, and the reflections of the seismic waves from various boundaries between the layers of rock and oil are analyzed to determine the constitution of the ground under the explosion and nearby. Of course, an explosion sets up a spherical wave in the ground rather than a plane wave, but we shall treat the seismic wave here as if it were a plane wave. (Spherical waves are discussed in Chap. 10.)

Assume that a plane seismic pulse is to be used to determine the depth of an interface between two strata of rock. The wave is started in a stratum of Sudbury norite, and a pickup at the same depth as the starting explosion detects a return pulse in 30 msec. What is the depth of the interface between the two strata? When does the shear-wave pulse (which was set up at the same time as the longitudinal wave) arrive?

For Sudbury norite, Young's modulus is given approximately by

$$Y_B \approx 9 \times 10^{10}\dagger$$
$$\rho_v = 1.67 \times 10^3 \text{ kg/m}^3$$
$$\mu = 3.6 \times 10^{10} \text{ newtons/m}^2$$

For the longitudinal wave we have

$$v_p = \sqrt{\frac{Y_B}{\rho_v}} = \sqrt{\frac{9 \times 10^{10}}{0.167 \times 10^4}} = \sqrt{54 \times 10^6}$$
$$= 7.35 \times 10^3 \text{ m/sec}$$

The first pulse returns from the boundary after a traveltime corresponding to the distance down and back, so that, if d is the depth, we have

$$30 \times 10^{-3} v_p = 2d$$
$$d = 15 \times 10^{-3} \times 7.35 \times 10^3 = 110 \text{ m}$$

The velocity for the shear wave is lower and is given by

$$v_p = \sqrt{\frac{\mu}{\rho_v}} = \sqrt{\frac{3.6 \times 10^{10}}{1.67 \times 10^3}} = 4.65 \times 10^3 \text{ m/sec}$$

Hence, the time of travel for this is given by the distance divided by the velocity:

$$T = \frac{2.2 \times 10^2}{4.65 \times 10^3} = 47.5 \text{ msec}$$

There are many interesting wave problems in seismology. Some of these are described by Leet[1] and Heiland.[2]

3-5. Thermal Conduction

In heat transfer three types of processes are ordinarily discussed: conduction, convection, and radiation. Although waves may occur in the fluid-dynamic processes of convection, they are not treated here. Radiation of heat is, of course, a wave process at a very high frequency, but it is not ordinarily treated as such. Hence, in this book we restrict ourselves to the discussion of *heat conduction.*

Heat can be conducted through gases and liquids as well as through solids, but since this involves the process of setting the masses in motion,

† L. D. Leet, "Practical Seismology and Seismic Prospecting," p. 97, Appleton-Century-Crofts, Inc., New York, 1938.

[1] *Ibid.*

[2] C. A. Heiland, "Geophysical Exploration," Prentice-Hall, Inc., Englewood Cliffs, N.J., 1940.

in addition to the wave process of conduction, it is not extensively discussed; rather, attention is largely restricted to conduction through solids. Many of the principles involved are also applicable to conduction through liquids and gases when the appropriate thermodynamic and fluid-dynamic processes are also considered.

In the preceding sections we have discussed waves that are essentially lossless; in fact, we have neglected losses altogether in our calculations. Thermal waves, however, correspond to a very lossy type of transmission line and obey the same relations as electromagnetic waves passing through a conductor which is so good that displacement currents may be neglected.

The basic experimental law for heat transfer through solids is due to Fourier. It states that the quantity of heat flowing through a surface per second is proportional to the temperature gradient normal to the surface. Mathematically, we express this, using the thermal conductivity k, as

$$\frac{\partial Q_h}{\partial t} = -k\,\Delta A\,\frac{\partial \tau}{\partial x}$$

Here Q_h = quantity of heat, joules
τ = temperature, °C
t = time
k = conductivity, watts/m-°C
and the geometric quantities are indicated in Fig. 3-3.

In dealing with quantities analogous to the electrical quantities, it is convenient to talk about a rate of flow per unit area, this being a flux density which corresponds directly to electrical current density. We define this rate of flow per unit area as

$$\mathbf{q} = \lim_{\Delta A \to 0} \frac{1}{(\Delta A)^2} \frac{\partial Q_h}{\partial t} \mathbf{\Delta A}$$

Like other flux densities, $\mathbf{q}$ is a vector. As in the preceding derivations, we shall here assume a plane wave, so that we need only one component of $\mathbf{q}$. Since the heat flow is in the direction of the wave propagation, the wave associated with heat is a longitudinal one, and for propagation in the x direction we shall use q_x. The above equation thus becomes

$$q_x = -k\,\frac{\partial \tau}{\partial x}$$

Rewriting in the appropriate form for its analogy, we have

Transmission-line analogy

(3-36) $$\frac{\partial \tau}{\partial x} = -\frac{1}{k} q_x \qquad\qquad \frac{\partial v}{\partial x} = -Ri$$

When we add the components of $\mathbf{q}$ in the various directions, this equation becomes

$$\nabla\tau = -\frac{1}{k}\mathbf{q} \tag{3-37}$$

It can be seen that this corresponds to the transmission-line current equation in which series inductance is neglected. The second "telegrapher's equation" for heat flow, however, corresponds to the current equation for the transmission line in which the shunt conductance is neglected, so that energy storage is involved, as in the phenomena previously treated.

Consider Fig. 3-3. If heat is being stored or removed from the volume indicated, we should be able to write an equation for this volume comparable with the one having to do with storage or removal of mass in the

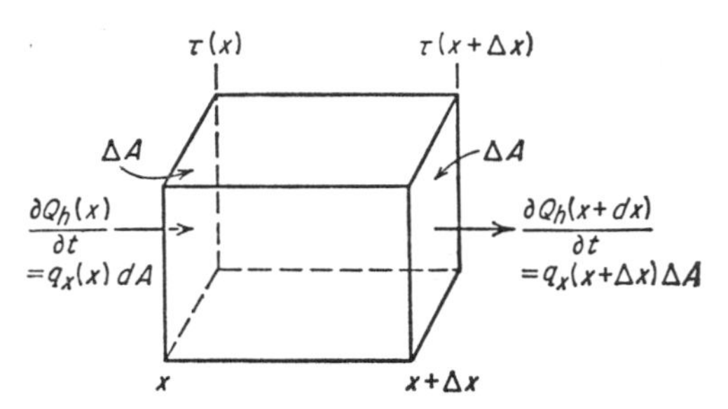

FIG. 3-3. Heat-conduction parameters.

case of sound waves. The amount of heat flowing into the small volume from the left is given by

$$\frac{dQ_h}{dt}(x) = q_x(x)\Delta A$$

The heat flowing out the other side is given by

$$\frac{dQ_h}{dt}(x + \Delta x) = q_x(x + \Delta x)\Delta A$$

The difference between these two quantities represents the net amount of heat which is being removed from the volume. Before writing the equation for this, we must introduce a new term, the *specific heat*. This is defined as a constant relating the change in the heat stored within a given mass to the change of temperature. Mathematically, the definition for a given volume ΔV is

$$dQ_h = S\rho_v\,\Delta V\,d\tau$$

Here S is the specific heat per unit of mass and ρ_v is the density in kilograms per cubic meter. Hence, we may write the rate of change of stored heat with time as

$$\begin{aligned}\frac{dQ_h}{dt}(x + \Delta x) - \frac{dQ_h}{dt}(x) &= [q_x(x + \Delta x) - q_x(x)]\Delta A \\ &= -S\rho_v\,\Delta V\,\frac{\partial \tau}{\partial t} \\ &= -S\rho_v\,\Delta A\,\Delta x\,\frac{\partial \tau}{\partial t}\end{aligned}$$

Dividing both sides of this equation by Δx and passing to the limit, we obtain

Transmission-line analogy

(3-38) $$\frac{\partial q_x}{\partial x} = -S\rho_v\frac{\partial \tau}{\partial t} \qquad \frac{\partial i}{\partial x} = -C\frac{\partial v}{\partial t}$$

The three-dimensional version of this is simply

Electromagnetic-wave analogy

(3-39) $$\nabla \cdot \mathbf{q} = -S\rho_v\frac{\partial \tau}{\partial t} \qquad \nabla \times \mathbf{H} = -\epsilon\frac{\partial \mathbf{E}}{\partial t}$$

It can be seen that (3-36) and (3-38) correspond to the "telegrapher's equations" and that (3-37) and (3-39) correspond in a rough way to Faraday's and Ampere's laws. Of course, there is no term in Faraday's law comparable to the series-resistance term, so a direct analogy between the electromagnetic wave in space and the thermal wave is not possible with temperature corresponding to electric field and flow rate corresponding to magnetic field. As discussed in Sec. 3-8, an alternative analogy is possible in which flow rate corresponds to electric field and temperature to magnetic field.

As usual, we may combine the telegrapher's equations to obtain a second-order differential equation. Thus, combining (3-36) and (3-38), we have, after differentiating (3-36) with respect to x,

Transmission-line analogy

(3-40a) $$\frac{\partial^2 \tau}{\partial x^2} = \frac{S\rho_v}{k}\frac{\partial \tau}{\partial t} \qquad \frac{\partial^2 v}{\partial x^2} = RC\frac{\partial v}{\partial t}$$

which is sometimes written

(3-40b) $$\frac{\partial^2 \tau}{\partial x^2} = \frac{1}{D_t}\frac{\partial \tau}{\partial t}$$

Here D_t is called the thermal diffusivity as a recognition that heat conduction follows the same mathematical relations as chemical diffusion (which is discussed in Sec. 3-6). A similar derivation left as a problem results in the comparable equation

(3-41) $$\frac{\partial^2 q_x}{\partial x^2} = \frac{S\rho_v}{k}\frac{\partial q_x}{\partial t} = \frac{1}{D_t}\frac{\partial q_x}{\partial t}$$

The three-dimensional temperature equation may be obtained by substituting for **q** from (3-37) into (3-39):

Electromagnetic-wave analogy

(3-42) $$\nabla^2\tau = \frac{S\rho_v}{k}\frac{\partial\tau}{\partial t} = \frac{1}{D_t}\frac{\partial\tau}{\partial t} \qquad \nabla^2\mathbf{E} = \mu\sigma\frac{\partial\mathbf{E}}{\partial t}$$

When we deal with the steady-state alternating flow of heat with angular frequency ω, (3-40) becomes

Transmission-line analogy

(3-43) $$\frac{\partial^2 T}{\partial x^2} = \frac{j\omega T}{D_t} = \frac{j\omega S\rho_v}{k}T \qquad \frac{\partial^2 V}{\partial x^2} = j\omega RCV$$

where $\tau = \text{Re}\, Te^{j\omega t}$. From this it can be seen that

Transmission-line analogy

$$\gamma = \sqrt{\frac{j\omega S\rho_v}{k}} = \sqrt{\frac{\omega S\rho_v}{2k}}(1+j) \qquad \gamma = \sqrt{j\omega CR} = \sqrt{\frac{\omega CR}{2}}(1+j)$$

$$= \sqrt{\frac{\omega}{2D_t}}(1+j)$$

Hence

(3-44) $$\alpha = \beta = \sqrt{\frac{\omega S\rho_v}{2k}} = \sqrt{\frac{\omega}{2D_t}} \qquad \alpha = \beta = \sqrt{\frac{\omega CR}{2}}$$

(3-45) $$v_p = \frac{\omega}{\beta} = \sqrt{\frac{2\omega k}{S\rho_v}} = \sqrt{2\omega D_t} \qquad v_p = \sqrt{\frac{2\omega}{CR}}$$

and

(3-46) $$\lambda = \frac{v_p}{f} = \sqrt{\frac{4\pi k}{fS\rho_v}} = \sqrt{\frac{4\pi D_t}{f}} \qquad \lambda = \sqrt{\frac{4\pi}{fCR}}$$

It is most important to note the implications of the values of α and β given above. Thus

α in nepers per unit distance $= \beta$ in radians per unit distance

This is a very large attenuation indeed—55 db in 1 wavelength. This same attenuation is found for an electromagnetic wave in a conductor. To see what it means in terms of temperature ratios, consider Table 3-6, where $T_1 = T_0 e^{-\alpha x}$.

Another interesting result is that the velocity of propagation is a function of frequency, whereas in the lossless waves treated heretofore it is

TABLE 3-6. ATTENUATION OF ALTERNATING TEMPERATURE

x	αx	T_1/T_0
$\lambda/4$	$\pi/2$	0.207
$\lambda/2$	π	0.0401
λ	2π	0.0019

independent of frequency (except for some of the transmission-line examples in Chap. 2). This means that the thermal wave is highly distorted, because both its velocity and its attenuation are functions of frequency. An example of the degree of this distortion is given in Chap. 4, where distortion is studied in the traveling wave caused by a sudden change of surface temperature. Sinusoidal waves are, of course, not distorted.

The characteristic impedance for the thermal wave is defined as the ratio of temperature to heat flow and is given by

$$Z_0 = \frac{T}{Q} = \sqrt{\frac{1}{j\omega k S\rho_v}} = \sqrt{\frac{1}{2\omega k S\rho_v}}\,(1 - j) \tag{3-47}$$

where $q = \text{Re}\ Qe^{j\omega t}$. From this it can be seen that the temperature and flow of heat are always 45° out of phase, with the temperature lagging the heat flow.

The analogies between the transmission line and the heat flow are shown in Table 3-7.

TABLE 3-7. HEAT–TRANSMISSION-LINE ANALOGIES

Electrical quantity	*Thermal quantity*
v	τ
i	q_x
R	$1/k$
C	$S\rho_v$
L	0
G	0

For the flow of heat through thin rods, it is more convenient to use as an analogy for current the total flow of heat along the rod than the flow per unit area. Thus, if

$$Q'_h = \frac{dQ_h}{dt} = qA$$

then (3-36) may be written for the rod of cross section A

$$\frac{\partial \tau}{\partial x} = -\frac{Q'_h}{kA} \tag{3-48}$$

Likewise, (3-38) becomes

$$\frac{\partial Q'_h}{\partial x} = -S\rho_v A\,\frac{\partial \tau}{\partial t} \tag{3-49}$$

On a thin rod, of course, there may be heat loss due to radiation and convection at the surface. Where the difference in temperature between the rod and the surrounding air is small, this loss is proportional to the temperature difference and to the area of the surface:

$$\Delta Q'_h = h(\tau - \tau_0)\Delta S$$

Here h is the heat-transfer coefficient, or surface emissivity, in watts per square meter–degree centigrade, τ_0 is the ambient temperature, and ΔS is the area of the surface between the solid and its surroundings. For a rod of circumference g, the heat flow in a length Δx is given by

$$\Delta Q'_h = h(\tau - \tau_0)g\,\Delta x$$

When this quantity is inserted into the derivation of (3-38) and (3-49), (3-49) becomes

$$\frac{\partial Q'_h}{\partial x} = -S\rho_v A\,\frac{\partial \tau}{\partial t} - gh(\tau - \tau_0) \tag{3-50}$$

This means that for the rod the wave equation is, instead of (3-40),

Transmission-line analogy

$$\frac{\partial^2 \tau}{\partial x^2} = \frac{S\rho_v}{k}\frac{\partial \tau}{\partial t} + \frac{gh(\tau - \tau_0)}{kA} \qquad \frac{\partial^2 v}{\partial x^2} = RG(v - v_0) + RC\,\frac{\partial v}{\partial t} \tag{3-51}$$

When we refer temperatures in the solid bar or rod to the ambient temperature (that is, make $\tau_0 = 0$ by translating our temperature scale), we find

$$\gamma = \sqrt{\frac{gh}{kA} + \frac{j\omega S\rho_v}{k}}$$

When the amount of surface loss is *small*, we may write

$$\begin{aligned}
\gamma &= \sqrt{\frac{j\omega S\rho_v}{k}}\sqrt{1 - j\,\frac{gh}{\omega S\rho_v A}} \\
&\approx \sqrt{\frac{j\omega S\rho_v}{k}}\left(1 - j\,\frac{gh}{2\omega S\rho_v A}\right) \\
&= \sqrt{\frac{\omega S\rho_v}{2k}}\left[\left(1 + \frac{gh}{2\omega S\rho_v A}\right) + j\left(1 - \frac{gh}{2\omega S\rho_v A}\right)\right]
\end{aligned}$$

Some examples of the constants involved in thermal calculations are given in Table 3-8. It should be noted that the units in Table 3-8 are mks units, whereas the units ordinarily given in tables are cgs or English units. The conversion factors for the appropriate constants are

$$\begin{aligned}
k_{\text{mks}} &= k_{\text{cgs}} \times 418.5 \\
S_{\text{mks}} &= S_{\text{cgs}} \times 4.185 \times 10^3 \\
\rho_{v,\text{mks}} &= \rho_{v,\text{cgs}} \times 10^3 \\
D_{t,\text{mks}} &= D_{t,\text{cgs}} \times 10^{-4}
\end{aligned}$$

TABLE 3-8. TYPICAL THERMAL CONSTANTS†

Material	Temperature, °C	k, joules/sec-m-°C	S, joules/kg-°C	ρ_v, kg/m³	D_t, m²/sec
Silver	0	418	234	10.5×10^3	1.7×10^{-4}
Copper	0	387	380	8.94×10^3	1.14×10^{-4}
Aluminum	0	203	870	2.71×10^3	0.86×10^{-4}
Mild steel	0	44.6	460	7.85×10^3	1.24×10^{-5}
Mild steel	600	36.3	670	7.68×10^3	0.71×10^{-5}
Asbestos	0	0.15	1,046	0.58×10^3	2.48×10^{-7}
Brick masonry	20	0.628	838	1.7×10^3	4.4×10^{-7}
Concrete, dams		2.42	920	2.47×10^3	1.07×10^{-6}
Sandy clay, 15% moisture		0.92	1,380	1.78×10^3	3.7×10^{-7}
Some wet soils		1.25–3.35			$4\text{–}10 \times 10^{-7}$
Ordinary glass		0.88	670	2.6×10^3	5.1×10^{-7}

† Converted from a table in Leonard R. Ingersoll, Otto J. Zobel, and Alfred C. Ingersoll, "Heat Conduction," pp. 241–245, McGraw-Hill Book Company, Inc., New York, 1948.

Example 3-9. In a gasoline engine it is found that the surface temperature difference of the cylinder wall over the duration of a cycle (2 revolutions) is 2.5°C at 2,400 rpm. Assume a plane wave penetrating into the mild-steel cylinder wall (this is not a bad assumption as the wavelength is so short that the curvature has a negligible effect). What is the temperature variation at depths of 1 mm and 5 mm into the wall? Determine all wave parameters. How much alternating heat flow is there into the wall?

$$f = \frac{1{,}200}{60} = 20 \text{ cps}$$

$$\alpha = \beta = \sqrt{\frac{\omega S \rho_v}{2k}} = \sqrt{\frac{2\pi \times 20 \times 460 \times 7.85 \times 10^3}{2 \times 44.6}}$$
$$= 2.25 \times 10^3 \text{ nepers/m, or radians/m}$$

At 1 mm,

$$\Delta T = \Delta T_0 e^{-2.25} = 2.5e^{-2.25} = 0.263°\text{C}$$

At 5 mm,

$$\Delta T = 2.5e^{-11.25} = 3.23 \times 10^{-5}\,°\text{C}$$

$$v_p = \frac{\omega}{\beta} = \frac{6.28 \times 20}{2.25 \times 10^3} = 0.0558 \text{ m/sec}$$

Note how much slower the heat wave is than the others that have been studied. The wavelength is quite short and is given by

$$\lambda = \frac{2\pi}{\beta} = 2.79 \times 10^{-3} \text{ m} = 2.79 \text{ mm}$$

The impedance is given by

$$Z_0 = \frac{1-j}{\sqrt{2\omega k S \rho_v}} = \frac{1-j}{\sqrt{4\pi \times 20 \times 44.6 \times 460 \times 7.85 \times 10^3}}$$
$$= 4.96 \times 10^{-6}(1-j) \qquad {}^\circ\text{C-m}^2/\text{watt}$$

The peak amount of power flowing into the wall per square meter is given by

$$Q = \frac{T}{|Z_0|} = \frac{2.5/2 \times 10^6}{4.96\sqrt{2}} = 1.26 \times 10^5 \text{ watts/m}^2$$

Here the peak value is shown as 2.5/2, or one-half the peak-to-peak value. To give this more meaning, consider a cylinder whose radius is $10/\pi$ cm and whose length is 10 cm. The cross-sectional area for a flow of heat is therefore 200 cm², or 2×10^{-2} m². The peak amount of heat which flows into and out of the wall is therefore 2.52×10^3 watts, or 2.52 kw.

Example 3-10. In a certain city the fundamental annual component resulting from Fourier analysis of the temperature variation has a minimum of 20°F and a maximum of 90°F. It is desired to find the depth to which water pipes must be buried if they are not to encounter freezing temperatures.

The temperature range is from -6.7 to $+32.2$°C. The peak amplitude of the sinusoidal wave is therefore 38.9/2, or 19.45°C. If the temperature is not to go below freezing, its peak swing about the mean must be reduced from 19.45 to (19.45 − 6.7)°C, or 12.75°C. In terms of the attenuation of the wave, we therefore have

$$e^{-\alpha x} = \frac{12.75}{19.45}$$

or

$$\alpha x = 0.42$$

If we examine Table 3-8 and use $D_t = 4 \times 10^{-7}$ for the soil involved, we may write, from (3-44),

$$\alpha = \sqrt{\frac{\omega}{2D_t}}$$

Here ω is the angular frequency for a period of 1 year and is therefore given by

$$\frac{2\pi}{365 \times 24 \times 3{,}600} = 1.99 \times 10^{-7}$$

Hence,

$$\alpha = \sqrt{\frac{1.99 \times 10^{-7}}{2 \times 4 \times 10^{-7}}} = 0.5 \text{ neper/m}$$

Therefore

$$0.5x = 0.42$$

or

$$x = 0.84 \text{ m} = 84 \text{ cm}$$

3-6. Diffusion Processes

In diffusion processes, random thermal motion results in the spread of particles throughout a volume, so that any initial nonuniformities in the distribution of a particular constituent tend to be smoothed out. Thus, if one introduces cream into a cup of coffee, it gradually tends to disperse throughout the coffee. As a result, the mixture eventually becomes uniform.

Diffusion processes are much the same whether we are talking of the diffusion of one gas into another, of one liquid into another, of a liquid into a solid, of a solid into another solid, or of ions, holes, electrons, molecules, or neutrons in a reactor. Whereas it is easy to see the diffusion of gases into gases and liquids into liquids, the idea of diffusion of solids into solids may be unfamiliar. Nevertheless, this process is widely used in the manufacture of transistors, where the collector and emitter materials are deposited on either side of the base, and is essential to the operation of transistors.

Actually, the processes of molecular diffusion in gases and liquids are usually overshadowed by turbulent mixing; and the more practical applications involve diffusion through solids.

Turbulent mixing involves *eddy diffusion*, which can be described mathematically in the same manner as molecular diffusion. This description is of little practical importance, however, for "constants" in the molecular-diffusion process are functions of concentration and position in the eddy-diffusion process. The resulting nonlinear equations are difficult both to establish and to solve, and they are not treated here, even though eddy diffusion is an important physical process.

Diffusion processes follow Fick's law.[1] In general, when two different

[1] E. R. G. Eckert, "Introduction to the Transfer of Heat and Mass," pp. 236–238, McGraw-Hill Book Company, Inc., New York, 1950; and W. Jost, "Diffusion in Solids, Liquids, and Gases," p. 2, Academic Press, Inc., New York, 1952.

substances are free to diffuse, they diffuse into each other. That is, if we take a box containing substance A and another box containing substance B and suddenly remove the barrier between them, A will tend to move into the box occupied by B, and B will tend to move into the box occupied by A, until the concentration of A and B is the same throughout the entire double box. Here, however, we are concerned with the case in which only one material is diffusing and the other is stationary.

Moisture evaporating into the air from a wet surface is an example of an essentially one-way diffusion process. There can be no significant diffusion of the air into the wet surface. An example in a liquid might be the diffusion of a dilute acid solution into water. Since the concentration of the ions associated with the water would be essentially the same in the weak solution and in the water into which the solution is diffusing, only the diffusion of the acid would be important. Another example is the diffusion of a gaseous impurity into a solid, a process sometimes used in transistor manufacture.

In chemical engineering, the diffusion situation is often complicated by the fact that chemical reactions take place along with diffusion. This may have the effect of providing a distributed source of ions to diffuse, or conversely, a distributed sink into which they may vanish as they are diffusing. A similar effect occurs in semiconductors, because of the generation and recombination of carriers.

Fick's first law may be stated as

$$\frac{dM_1}{dt} = -D_{12}\frac{dc_1}{dx}\Delta A$$

where M_1 = mass of material, kg, crossing surface ΔA

ΔA = area of the surface

c_1 = mass concentration of substance 1, kg/m^3

D_{12} = constant known as the diffusivity of substance 1 in substance 2

The unit for flux density of mass flow in diffusion is the mass transfer per unit area in a second. Thus we may rewrite Fick's law as

$$m_{vx} = -D\frac{\partial c}{\partial x} \qquad \text{kg/m}^2\text{-sec}$$

where m_{vx} is the mass transfer per unit area in a second in the x direction. Here the subscripts 1 and 2 have been dropped from the equation, since we are talking about diffusion of one material into the other. It should be recognized that a comparable equation can be written for each constituent in a mixture (in this sense, a solution is considered a mixture in which each type of ion is considered a separate constituent).

Fick's law may also be written in terms of number of particles instead of mass.

Fick's law itself is one of the telegrapher's equations for the diffusion process. It is more recognizable as

Transmission-line analogy

$$\frac{\partial c}{\partial x} = -\frac{1}{D} m_{vx} \qquad \frac{\partial v}{\partial x} = -Ri \tag{3-52}$$

It is not surprising that this has the same form as the Fourier heat-conduction equation (3-36), for both are due to random (thermal) agitation of molecules.

The second telegrapher's equation is simply the continuity equation for mass. Consider the volume shown in Fig. 3-4. The net amount of

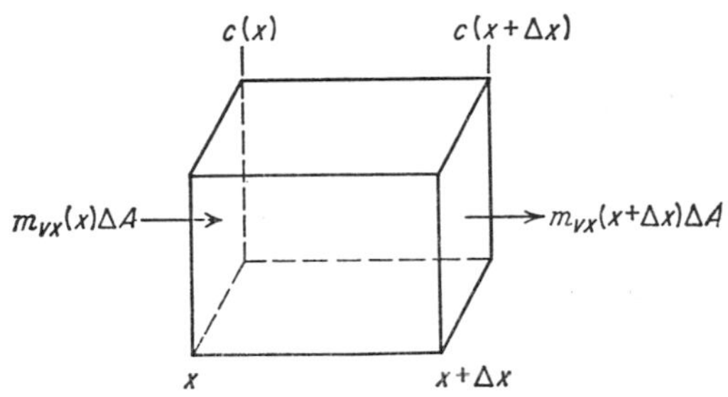

FIG. 3-4. Mass transfer by diffusion.

mass leaving the volume on the right is a measure of the rate of mass decrease in the volume. Thus

Mass leaving − mass entering = −(rate of mass increase)

In the symbols of Fig. 3-4, this is

$$\Delta A\,[m_{vx}(x + \Delta x) - m_{vx}(x)] = -\frac{dM}{dt} = -\frac{d(c\,\Delta x\,\Delta A)}{dt}$$

Canceling out the ΔA's, dividing through by Δx, and passing to the limit as usual, this becomes

Transmission-line analogy

$$\frac{\partial m_{vx}}{\partial x} = -\frac{\partial c}{\partial t} \qquad \frac{\partial i}{\partial x} = -C\frac{\partial v}{\partial t} \tag{3-53}$$

The two equations may be combined in the usual fashion to obtain the wave equation for concentration in the diffusion process:

Transmission-line analogy

$$\frac{\partial^2 c}{\partial x^2} = \frac{1}{D}\frac{\partial c}{\partial t} \qquad \frac{\partial^2 v}{\partial x^2} = RC\frac{\partial v}{\partial t} \tag{3-54}$$

Three-dimensional forms of (3-52) to (3-54) are readily seen to be

(3-52a) $$\nabla c = -\frac{1}{D}\mathbf{m}_v$$

(3-53a) $$\nabla \cdot \mathbf{m}_v = -\frac{\partial c}{\partial t}$$

Electromagnetic-wave analogy

(3-54a) $$\nabla^2 c = \frac{1}{D}\frac{\partial c}{\partial t} \qquad \nabla^2 \mathbf{E} = \mu\sigma \frac{\partial \mathbf{E}}{\partial t}$$

This equation is appropriate in this form for uncharged particles in liquids and solids. With charged particles, as in diffusion of charge carriers through semiconductors in diodes and transistors, other considerations must be taken into account.

Fick's law may be written for charge carriers in terms of current density as

$$\frac{\partial(q_i n_i)}{\partial x} = -\frac{1}{D_i}\mathcal{J}_{ix}$$

where q_i = charge on each particle of the ith constituent
n_i = number of such particles per unit volume
$\mathcal{J}_{ix}$ = current density for that constituent, in the x direction
D_i = diffusion coefficient for the ith constituent

This may be written as

Transmission-line analogy

(3-55) $$\frac{\partial n_i}{\partial x} = -\frac{1}{D_i q_i}\mathcal{J}_{ix} \qquad \frac{\partial v}{\partial x} = -Ri$$

The diffusion coefficient may be determined from the charge mobility, using the Nernst-Einstein relation:[1]

(3-56) $$D_i = \frac{kT}{q_i}\mu_i$$

where k = Boltzmann's constant (1.23×10^{-23} joule/deg)
T = temperature, °K
μ_i = mobility of the ith constituent

In semiconductors the charge-carrying constituents designated by i are electrons and "holes." They both have the same charge, 1.60×10^{-19} coulomb, the electron being charged negatively and the hole being charged positively. The mobility is defined as the average velocity of a particle in a unit electric field.

In the situation where the total number of charge carriers is con-

[1] Adrianus J. Dekker, "Electrical Engineering Materials," p. 177, Prentice-Hall, Inc., Englewood Cliffs, N.J., 1959.

served, that is, where no charge carriers are created or destroyed, the derivation of the second telegrapher's equation is almost identical with the derivation of (3-53), with $\mathcal{J}_{ix}$ replacing m_{vx} and $q_i n_i$ replacing c. Hence, the result may be written at once as

Transmission-line analogy

(3-57) $$\frac{\partial \mathcal{J}_{ix}}{\partial x} = -q_i \frac{\partial n_i}{\partial t} \qquad \frac{\partial i}{\partial x} = -C \frac{\partial v}{\partial t}$$

In a semiconductor, generation and recombination of charge carriers take place, so that the number does not remain constant. If the density is increased above the equilibrium value, it decays exponentially back to equilibrium, because of recombination. If it is reduced, it increases exponentially, because of the thermal generation of carriers. The resulting expression for density, in the absence of current flow, is[1]

$$\frac{\partial n_i}{\partial t} = \frac{n_{i0} - n_i}{\tau_i}$$

where n_{i0} = equilibrium density at the temperature involved
τ_i = lifetime for the particular constituent and temperature

The total time rate of change of density at a point is therefore the sum of this contribution and the one due to current flow, given by (3-57). Thus the modified telegrapher's equation is

$$\frac{\partial \mathcal{J}_{ix}}{\partial x} = q_i \frac{n_{i0} - n_i}{\tau_i} - q_i \frac{\partial n_i}{\partial t}$$

If the semiconductor is homogeneous and if the temperature is uniform, n_{i0} is a function of neither position nor time. Therefore the density may be expressed in terms of the deviation from this value, with derivatives of n_i and the difference being the same. Let this difference be

$$\nu_i = n_i - n_{i0}$$

Then the equation becomes

Transmission-line analogy

(3-58) $$\frac{\partial \mathcal{J}_{ix}}{\partial x} = -q_i \frac{\nu_i}{\tau_i} - q_i \frac{\partial \nu_i}{\partial t} \qquad \frac{\partial i}{\partial x} = -Gv - C \frac{\partial v}{\partial t}$$

In these terms, (3-55) becomes

Transmission-line analogy

(3-59) $$\frac{\partial \nu_i}{\partial x} = \frac{-1}{D_i q_i} \mathcal{J}_{ix} \qquad \frac{\partial v}{\partial x} = -Ri$$

[1] *Ibid.*, pp. 177–180.

Combining, we have the diffusion equation

Transmission-line analogy

$$(3\text{-}60) \qquad \frac{\partial^2 \nu_i}{\partial x^2} = \frac{\nu_i}{D_i\tau_i} + \frac{1}{D_i}\frac{\partial \nu_i}{\partial t} \qquad \frac{\partial^2 v}{\partial x^2} = RGv + RC\frac{\partial v}{\partial t}$$

This is the form of the wave equation involved in carrier diffusion in semiconductors. It can be solved in the usual manner.

It should be noted that the assumption was made that no electric field was acting. When this assumption cannot be made, Fick's law is replaced by

$$(3\text{-}61) \qquad \mathcal{J}_{ix} = n_i q_i \mu_i E_x - q_i D_i \frac{\partial n_i}{\partial x}$$

This equation is such that the wave analogies are not possible—nothing in the wave equations is analogous to the first term on the right-hand side of this equation. Of course, approximations can be made assuming nonlinear elements or a nonconstant diffusivity D_i.†

In a transistor, the equations developed above prior to (3-61) may be used to describe the minority carrier density and current when small voltage is impressed across the junction. In such a situation the *drift current* indicated by the first term on the right-hand side of (3-61) may be neglected.

A summary of the analogies for diffusion is given in Table 3-9.

TABLE 3-9. DIFFUSION–TRANSMISSION-LINE ANALOGIES

Electrical quantity	Molecular diffusion	Charge-carrier diffusion
v	c	n_i or ν_i
i	m_{vx}	$\mathcal{J}_{ix}$
R	$1/D$	$1/D_i q_i$
L	0	0
G	0	q_i/τ_i
C	1	q_i

"Alternating-current" diffusion is important in describing current flow in the transistor, but not in general for other processes, since most diffusion sources are transient. Nevertheless, it is instructive to consider it even for the other processes, for the transients encountered there may be resolved by Fourier-integral methods into components at different frequencies, and insight into the results may be obtained from the impedance, attenuation, and velocity expressions.

For diffusion not involving creation or recombination of the diffusing

† F. M. Smits, Formation of Junction Structures by Solid-state Diffusion, *Proc. IRE*, vol. 46, pp. 1049–1061, 1958.

particles, the impedance may be written, after definition, by analogy to the electrical case. Thus,

Transmission-line analogy

(3-62) $$Z_0 = \frac{C}{M_{vx}} = \sqrt{\frac{1}{j\omega D}} = \sqrt{\frac{1}{2\omega D}}(1 - j) \qquad\qquad Z_0 = \sqrt{\frac{R}{j\omega C}}$$

where $m_{vx} = \text{Re } M_{vx}e^{j\omega t}$ and $c = \text{Re } Ce^{j\omega t}$. The expressions for attenuation, velocity, and wavelength are the same as for the thermal wave and are given by Eqs. (3-44) to (3-46).

TABLE 3-10. TYPICAL DIFFUSION COEFFICIENTS

Type of diffusion	Temperature T, °C	Diffusion coefficient D	Lifetime τ_i, μsec
Water to air†	8.0	0.329×10^{-4} m²/sec	
Hydrogen to air†	0.0	0.634×10^{-4} m²/sec	
Hydrochloric acid to water†	12	2.21×10^{-4} m²/day	
Sugar to water†	12	0.254×10^{-4} m²/day	
Indium to germanium‡	700	10^{-18} m²/sec	
Antimony to germanium‡	875	10^{-14} m²/sec	
Boron to silicon‡	1,140	10^{-16} m²/sec	
Gaseous carburizing agent to iron§	1,000	2×10^{-11} m²/sec	
Silver to aluminum§	500	$2\text{–}1.1 \times 10^{-18}$ m²/sec	
Holes (+) to germanium¶	30	4.4×10^{-3} m²/sec	100–1,000
Electrons (−) to germanium¶	30	9.3×10^{-3} m²/sec	100–1,000
Holes to silicon¶	30	6.5×10^{-4} m²/sec	50–500
Electrons to silicon¶	30	3.1×10^{-3} m²/sec	50–500

† C. D. Hodgman, "Handbook of Chemistry and Physics," p. 1191, Chemical Rubber Publishing Company, Cleveland, 1935.

‡ L. P. Hunter, "Handbook of Semiconductor Electronics," pp. 7–14, 15, McGraw-Hill Book Company, Inc., New York, 1956.

§ W. Jost, "Diffusion in Solids, Liquids, and Gases," chap. 5, Academic Press, Inc., New York, 1952.

¶ R. D. Middlebrook, "An Introduction to Junction Transistor Theory," p. 292, John Wiley & Sons, Inc., New York, 1957.

For charge diffusion in semiconductors, the wave equation (3-60) is the same as that for heat conduction along a thin rod [Eq. (3-51)], and the wave parameters take similar form. Thus,

Transmission-line analogy

(3-63) $$Z_0 = \frac{N_i}{J_i} = \frac{1}{q_i}\sqrt{\frac{1}{j\omega D_i(1 + 1/j\omega\tau_i)}} \qquad\qquad Z_0 = \sqrt{\frac{R}{G + j\omega C}}$$

where $\mathcal{J}_i = \text{Re } J_i e^{j\omega t}$ and $\nu_i = \text{Re } N_i e^{j\omega t}$.

Transmission-line analogy

(3-64) $$\gamma = \sqrt{\frac{1/\tau_i + j\omega}{D_i}} \qquad \gamma = \sqrt{R(G + j\omega C)}$$

Diffusion also takes place in gases, although intergaseous diffusion without mass transport by other means is unusual. The diffusion equation is the same as (3-54), but it is frequently desirable to write it in terms of partial pressure, in which case it takes the same form, with concentration replaced by partial pressure, and a different definition for the diffusion coefficient.

Some typical examples of diffusion coefficients are given in Table 3-10.

Example 3-11. *Diffusion of Water to Air.* Diffusion among two gases and two liquids is so slow that other processes, like turbulent mixing, overshadow it. Further, periodic excitation is unusual with such processes. Nonetheless, it is interesting to examine velocities and attenuations for some periodic excitations which might be significant components in Fourier analysis of a more likely transient excitation. The special case of a step transient is treated in Chap. 4.

Consider excitation with periods of 100 sec, 10,000 sec (almost 3 hr), and 10^6 sec (almost 12 days), for water vapor diffusing into air:

$$\text{Attenuation per meter} = \alpha = \sqrt{\frac{\omega}{2D}}$$

$$D = 0.329 \times 10^{-4} \text{ m}^2/\text{sec}$$

So

$$\alpha = 10^2 \sqrt{\frac{\pi f}{0.329}} = 310 \sqrt{f}$$

For a 100-sec period, $f = 10^{-2}$, $\alpha = 31$ nepers/m, and the water concentration is down to 37 per cent of maximum at a distance of $\frac{1}{31}$m , or only 3.22 cm. For a 10^4-sec period, $\alpha = 3.10$ nepers/m, so concentration is down to 37 per cent in 32 cm. Only for the 12-day period is the 37 per cent point out at 3.2 m. The velocity for a 100-sec period is

(3-45) $$\begin{aligned} v_p &= \sqrt{2\omega D} \\ &= \sqrt{4\pi f \times 0.329 \times 10^{-4}} = 0.0204 \sqrt{f} \\ &= 2.04 \times 10^{-3} \text{ m/sec} = 2.04 \text{ mm/sec} \end{aligned}$$

For 10^6 sec, it is a mere 2.04×10^{-2} mm/sec, or 7.3 cm/hr.

Example 3-12. *Diffusion of Hydrochloric Acid to Water.* Although in Example 3-11 velocities for gaseous diffusion seemed low and attenuation seemed high, gaseous diffusion is very *rapid* compared with liquid diffusion. Consider the case of HCl diffusing into

H_2O. Here $\alpha = 10^2 \sqrt{\pi f/2.21} = 119 \sqrt{f}$ nepers/m, where f is in cycles per *day*. Since there are 86,400 sec/day, a period of 100 sec is 864 cycles/day; a period of 10,000 sec, 8.64 cycles/day; and a period of 10^6 sec, 0.0864 cycle/day. Therefore, $\alpha = 3{,}500$ nepers/m, 350 nepers/m, and 35 nepers/m, respectively. Thus the concentration is $1/e$ (37 per cent) times the maximum at distances of 0.285 mm, 2.85 mm, and 2.85 cm, respectively. The velocities are

$$v_p = 10^{-2} \sqrt{4\pi \times 2.21 f} = 5.27 \times 10^{-2} \sqrt{f} \qquad \text{m/day}$$

This comes out as 1.52 m/day, 15.2 cm/day, and 1.52 cm/day—very low velocities indeed! But even these velocities are high compared with intermetallic diffusion.

Example 3-13. *Hole Diffusion in a Diode.* Consider a greatly idealized situation in which an rms hole current of 10^3 amp/m^2 at a frequency of 10^4 cps is injected into the N region of a silicon $P - N$ diode, under conditions such that the electric field, and therefore the drift current, may be neglected. Find the attenuation factor and diffusion impedance in the silicon. Assume a lifetime of 100 μsec.

Using (3-64), we find

$$\begin{aligned}\gamma &= \sqrt{\frac{1/10^{-4} + j2\pi \times 10^4}{6.5 \times 10^{-4}}} = 10^4 \sqrt{\frac{1 + j2\pi}{6.5}} \\ &= 10^4 \sqrt{\frac{6.37\underline{/81^\circ}}{6.5}} \\ &= 0.98 \times 10^4\underline{/40.5^\circ} = 7.45 \times 10^3 + j6.37 \times 10^3\end{aligned}$$

Hence $\alpha = 7.45$ nepers/mm, and the diffusion current damps out to a negligible value in much less than a millimeter. It is interesting to note that the wavelength is only

$$\lambda = \frac{2\pi}{6.37} = 0.98 \text{ mm}$$

in the silicon.

The diffusion impedance is

$$\begin{aligned}Z_0 &= \frac{1}{1.60 \times 10^{-19} \sqrt{j2\pi \times 10^4 \times 6.5 \times 10^{-4}[1 + 1/(j2\pi \times 10^4 \times 10^{-4})]}} \\ &= \frac{1}{1.60 \times 10^{-19} \sqrt{(40.7\underline{/90^\circ})(1 - j0.16)}} \\ &= \frac{1}{1.60 \times 10^{-19} \sqrt{41.2\underline{/80.9^\circ}}} \\ &= 9.85 \times 10^{17}\underline{/-40.45^\circ} \text{ (amp-m)}^{-1}\end{aligned}$$

Hence, the rms concentration of holes at the surface is

$$N_h = Z_0 J_x = 9.85 \times 10^{17} \times 10^3 = 9.85 \times 10^{20} \text{ holes/m}^3$$

with decrease into the N region as $\exp(-\alpha x)$.

3-7. Other Waves

Many other wave equations could be analyzed. Two of the most important are Schrödinger's equation of wave mechanics and the water-wave equation. One might also discuss such things as the wave equation for traffic starting up at a light or that for signals traveling along a nerve.

Schrödinger's equation[1] deals with the potential function of wave mechanics, a quantity which describes the wave nature of matter. It is normally operated on by a differential operator, and the combination represents momentum squared, although sometimes it represents other quantities in wave mechanics.

Schrödinger's equation is simply the familiar wave equation

$$\frac{\partial^2 \psi}{\partial x^2} = \frac{1}{v_p^2} \frac{\partial^2 \psi}{\partial t^2} \tag{3-65a}$$

or, in three-dimensional form,

$$\nabla^2 \psi = \frac{1}{v_p^2} \frac{\partial^2 \psi}{\partial t^2} \tag{3-65b}$$

$\psi(x,t)dx\,dt$ represents the probability that the space between x and $x + dx$ is occupied between time t and $t + dt$. In this case, the velocity of propagation is the "velocity of light." Many of the techniques of transmission-line waves may be applied to solutions of Schrödinger's equation.

Water waves follow a similar equation:

$$\frac{\partial^2 u_w}{\partial x^2} = \frac{1}{v_p^2} \frac{\partial^2 u_w}{\partial t^2} \tag{3-66}$$

where u_w is the vertical particle velocity. This is the same equation as for the vibrating membrane. Here the membrane is the surface of the water, and surface tension tends to hold it together like a membrane. There are several versions of v_p for water waves. The rather complex expression for deep water involves both the gravitational attraction and the surface tension. For shallow water the velocity is given by

$$v_p = \sqrt{gh} \tag{3-67}$$

where g is the gravitational acceleration and h is the depth of the water.

[1] D. C. Peaslee and H. Mueller, "Elements of Atomic Physics," chap. 10, Prentice-Hall, Inc., Englewood Cliffs, N.J., 1955.

3-8. Summary

Many traveling-wave phenomena may be treated in a manner directly parallel to that used for electric transmission lines. The analogy has been developed here for the vibrating string; for sound in gases, liquids, and solids; for shear and torsion in solids; for heat; and for diffusion.

The technique used for each phenomenon is to develop a set of *telegrapher's equations* which apply to the type of wave in question. Combining these gives a second-order differential *wave equation* which may be solved in the usual manner. For each set of telegrapher's equations, variables and parameters are found which correspond to the voltage and current variables and the resistance, inductance, conductance, and capacitance parameters of the transmission line. Although it would be possible in each case to determine the propagation constant, the attenuation and phase-shift constants, the velocity of propagation, the wavelength, and the characteristic impedance in the same way that they were determined for the transmission line in Chap. 2, the similarity of the equations makes possible the simpler approach of writing these down directly in terms of their transmission-line analogs.

The wave equations derived in this chapter fall into two forms, a lossless one and a very lossy (diffusion) equation. The lossless-wave equation in every case is analogous to the transmission-line-wave equation (2-3) with the terms involving R and G omitted. Using v as a sort of universal variable, this equation takes the form

$$\frac{\partial^2 v}{\partial x^2} = \frac{1}{v_p^2}\frac{\partial^2 v}{\partial t^2} \tag{3-68}$$

For the transmission-line equation

$$v_p = \sqrt{\frac{1}{LC}}$$

For the various other waves governed by the lossless-wave equation, the velocity of propagation may be obtained by substituting those quantities which are analogous to L and C.

In general, the *diffusion equation* (which was applied to both heat and diffusion) is in the form

$$\frac{\partial^2 v}{\partial x^2} = \frac{1}{D}\frac{\partial v}{\partial t} \tag{3-69}$$

This corresponds to the transmission-line equation (2-3) with the first and third terms of the right-hand side dropped and one of the two terms involving the first time derivative retained. For a transmission line it corresponds to a line with only series resistance and shunt capacitance or one with only shunt conductance and series inductance. In heat conduction on thin rods and carrier diffusion in semiconductors, a term in v (undifferentiated) appears also.

Lossless waves are treated again in the first part of Chap. 4 and in Chaps. 5, 9, and 10. Lossy waves of the diffusion type are treated in the second part of Chap. 4 and to some extent in Chap. 6. Lossy waves in general are treated in various other places.

It is sometimes easier to see the parallelism between the transmission-line equations and the nonelectrical telegrapher's equations when they are all written together. This is done in Table 3-11 for some of the various types of waves treated here. The analogies which are apparent from examination of Table 3-11 are listed in Table 3-12. Two special cases are shown for the transmission line—the lossless LC line and the RC line which corresponds to diffusion. When more than two parameters (L, C, R, G) are present, the form is complicated and is not listed.

TABLE 3-11. TELEGRAPHER'S-EQUATION ANALOGS—A

	First equation	Type of wave	Second equation	
(2-1)	$\dfrac{\partial v}{\partial x} = -Ri - L\dfrac{\partial i}{\partial t}$	Transmission line	$\dfrac{\partial i}{\partial x} = -Gv - C\dfrac{\partial v}{\partial t}$	(2-2)
(3-2)	$\dfrac{\partial u_w}{\partial x} = \dfrac{1}{T}\dfrac{\partial f_w}{\partial t}$	Vibrating string	$\dfrac{\partial f_w}{\partial x} = \rho_L \dfrac{\partial u_w}{\partial t}$	(3-1)
(3-14)	$\dfrac{\partial u_x}{\partial x} = -\dfrac{1}{\gamma_g p_a}\dfrac{\partial p}{\partial t}$	Sound in gas	$\dfrac{\partial p}{\partial x} = -\rho_v \dfrac{\partial u_x}{\partial t}$	(3-15)
(3-24)	$\dfrac{\partial u_x}{\partial x} = -K\dfrac{\partial p}{\partial t}$	Sound in liquid	$\dfrac{\partial p}{\partial x} = -\rho_v \dfrac{\partial u_x}{\partial t}$	(3-15)
(3-29)	$\dfrac{\partial u_x}{\partial x} = -\dfrac{1}{Y_B}\dfrac{\partial p_x}{\partial t}$	Sound in solid	$\dfrac{\partial p}{\partial x} = -\rho_v \dfrac{\partial u_x}{\partial t}$	(3-15)
(3-36)	$\dfrac{\partial \tau}{\partial x} = -\dfrac{1}{k} q_x$	Thermal conduction	$\dfrac{\partial q_x}{\partial x} = -S\rho_v \dfrac{\partial \tau}{\partial t}$	(3-38)
(3-52)	$\dfrac{\partial c}{\partial x} = -\dfrac{1}{D} m_{vx}$	Diffusion in general	$\dfrac{\partial m_{vx}}{\partial x} = -\dfrac{\partial c}{\partial t}$	(3-53)
(3-59)	$\dfrac{\partial \nu_i}{\partial x} = \dfrac{-1}{D_i q_i} \mathcal{J}_{ix}$	Charge-carrier diffusion	$\dfrac{\partial \mathcal{J}_{ix}}{\partial x} = -q_i \dfrac{\nu_i}{\tau_i} - q_i \dfrac{\partial \nu_i}{\partial t}$	(3-58)

The analogies used in this chapter are those indicated by Tables 3-11 and 3-12. It is not necessary, however, to associate the transmission-line voltage equation with the particular telegrapher's equation that has been used here for each type of wave. There is nothing to prevent us from associating, for the vibrating string, (3-1) with (2-1) and (3-2) with (2-2). Heretofore the association has been the reverse of this.

TABLE 3-12. SUMMARY OF ANALOGS—A

Type of wave	Variables		Parameters				v_p	Z_0
Transmission line	v	i	R	L	G	C	$\frac{1}{\sqrt{LC}}$ (lossless) $\sqrt{\frac{2\omega}{RC}}$ (diffusion)	$\sqrt{\frac{R + j\omega L}{G + j\omega C}}$
Vibrating string	u_w	f_w	0	$-\frac{1}{T}$	0	$-\rho_L$	$\sqrt{\frac{T}{\rho_L}}$	$\sqrt{\frac{1}{\rho_L T}}$
Sound in gas	u_x	p	0	$\frac{1}{\gamma_g p_a}$	0	ρ_v	$\sqrt{\frac{\gamma_g p_a}{\rho_v}}$	$\frac{1}{\sqrt{\gamma_g \rho_v p_a}}$
Sound in liquid	u_x	p	0	K	0	ρ_v	$\sqrt{\frac{1}{K\rho_v}}$	$\sqrt{\frac{K}{\rho_v}}$
Sound in solid	u_x	p_x	0	$\frac{1}{Y_B}$	0	ρ_v	$\sqrt{\frac{Y_B}{\rho_v}}$	$\sqrt{\frac{1}{Y_B \rho_v}}$
Shear in solid	u_y	p_y	0	$\frac{1}{\mu}$	0	ρ_v	$\sqrt{\frac{\mu}{\rho_v}}$	$\frac{1}{\sqrt{\mu \rho_v}}$
Torsion in solid	ω	τ	0	†	0	†	$\sqrt{\frac{C}{\rho_v}}$	†
Heat	τ	q_x	$\frac{1}{k}$	0	0	$S\rho_v$	$\sqrt{2\omega D_t}$	$\sqrt{\frac{1}{j\omega k S \rho_v}}$
Diffusion in general	c	m_{vx}	$\frac{1}{D}$	0	0	1	$\sqrt{2\omega D}$	$\frac{1}{\sqrt{j\omega D}}$
Charge-carrier diffusion	n_i or ν_i	$\mathcal{J}_{ix}$	$\frac{1}{D_i q_i}$	0	$\frac{q_i}{\tau_i}$	q_i	$\frac{\omega}{\text{Re}\sqrt{(1/\tau_i + j\omega)/D_i}}$	$\frac{1}{q_i\sqrt{D_i(1/\tau_i + j\omega)}}$
Wave mechanics	ψ	†	†	†	†	†	c	†

† These parameters have not been defined here.

In fact, it is possible to interchange (2-1) and (2-2) as the column heads of Table 3-11. When this is done, a whole new set of analogies is developed in which velocity, temperature, and concentration become analogous to current; and force, pressure, stress, heat flux, and mass-transfer rate become analogous to voltage. When this set of analogies is used, Table 3-12 is replaced by Table 3-13. Here only a few of the types of waves treated in Table 3-12 are shown, since the purpose of Table 3-13 is only to indicate the changes that occur.

Notice that the velocities are unchanged. The reason for this is that they are related directly to the wave equations (3-68) and (3-69) and do not depend on the telegrapher's equations. The impedances, on the other hand, are reciprocal to what they were with the analogies of Tables 3-11 and 3-12.

For analysis of the various nonelectric waves it does not matter whether one set of analogies or another is used. The acoustic literature commonly uses the B analogies, like those of Table 3-12.

TABLE 3-13. SUMMARY OF ANALOGS—B

Wave type	Variables		Parameters				v_p	Z_0
Transmission line	v	i	R	L	G	C		$\sqrt{\dfrac{R + j\omega L}{G + j\omega C}} = \dfrac{V}{I}$
Vibrating string	f_w	u_w	0	$-\rho_L$	0	$-\dfrac{1}{T}$	$\sqrt{\dfrac{T}{\rho_L}}$	$\sqrt{T\rho_L} = \dfrac{F_w}{U_w} = \dfrac{f_w}{u_w}$
Sound in solid	p_x	u_x	0	ρ_v	0	$\dfrac{1}{Y_B}$	$\sqrt{\dfrac{Y_B}{\rho_v}}$	$\sqrt{\rho_v Y_B} = \dfrac{P_x}{U_x} = \dfrac{p_x}{u_x}$
Heat	q_x	τ	0	$S\rho_v$	$\dfrac{1}{k}$	0	$\sqrt{2\omega D_t}$	$\sqrt{j\omega k S\rho_v} = \dfrac{Q_x}{T}$
Diffusion in general	m_{vx}	c	0	1	$\dfrac{1}{D}$	0	$\sqrt{2\omega D}$	$\sqrt{j\omega D} = \dfrac{M_{vx}}{C}$

It is frequently convenient to model a nonelectrical problem by an electrical analog on which voltages and currents may be measured more easily than pressures and velocities or temperatures and heat fluxes. Likewise it is sometimes convenient to model one nonelectric wave with another. In each case we have a choice between two sets of analogies when we use these models for the waves being studied. Sometimes one is more convenient than the other, and it is desirable to keep in mind that both are available.

Lumped-constant models are sometimes used for the various types of waves, as discussed in Chap. 7, but it is not necessary to use them. For example, acoustic waves in water are used to model electromagnetic waves in air for training radar operators, and a lossy transmission line can be used to model a heat-flow problem.

The immediately preceding discussion on analogies has dealt strictly with plane waves. At various points in this chapter, however, the general three-dimensional waves have been discussed. Analogies between electric and nonelectric waves and between various nonelectric waves are also possible in the general case. Here, however, the fact that some waves are vector waves and others deal with scalar quantities makes the analogies less nearly perfect than for plane waves. The analogies also differ because some waves are longitudinal and others are transverse. Nevertheless, these analogies are frequently very useful.

PROBLEMS

3-1. A steel wire is used in a musical instrument, and 1 wavelength is 80 cm (only a half-wavelength is used in the instrument). This wire has a diameter of 0.5 mm and a mass of 1.53 g/m. If it is stretched to a tension of 90 newtons (about 20 lb), what will be its vibration frequency?

3-2. A 500,000-circular mil hard-drawn copper cable is strung between towers spaced 200 ft apart. The towers are 80 ft high, and the wire may come within 60 ft of the ground. The wire weighs 8,151 lb/mile. Calculate the velocity of a vibration wave on this cable, and determine the wavelength for a period of 5 sec. Determine the frequency for which the distance between towers is a half-wavelength. Note that a useful empirical formula for determining sag on a transmission line is

$$d = \frac{S^2 \rho_L}{8T_h}$$

where d = sag, ft
S = spacing between supports, ft
ρ_L = density, lb/ft
T_h = horizontal component of tension, lb

3-3. A certain loudspeaker may be considered to generate a plane wave in air at 3,000 cps. The speaker diaphragm moves a distance of 0.01 mm peak to peak at this frequency, and the particle velocity in the wave is the same as the diaphragm velocity. Calculate the pressure in the wave, and the wavelength.

3-4. Derive an expression for the sound intensity (power transmitted per unit area) in a plane sound wave. Use this expression to determine the intensity for the loudspeaker of Prob. 3-3.

3-5. A sound wave is generated in alcohol, for which the velocity of propagation is quoted as being 1,240 m/sec. The density is given in tables as 0.80 g/cm^3. Find the compressibility and characteristic impedance for the wave. What is the wavelength for a frequency of 10 kc?

3-6. Derive Eqs. (3-16) and (3-17).

3-7. Compare the characteristic impedance for copper, aluminum, granite, and rubber for longitudinal waves.

3-8. Using the result of Prob. 3-4, determine the pressure and particle velocity for a sound wave in water with an intensity of 100 watts/m^2.

3-9. Derive the wave equation for shear waves.

3-10. In an automobile-crash study, the time of a crash is determined by transmission of a signal from a crystal transducer mounted 20 cm back on the frame. If the velocity prior to crash is 30 m/sec, how long does it take for the impact shock to reach the crystal through the steel frame? If the crystal requires 1 msec to operate after receipt of the initial pulse, does it have time to function before it arrives at the barrier into which the auto crashes (assuming it continues at the inital velocity)?

3-11. Equation (3-35) describes transverse vibrations of a bar of circular cross section. Determine a solution of this equation for sinusoidal time variation.

3-12. A plane vibration is established in a large piece of aluminum. If the velocity of the wave is determined from its displacement, calculate the pressure at a frequency of 100 cps for a wave with an rms displacement of 1 mm. Calculate the wavelength in the aluminum.

3-13. The daily range of temperature outside a masonry building (brick) is 25°C. Find the range at a depth in the wall of 1 cm, of 10 cm. Calculate the heat flow at the surface and at 10 cm depth.

3-14. Derive Eq. (3-48).

3-15. A copper rod is of circular cross section and 1 cm in diameter. In still air, the surface emissivity may be taken as 10 watts/m^2-°C. Determine the propagation constant in the rod for a sinusoidal variation with a period of 500 sec. Compare the values of α obtained when surface loss is considered and when it is neglected.

3-16. Consider a germanium junction diode so biased that the electric field in the P region may be neglected. If the cross section is 1 mm^2 and an electron current of 50 rms mamp is injected into the P region, find α, the wavelength, and the electron density at 1 mm depth. Assume a frequency of 1 kc.

3-17. Repeat Prob. 3-16 for a silicon diode.

3-18. Determine the depth in aluminum at 500°C at which the silver concentration would be $1/e$ times its surface value for a component of a periodic exposure for which the period is 1 hr and one for which it is 1 day. Calculate the pertinent velocities of propagation.

3-19. Repeat Prob. 3-18 for hydrogen to air, adding a 100-sec period.

3-20. Derive for current density an equation analogous to (3-60).

3-21. Solve (3-60) for sinusoidal excitation, finding N_i from $n_i = \text{Re } N_i e^{j\omega t}$. Determine J_{ix} from this.

3-22. Derive the wave equation which results from the form of Fick's law shown in Eq. (3-61).

3-23. Derive Eq. (3-54a) and the comparable equation for m_v.

3-24. A home gas-fired furnace operates under certain heating loads with a cycle of 10 min on and 10 min off. The temperature at the furnace wall fluctuates from 175 to 90°F. The furnace wall is attached to mild-steel mounting brackets which make contact with wooden supports at a distance of 2 in. from the furnace. Neglecting radiation and convection losses from the bracket, determine the temperature range at the nearest point of the wood. Neglect reflection.

3-25. Show that Eq. (3-39) follows from Eq. (3-38).

3-26. Derive the equation for q comparable with (3-42).

3-27. Prepare a table like Table 3-12 using heat as the basis, rather than a transmission line, and making the following quantities analogous to temperature: force for vibrating string, sound velocity in gas, sound pressure in liquid, sound pressure in solid, shear velocity, torque, mass-flow rate.

3-28. Complete Table 3-13.

3-29. Add an "attenuation" column to Table 3-12.

BIBLIOGRAPHY

Numbers in parentheses refer to sections of this chapter related to the reference.

Dekker, Adrianus J.: "Electrical Engineering Materials," chaps. 5–7, Prentice-Hall, Inc., Englewood Cliffs, N.J., 1959. (6)

Ewing, W. M., W. S. Jardetsky, and Frank Press: "Elastic Waves in Layered Media," McGraw-Hill Book Company, Inc., New York, 1957. (4)

Guy, Albert G.: "Elements of Physical Metallurgy," Addison-Wesley Publishing Company, Reading, Mass., 1951. (6)

Hausmann, Erich, and Edgar P. Slack: "Physics," 2d ed., D. Van Nostrand Company, Inc., Princeton, N.J., 1939. (2, 3)

Heiland, C. A.: "Geophysical Exploration," Prentice-Hall, Inc., Englewood Cliffs, N.J., 1940. (4)

Hix, C. F., Jr., and R. P. Alley: "Physical Laws and Effects," John Wiley & Sons, Inc., New York, 1958. (3–7)

Hueter, Theodore F., and Richard H. Bolt: "Sonics," John Wiley & Sons, Inc., New York, 1955. (2–4)

Hunter, Lloyd P.: "Handbook of Semiconductor Electronics," McGraw-Hill Book Company, Inc., New York, 1956. (6)

Ingersoll, Leonard R., Otto J. Zobel, and Alfred C. Ingersoll: "Heat Conduction," McGraw-Hill Book Company, Inc., New York, 1948. (5)

Jost, W.: "Diffusion in Solids, Liquids, and Gases," Academic Press, Inc., New York, 1952. (5, 6)

Leet, L. D.: "Practical Seismology and Seismic Prospecting," Appleton-Century-Crofts, Inc., New York, 1938. (4)

Lo, Arthur W., Richard O. Endres, Jakob Zawels, Fred D. Waldhauer, and Chung-Chih Cheng: "Transistor Electronics," Prentice-Hall, Inc., Englewood Cliffs, N.J., 1955. (6)

Mickley, Harold S., Thomas K. Sherwood, and Charles E. Reed: "Applied Mathematics in Chemical Engineering," 2d ed., McGraw-Hill Book Company, Inc., New York, 1957. (5, 6)

Middlebrook, R. D.: "An Introduction to Junction Transistor Theory," John Wiley & Sons, Inc., New York, 1957. (6)

Moore, Richard K.: "Wave and Diffusion Analogies," McGraw-Hill Book Company, Inc., New York, in preparation. (1–8)

Officer, C. B.: "Introduction to the Theory of Sound Transmission," McGraw-Hill Book Company, Inc., New York, 1958. (4)

Olson, Harry F.: "Acoustical Engineering," D. Van Nostrand Company, Inc., Princeton, N.J., 1957. (2–4)

———: "Dynamical Analogies," D. Van Nostrand Company, Inc., Princeton, N.J., 1943. (3, 4)

Peaslee, D. C., and H. Mueller: "Elements of Atomic Physics," Prentice-Hall, Inc., Englewood Cliffs, N.J., 1955. (7)

Rayleigh, Lord: "The Theory of Sound," 2d ed., vols. I and II, Dover Publications, New York, 1945. (2–4)

"Reference Data for Radio Engineers," 4th ed., International Telephone and Telegraph Corp., New York, 1956. (2–4, 7)

Schneider, P. J.: "Conduction Heat Transfer," Addison-Wesley Publishing Company, Reading, Mass., 1955. (5)

Timoshenko, S., and J. N. Goodier: "Theory of Elasticity," 2d ed., chap. 15, McGraw-Hill Book Company, Inc., New York, 1951. (4)

4. Transient Traveling Waves

Many important traveling-wave applications involve *transients* which propagate down a transmission line or through some sort of medium. The general wave equation was developed in Chap. 2 for the transmission line, and comparable equations were developed in Chap. 3 for nonelectric waves. The solutions given in these chapters, however, were the steady-state a-c solutions.

Transient problems on transmission lines are of many types. Transients on transmission lines are used to make rectangular pulses in pulse generators. The earliest transients studied on transmission lines were the rectangular pulses of telegraphy. Later transient studies in the transmission-line field had to do with faults on power lines. When lightning strikes a power line, a large surge voltage is induced in the local area, and this propagates as a transient to other parts of the line. When a power line is suddenly short-circuited, the transient associated with this short circuit takes a finite time to reach the circuit breakers, which permit the shorted section to be disconnected from the rest of the power system.

Transients are not of great importance on strings, with the possible exception of those which support suspension bridges. The moving of a large vehicle across a bridge represents transient loading of the supporting cables. This problem is much more complex than the simple string treated in Chap. 3.

Transient sound waves in air are frequently important. The thunderclap is the first example that comes to mind. Often, the time of propagation of the sound of the firing of a gun is significant, and the sound wave emitted from an atomic blast provides one of the methods for detecting a blast at a long distance. Sound waves generated by jet aircraft create "sonic booms."

Sound waves in liquids are used in *sonar* devices for locating submerged submarines and also for determining the depth of the bottom of the sea beneath a ship. Transient modulated-carrier pulses are used for this purpose.

Transient pulses are used in solids for flaw detection. In the earth, of course, they are used for seismic prospecting. Transient pulses are important wherever sudden loading occurs on a machine, whether it is a pile driver or an automobile engine.

Steady-state alternating heat flow is much less common than transient heat flow, and nearly all diffusion problems are transient problems.

In this chapter two special types of transients are treated: those for which losses may be neglected [which follow the lossless-wave equation (3-68)] and those which follow the diffusion equation (3-69). Section 4-5 discusses a general approach to transients, but the treatment is not detailed, since the general solution is considerably more complicated than the special cases.

4-1. Transients on Lossless Lines and Media

In the general case, the velocity of transmission is a function of frequency. As indicated in Sec. 4-5, the steady-state type of solution may be superimposed after Fourier analysis when the velocity is a function of frequency.

For many important problems, the velocity may be considered independent of frequency. With the exception of a special case treated in Chap. 8 (the "distortionless line"), such waves are restricted to those situations in which loss may be neglected.

The velocity of propagation may be written, using (2-20) and (2-16), as

$$v_p = \frac{\omega}{\beta} = \frac{\omega}{\operatorname{Im} \sqrt{(R + j\omega L)(G + j\omega C)}}$$

where Im means "imaginary part of." This may be rewritten as

$$v_p = \frac{\omega}{\operatorname{Im} (j\omega \sqrt{LC}) \sqrt{(1 + R/j\omega L)(1 + G/j\omega C)}}$$

to show the effect of small R and G. If R and G are zero, the velocity is completely independent of frequency, as the ω's cancel out in the numerator and denominator. Since a practical line must have some loss, this situation may only be approximated in practice, but the approximation is a good one whenever

$$R \ll \omega L \qquad G \ll \omega C$$

When this is true, we have the low-loss value for velocity of propagation

$$v_p \approx \frac{1}{\sqrt{LC}} \tag{2-34}$$

This approximation is valid for most radio-frequency transmission lines; for electromagnetic waves through nonconducting space; for acoustic waves in gases, liquids, and most solids; and for waves on a string. This assumption corresponds to rewriting the wave equation neglecting the terms involving R and G. Hence, (2-3) becomes

$$\frac{\partial^2 v}{\partial x^2} = LC\frac{\partial^2 v}{\partial t^2} = \frac{1}{v_p^2}\frac{\partial^2 v}{\partial t^2} \tag{4-1}$$

Note that this is also the form of Eq. (3-68).

The traveling-wave solution of this equation takes the general form

$$v = f(x - v_p t) + g(x + v_p t) \tag{4-2}$$

It can be shown that both the f and the g terms satisfy (4-1) independently and, therefore, that their sum must satisfy it, since (4-1) is a linear equation. Note that f is any function of the variable in its parenthesis and g is any function of the variable in its parenthesis. As long as x is increased with velocity v_p, the value of the function f remains the same. To keep the value the same for g, it is necessary that x be decreased at a rate v_p as t is increased, thus keeping the sum in parenthesis the same.

To verify that this kind of solution does satisfy Eq. (4-1), consider the situation when $g = 0$ (wave traveling only in the direction of increasing x). Substituting (4-2) in (4-1), we have

$$\frac{\partial v}{\partial x} = \frac{\partial f}{\partial(x - v_p t)}\frac{\partial(x - v_p t)}{\partial x} = \frac{\partial f}{\partial(x - v_p t)}$$

$$\frac{\partial v}{\partial t} = \frac{\partial f}{\partial(x - v_p t)}\frac{\partial(x - v_p t)}{\partial t} = -v_p\frac{\partial f}{\partial(x - v_p t)}$$

$$\frac{\partial^2 v}{\partial x^2} = \frac{\partial^2 f}{\partial(x - v_p t)^2}\frac{\partial(x - v_p t)}{\partial x} = \frac{\partial^2 f}{\partial(x - v_p t)^2}$$

$$\frac{\partial^2 v}{\partial t^2} = -v_p\frac{\partial^2 f}{\partial(x - v_p t)^2}\frac{\partial(x - v_p t)}{\partial t} = v_p^2\frac{\partial^2 f}{\partial(x - v_p t)^2}$$

Substituting these results in (4-1) shows that the equation is indeed satisfied by any function $f(x - v_p t)$.

Since there is a current equation which corresponds directly to Eq. (4-1), its solution must be of the same form. Thus,

$$i = Mf(x - v_p t) + Ng(x + v_p t)$$

Again considering only a wave traveling in the increasing x direction ($g = 0$), we may apply the telegrapher's equation (2-1) with $R = 0$:

$$\frac{\partial v}{\partial x} = -L\frac{\partial i}{\partial t}$$

Using the values for the derivative given above and substituting into this equation,

$$\frac{\partial f(x - v_p t)}{\partial(x - v_p t)} = -LM\left[-v_p\frac{\partial f(x - v_p t)}{\partial(x - v_p t)}\right]$$

Canceling out the equal derivatives, we have

$$M = \frac{1}{Lv_p} = \frac{\sqrt{LC}}{L} = \sqrt{\frac{C}{L}} = \frac{1}{Z_0}$$

This expression for Z_0 is the one which applies for low loss and was given in (2-32).

For the wave traveling in the increasing x direction, the ratio of voltage to current is given by

(4-3) $$\frac{v}{i} = \frac{f}{Mf} = Z_0$$

Thus, for *any* such wave on a lossless line, the voltage-to-current ratio is the characteristic impedance, the same as it is for the steady-state *sine wave* discussed in Chap. 2. The characteristic impedance is written as an impedance Z rather than a resistance R; but, in fact, for the lossless line, it is a resistance and must be so if the instantaneous values of voltage and current are to be related by (4-3).

It can be shown by a similar technique that the voltage-to-current ratio for a wave traveling to the left is given by

(4-4) $$\frac{v}{i} = \frac{g}{Ng} = -Z_0$$

The minus sign for the wave traveling in the decreasing x direction may be better understood by considering the following: if the voltage is in the same direction for both waves, it is necessary that the current flow in the positive x direction if power flows in that direction and that the current flow in the negative x direction if the power is to flow in that direction.

We are now prepared to answer a very important question regarding the transmission line. Suppose that a battery with internal emf E and internal resistance R_g is suddenly connected to one end of an uncharged transmission line, as indicated in Fig. 4-1. An important question which must be answered is, With this known supply, what voltage and current establish themselves immediately on the line? This is easily seen, since (4-2) is the ratio of voltage to current for a wave traveling in the increasing x direction. When the battery is first connected to the line and the line is uncharged, there can be no wave toward the battery, so the wave toward the right must be the one which is set up. If its voltage-to-current ratio is known to be a certain resistance, the emf of the source divides between this resistance and the source resistance in the usual circuit manner. Thus, the equivalent circuit for this purpose is as shown in Fig. 4-1*b*. The characteristic impedance is the resistance which the line initially presents to the source. *It is most important to realize that this is true only for the wave starting down the line and that it is not true for the final result in a steady-state situation.*

The circuit of Fig. 4-1 will eventually settle down to its d-c value. If the transmission line is assumed to be lossless, there will be no series

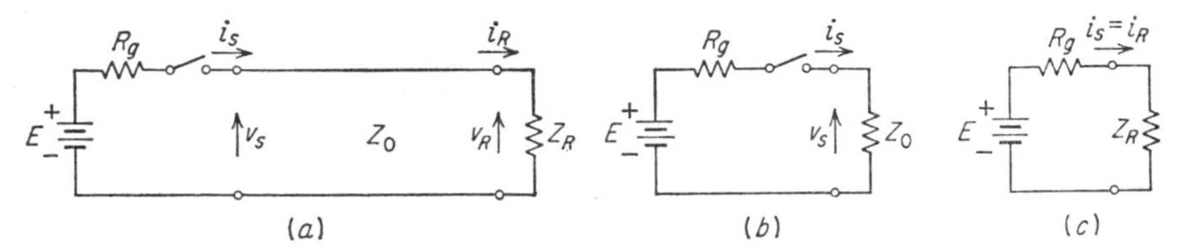

FIG. 4-1. Equivalent source circuit: (*a*) actual circuit; (*b*) initial equivalent circuit; (*c*) steady-current equivalent circuit.

drop in the line, and the result will be as if the load impedance Z_R were connected directly at the sending end of the line. This does not happen instantaneously, but it is the final steady value, even though Z_R may be quite a bit different from Z_0. The equivalent circuit appears in Fig. 4-1*c*.

For the initial pulse, then, the voltage at the sending end v_s is given by

$$v_s = \frac{Z_0}{Z_0 + R_g} E$$

Of course, a point down the line from the sending end "does not know immediately" that the switch has been closed, for it takes a finite time for the disturbance to propagate down the line, as was discussed in Chap. 2. In Chap. 2, it was assumed that the line was infinitely long or that it was terminated in its characteristic impedance. Here we are considering that the line is of finite length and is not necessarily terminated in its characteristic impedance.

Before discussing the implications of this finite-length line, it is desirable to study some of the types of excitation which may be set up on it and on its analogs. Frequently, both lines and other wave devices are excited by pulses whose duration is short rather than by the step functions discussed above. Such a pulse was shown in Chap. 2, but a simplified exciter for it is shown again in Fig. 4-2. The waveform at the sending end of the line associated with this is shown in Fig. 4-3. It is assumed that the line is initially uncharged. At time $t = 0$, the switch is thrown from the short-circuit position to the battery, so that current begins to flow into the line as shown in Fig. 4-1. After a time τ, the switch is thrown back to the short-circuit position, and the pulse ends as indicated in Fig. 4-3. It is frequently desirable to consider the solution for the pulse on the transmission line as if it were made up of two step functions at different times. This is indicated in Fig. 4-3, where the beginning of the pulse is shown as a step function which continues indefinitely. The end of the pulse is represented by superimposing a negative step equal and opposite to the initial positive step. Superimposing the equal

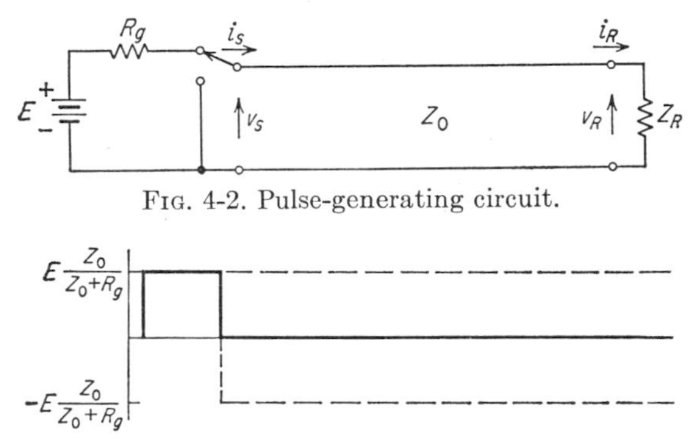

FIG. 4-2. Pulse-generating circuit.

FIG. 4-3. Pulse as superposition of steps. Solid line is resultant of two dash-line steps.

positive and negative voltages results in zero voltage from then on, as indicated in Fig. 4-3.

Actually, in many cases, the waveshapes are considerably more complicated than these. On a lossless line it is usually possible to sketch what happens even for the more complicated waveshapes without resolving the pulse into its Fourier components. On a lossy line, however, because of the different velocities and attenuations for the different Fourier components, the pulse must be resolved.

For the string, a pulse of constant force might be applied at some point. Probably, however, a pulse of constant velocity or (approximately) a pulse of constant displacement would be more likely.

For sound in air or water, continuous velocity, such as in the electrical equivalent of Fig. 4-1, is possible only when the object applying the velocity is moving continuously. Since it is difficult to accelerate such an object instantaneously, the step function (which is fairly easy to achieve on the electric transmission line) is quite difficult for either velocity or pressure in a sound wave. Of course, except for a slower rise, the motion of some sort of vehicle, like an airplane or car, would correspond to the battery voltage of Fig. 4-1. Pulses of pressure, however, are quite common. Thunder is simply a pressure pulse traveling from the expanding air about a lightning stroke. The air expanding in any kind of explosion creates a pressure pulse, which we hear as a boom. In sound-ranging devices, such as sonar and depth indicators, pulses are generated electrically and converted by transducers into mechanical motion pulses in the water.

Longitudinal step functions are also rather hard to achieve in a solid. An example of a step function in a solid is that due to the starting of a freight train. Initially, the engine begins to move forward, and the first car begins to move only after the slack in the coupling has been taken up. Hence, a sudden tension (and velocity) pulse is imposed on the frame of that car, and, because the engine continues to pull, is maintained. New steps occur as the load at the back of the car is increased when the next car begins to be pulled. A pressure rise of almost a step occurs in the frame of a rocket when the engine is turned on and its force begins to be felt through the shell of the rocket.

Pulses of sound in solids are quite common. In addition to the artificially generated ones used for locating flaws, there is a pulse of sound in a solid every time a solid is hit by a hammer. The pulse in a pile when hit by a pile driver is important to the boring in of the pile. Whenever a bullet hits a solid object, a pulse is sent through the object. Interesting studies have been made of the travel of this pressure pulse through both the solids and liquids present in the chest cavity of animals. Whenever two autos collide or one strikes a tree, a relatively short pressure pulse is transmitted to the frame, so that the back of the frame does not receive the pressure until after the front has been stopped.

4-2. Reflection

Our treatment of traveling waves would be almost over at this point if it were not for the phenomena of reflection. Desired reflections allow us to make resonant circuits out of transmission lines; enable us to produce musical instruments using strings, such as the violin and piano; play an important part in radar, sonar, and seismic prospecting; and have many other applications too numerous to mention.

Undesired reflections cause ghosts on television images, produce oscillation in telephone lines, create overvoltages on power lines, make it difficult to hear speakers in an auditorium, break supports for masses which are suddenly accelerated, and result in many other complications.

Reflections are treated here for the transient condition. *The treatment is perfectly general, however, and applies equally well for the steady state.* The details of the steady-state treatment are reserved for Chaps. 5 and 6.

For both the transient and the steady-state solution of the traveling-wave equation, there are two terms, one representing a wave traveling in the increasing x direction and the other representing a wave traveling in the decreasing x direction. In Chap. 2 we described these waves in terms of the value at the source and the distance from the transmitting end of the line, that is, the distance from the source. When reflected waves are present, it is more convenient to describe both the incident and the reflected waves in terms of their values at the reflection point and their distances from this point.

In the following discussion we use a superscript $+$ to represent a wave traveling from the source toward the load, called an incident wave, and a superscript $-$ to represent a wave reflected from the load and traveling toward the source, called the reflected wave. We let

v^+ = voltage traveling toward the load
v^- = voltage traveling toward the generator
i^+ = current traveling toward the load
i^- = current traveling away from the load and toward the generator

Thus, the voltage at the load (receiving end) is

$$v_R = v^+ + v^- \tag{4-5}$$

and

$$i_R = i^+ + i^- = \frac{v^+}{Z_0} - \frac{v^-}{Z_0} \tag{4-6}$$

The minus sign in (4-6) is that which was found in (4-4). Thus, if both voltage and current are positive, as indicated in (4-1), the power must be flowing toward the load; but if either is reversed, the power must be flowing from the load toward the generator.

To determine the relation between reflected and incident voltages and currents, it is necessary to apply at the receiving end of the line the *boundary conditions* placed there by the load impedance. Thus, the ratio of v_R to i_R is fixed by the load impedance, so that

$$\frac{v_R}{i_R} = Z_R = \frac{v^+ + v^-}{v^+/Z_0 - v^-/Z_0} = Z_0 \frac{v^+ + v^-}{v^+ - v^-}$$

Now let us define the *reflection coefficient* as the ratio of the reflected voltage to the incident voltage:

$$\Gamma_R = \frac{v^-}{v^+} \tag{4-7}$$

The reflection coefficient is a measure of voltage reflected in terms of the voltage incident. Thus, if there is no reflected voltage, the reflection coefficient is zero. If the reflected voltage is equal to the incident voltage and is in phase, the reflection coefficient is 1; if it is equal but is 180° out of phase, the reflection coefficient is -1. Because of the difference in signs between the reflected voltage and reflected current in (4-5) and (4-6), the reflection coefficient for current is the negative of the reflection coefficient for voltage.

To find the reflection coefficient in terms of the impedances, we rewrite the equation preceding (4-7) as

$$\frac{Z_R}{Z_0} = \frac{1 + \Gamma_R}{1 - \Gamma_R}$$

This may be solved for Γ_R, finally giving us

$$\Gamma_R = \frac{Z_R/Z_0 - 1}{Z_R/Z_0 + 1} = \frac{Z_R - Z_0}{Z_R + Z_0} \qquad (4\text{-}8)$$

This is a very important equation in traveling-wave work and it should be committed to memory.

Although this expression for the reflection coefficient is general and not based upon the lossless line, it can be applied to transients without Fourier analysis of the waveshape only if the line is lossless, and then only when the reflection coefficient is independent of frequency. This means that, if we are to calculate easily the reflections of pulses, the load impedance Z_R should be a resistance. If it is not, the reflection coefficient is complex and a function of frequency, so that the pulse must be resolved into its various Fourier components, each of which is reflected with a different magnitude and phase.

When a load is a resistance, the behavior of the reflection coefficient is as indicated in Fig. 4-4. Note that a short circuit ($Z_R = 0$) results in a reflection coefficient of -1, whereas an open circuit (Z_R infinite) results in a reflection coefficient of $+1$. These are easily justified by examination of (4-5) and (4-6), as well as by substitution in (4-8). Thus, for a short circuit, the voltage at the receiving end must be zero, so v^- must equal $(-v^+)$. Similarly, for an open circuit, the current must be zero. Hence, i^+ must equal $(-i^-)$, and the voltage reflection coefficient must be $+1$.

It can also be seen from the figure that the reflection coefficient is *zero* when the load impedance is *equal* to the characteristic impedance. Thus there is no reflected wave when the line is properly terminated. This was the basis of our discussions in Chaps. 2 and 3, in which proper termination was assumed.

To see what happens to a pulse when reflection is present, consider the circuit of Fig. 4-1 with the switch closed at $t = 0$ and with a reflection coefficient at the receiving end of the line of $+\frac{1}{2}$ ($Z_R = 3Z_0$). In Fig. 4-5*a* the pulse is shown approaching the end of the line for the first time. This picture is the same as has been shown previously. Figure 4-5*b* shows the condition shortly after the pulse reaches the end of the

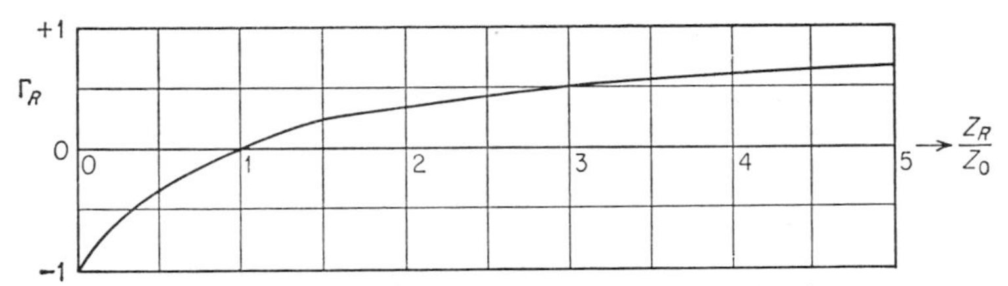

FIG. 4-4. Reflection coefficient for resistive load.

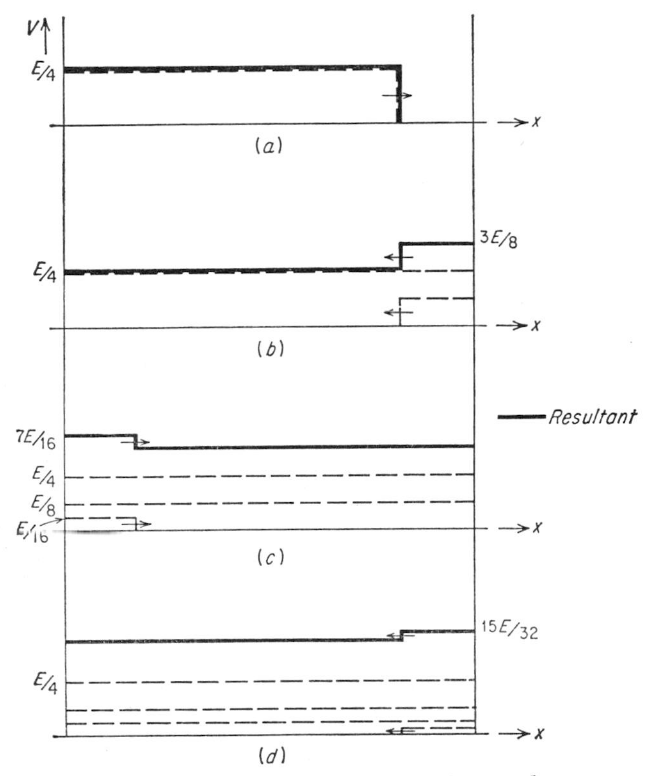

FIG. 4-5. Reflection on a line with $\Gamma_R = \frac{1}{2}$.

line. The reflected voltage wave has started back along the line. The resultant voltage between the end of the line and the end of the reflected pulse consists of the incident voltage plus the reflected voltage. Further back on the line it is just the incident voltage.

The next question which arises is, What happens when the reflected wave hits the battery? It can be easily shown that the reflection at the battery, for the wave which was initially reflected from the load, is governed by the reflection coefficient:

$$\Gamma_s = \frac{Z_g - Z_0}{Z_g + Z_0} \tag{4-9}$$

Thus, the reflection is the same as it would be if the emf were simply shorted out.

If in our example we assume that the battery internal resistance is also three times the characteristic impedance, the reflection coefficient at that end will also be $\frac{1}{2}$. The result will then appear as in Fig. 4-5*c* shortly after the reflection at the battery has taken place. The voltage on the line near the sending end now consists of the original incident voltage plus the original reflected voltage (one-half the original incident voltage) plus a new re-reflected voltage wave (going in the same direction as the original incident wave) whose magnitude is half the original reflected wave or one-fourth the original incident wave. When this wave has traveled to the end of the line and has been reflected there, the situation is as shown in Fig. 4-5*d*. Here the resultant has again gone up, but this time by one-eighth of the original incident voltage.

It is intuitively apparent that the resultant voltage must approach as a limit the voltage that would appear across the load if the line had been so short that traveling waves did not need to be considered at all. Thus, we expect that the steady state is just the same as would have been predicted by d-c circuit analysis. Of course, it takes an infinite time to achieve this result.

In order to keep track of the various reflections, it is convenient to use a "bounce diagram," on which the progress of the wave is plotted as a function of time and distance, with magnitudes of the various components indicated on the diagram numerically. It is usually desirable to make such a diagram for both the voltage and the current on the line. A bounce diagram is shown in Fig. 4-6, and its use is illustrated in Example 4-1.

Example 4-1. Consider a lossless line having

$$R_g = Z_R = \frac{Z_0}{2}$$

Thus,

$$\Gamma_R = \Gamma_S = -\tfrac{1}{3}$$

Initially, for a battery voltage of E, $2E/3$ appears across the line, and the current is $2E/3Z_0$. These are indicated on Fig. 4-6. Here T is the time for a wave to travel one length of the line. The

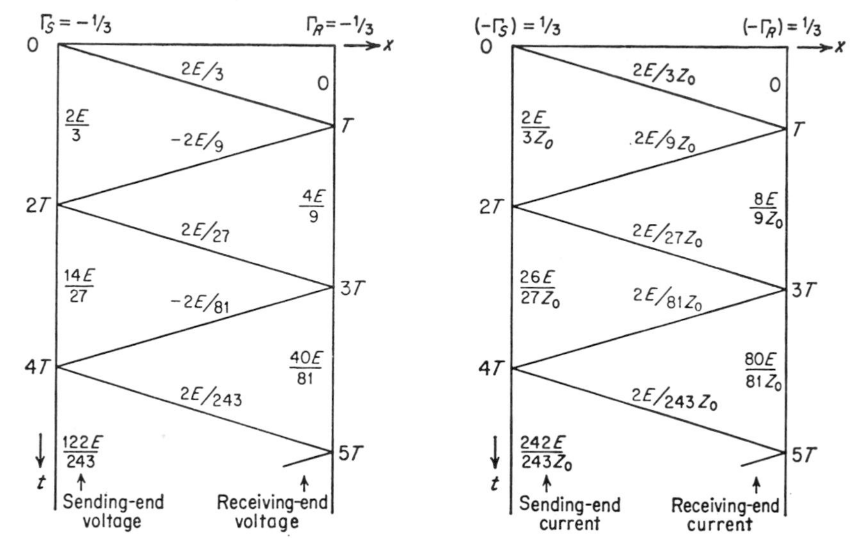

FIG. 4-6. Bounce diagrams for Example 4-1.

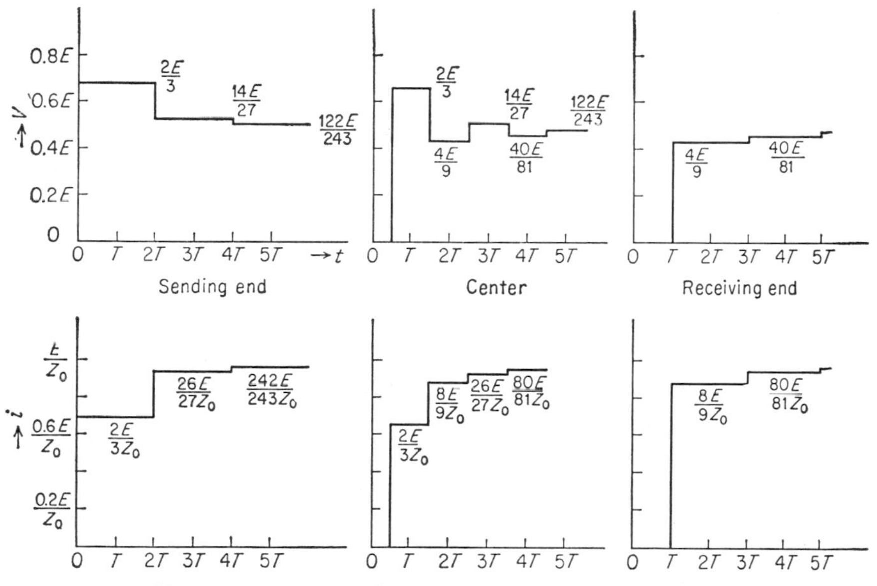

FIG. 4-7. Voltages and currents on line of Example 4-1.

voltage and current at three points on the line (the two ends and the middle) are plotted vs. time in Fig. 4-7. The data for plotting these were obtained directly from the bounce diagram.

On the bounce diagram, distance is measured horizontally and time vertically. The zigzag line traces the progress of the wavefront in space and time. The magnitude of each portion of wavefront is shown in the center, and the terminal voltages are shown at the sides. At any point x, the total voltage at a time t is the sum of all the voltages appearing above it on the diagram. Note that the voltages tend toward $E/2$ and the currents toward E/Z_0, the d-c values. This can also be seen from noting that v_R tends toward

$$\begin{aligned} v_R &= v^+[1 + (-\tfrac{1}{3}) + (\tfrac{1}{3})^2 - (\tfrac{1}{3})^3 + \cdots] \\ &= v^+ \frac{1}{1 + \frac{1}{3}} = \frac{2E}{3}\frac{3}{4} = \frac{E}{2} \end{aligned}$$

where v^+ represents the initial incident voltage. Similarly,

$$\begin{aligned} i_R &= i^+[1 + \tfrac{1}{3} + (\tfrac{1}{3})^2 + \cdots] \\ &= i^+ \frac{1}{1 - \frac{1}{3}} = \frac{2E}{3Z_0}\frac{3}{2} = \frac{E}{Z_0} \end{aligned}$$

where i^+ represents the initial incident current.

The implications of the reflection coefficient are now considered for the various nonelectric waves and the plane electromagnetic wave in space, with particular reference to the equivalents to short- and open-circuited transmission lines.

The reflection coefficient for a plane electromagnetic wave normally incident upon a boundary is just the ratio of the reflected electric field to the incident electric field. If this boundary is a perfect conductor, the reflection coefficient is -1, and the reflected electric field is equal and opposite to the incident electric field, since the net field in the conductor must be zero. Examination of Eq. (4-6) shows that, when this is true for a transmission line or a wave in space, the *current* (or *magnetic field*) in the reflected wave is *equal* to and in the *same direction* as the current in the incident wave, so that the resultant current or magnetic field at the load is twice its incident value. That is, the current in a short circuit or the magnetic field in a perfect-conductor boundary is twice the incident value. For an open circuit, the voltage is doubled, and the current is zero. An open circuit is hard to achieve for a plane electromagnetic wave because intrinsic impedances do not become infinite; however, the open-circuit situation is approached when a wave traveling through a medium with high dielectric constant, like water, is normally incident on a boundary with air.

With a vibrating string, it is normal to fix the ends. In musical instruments, of course, this is common; but even in the case of the power line described in Example 3-2, the line is reasonably well tied down at each tower. Because the string is considered limp, it is difficult to conceive of a situation where the end is loose. If the end is fixed, its velocity is zero, and this corresponds (in the analog labeled A in Chap. 3) to zero voltage. Hence, the fixed-end string corresponds to the short-circuited transmission line, and the reflection coefficient is $+1$ for the force or -1 for the velocity.

Because most gases have roughly the same acoustic impedance, the situation of a surface where a gas pressure drops essentially to zero (corresponding to an open-circuited transmission line) is not achievable. However, the kind of surface corresponding to a short-circuited transmission line is readily available, since this is just a hard surface for which the particle velocity must go to zero. Because of the big difference in pressures in a liquid and a gas for the same velocities (therefore a large difference in impedances), a sound wave arriving at a surface between a liquid and a gas from within the liquid experiences the equivalent of an open-circuit termination, since the pressure in a gas is essentially zero as compared with the pressure in the liquid for the same velocity.

Fixed ends are hard to achieve for shear or acoustic waves in metal or rock, but open ends occur when the wave in a solid strikes a liquid or gas.

The effect of reflections of sound in a bar from fixed and free ends is shown in the traced oscillograms of Fig. 4-8. These oscillograms were obtained experimentally by Dr. F. Ju, of the Mechanical Engineering Department, University of New Mexico.

TABLE 4-1. SOME EQUIVALENT TERMINATIONS

Type of wave	$\Gamma_R = 1$			$\Gamma_R = -1$		
	Name of termination	Zero quantity	Doubled quantity	Name of termination	Zero quantity	Doubled quantity
Transmission line	Open-circuit	i	v	Short-circuit	v	i
Plane electromagnetic	Open-circuit	H	E	Conductor	E	H
Vibrating string	Dangling-end	f_w	u_w	Fixed-end	u_w	f_w
Sound in gas				Hard-surface	u_x	p
Sound in liquid	Open-circuit	p	u_x	Hard-surface	u_x	p
Sound in solid	Open-end	p_x	u_x	Fixed-end	u_x	p_x
Shear in solid	Open-end	p_y	u_y	Fixed-end	u_y	p_y

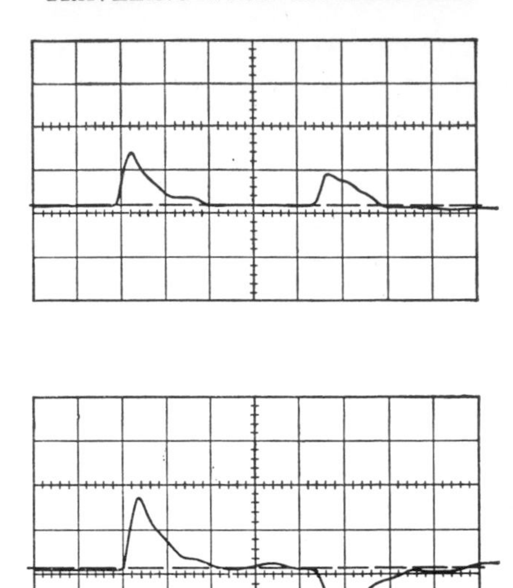

FIG. 4-8. Oscillograms of strain pulses—fixed and open ends. The top photograph shows incident and reflected strain pulses in a metal bar with a fixed end. The bottom photograph shows comparable pulses for an open-end bar. Since strain is proportional to stress, these photos also illustrate stress pulses. Note attenuation. For clarity, the photographs have been traced.

4-3. Some Practical Applications

The principles of transient traveling waves are best understood by reference to practical examples. Both transmission-line and acoustic examples are discussed.

Example 4-2. *Radar Modulator.* Transmission lines are frequently used to develop square pulses, particularly in high-power radar modulators. The duration of such pulses is determined by the time required for a wave to travel down and a reflection to travel back on a transmission line. In practice, it is more common to use an *artificial* line for this purpose. The circuit of such a pulse generator is shown in Fig. 4-9. The transmission line is charged slowly through the high resistance. When its charge is close to the supply voltage, a trigger is applied to the grid of the thyratron, causing it to act as a switch connecting the load across the line through the pulse transformer. It is customary to adjust the turns ratio of the transformer so that the load presented is the line's characteristic impedance. As shown in Fig. 4-10, the line delivers a constant current during the time required for the wave to travel down and back, after which the current is cut off and the thyratron automatically turns off. In practice, special provisions must be made to account for the fact that current in the inductance of the transformer cannot be stopped instantaneously.

One very important feature of this analysis is the equivalent circuit shown in Fig. 4-9 for discharging the line. Note that, in order to represent the discharge of the line properly, the voltage to which it is charged is considered as a battery which can determine the initial current flow and voltage on the line. This is the same technique that is used for determining the current which flows initially when a capacitor is being discharged. That is, a capacitor is represented as a battery whose voltage is equivalent to that appearing initially across the capacitor. The current which flows

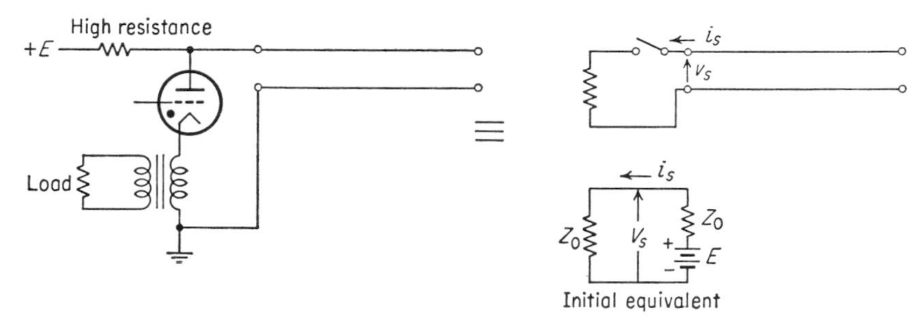

FIG. 4-9. Radar modulator.

initially is just that which would flow when this battery is connected in series with the resistance of the capacitor (actually Z_0) and the external resistance.

Here we are discussing the capacitance in the line and, for determining initial current, the capacitance is represented by the battery. The current in the line and the load must be the same. Hence, the ratio of voltage to current across the line for the wave which is set up in the line must be just the characteristic impedance. This acts like a series impedance. The external voltage-to-current ratio is the impedance of the load, which in this example is also set equal to the characteristic impedance. The voltage appearing across the terminals of the line is as indicated on the diagram. It is the value obtained when the voltage across the characteristic impedance due to the wave starting down the line is subtracted from the voltage of the battery which represents the charge on the capacitance in the line. On the bounce diagram, the initial charge of the line is represented by the horizontal line of voltage E at time zero.

This type of initial condition is common in other problems, such as the ballistic pendulum discussed in Example 4-4 and some types of heat and diffusion problems.

Once the wave has been established on the line, the results shown in Fig. 4-10 apply for the voltage at different positions on the line and at different times. The resultant voltage on the line is always the difference between the initial voltage to which the line was charged and the various pulse voltages reflecting back and forth

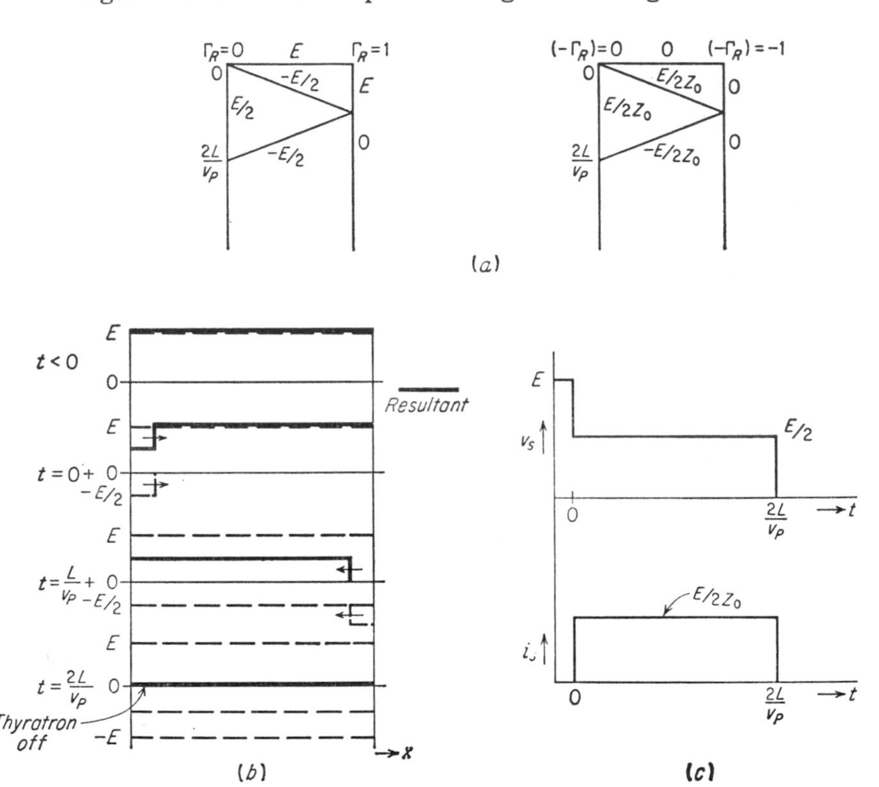

FIG. 4-10. Analysis of radar modulator: (*a*) left—bounce diagram for voltage; right—bounce diagram for current; (*b*) voltage on line; (*c*) sending-end conditions.

across the line. In this case, since the line is properly terminated by the load at the sending end, reflection occurs only from the far end of the line, and the complete pulse on the line stops when the out-of-phase current reflection returns to be absorbed in the load impedance.

Example 4-3. *Breaking Points for a Wire.* Figure 4-11 illustrates a mass being dropped and suddenly stopped by a wire or cable. It is interesting to note where the wire will break. When this experiment is performed with wires of the proper cross section (small enough to break, but large enough not to break at the point of load application), the breaking point is always at the point of connection of the wire to its support. The reason for this is that the acoustic wave in the wire travels up to the support, where it is reflected with zero velocity and double pressure (actually tension

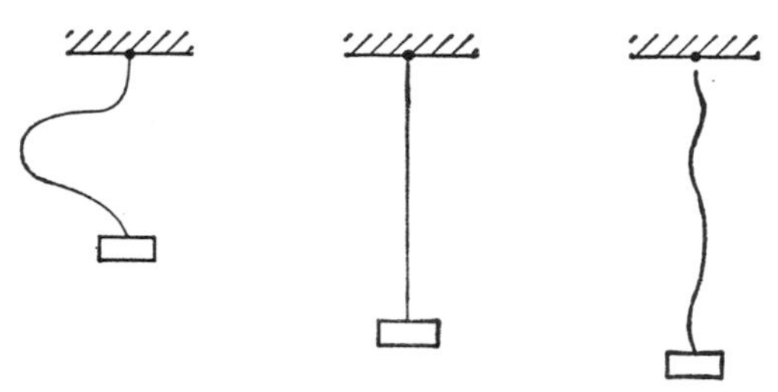

FIG. 4-11. Breaking point for a wire.

in this case). If this double tension is sufficient to cause the yield point of the wire to be exceeded, the wire breaks. Since the double tension occurs first at the fixed support, the breaking also occurs there.

Example 4-4. *The Ballistic Pendulum.* The problem of the ballistic pendulum is similar to various other problems of one object striking another. In particular, it is almost identical with the problems of a pile driver striking a pile, of a collision between two automobiles, and of an object dropping on a floor. The name *ballistic pendulum* comes from the use of a pendulum to measure the velocity of a projectile by its effect on the heavy pendulum.

A ballistic pendulum and its electric-circuit equivalent are shown in Fig. 4-12. For simplicity, it is assumed that both the main bar, which is struck, and the striker are of the *same cross section* and of the same material (steel), though this would not ordinarily be the case with a pendulum used with bullets. If they were not of the same cross section, spherical as well as plane waves would be set up after impact. If they were of different materials but the same cross section, the characteristic impedances involved would be different, and the distribution of velocities between the two objects would be different.

It is assumed that a bar 1 m long is suspended so that it is free to move almost horizontally. It is struck by a similarly suspended bar 10 cm long. Just before the time of impact, the striker (1) is moving with a velocity of 10 m/sec, and the struck bar (2) is stationary.

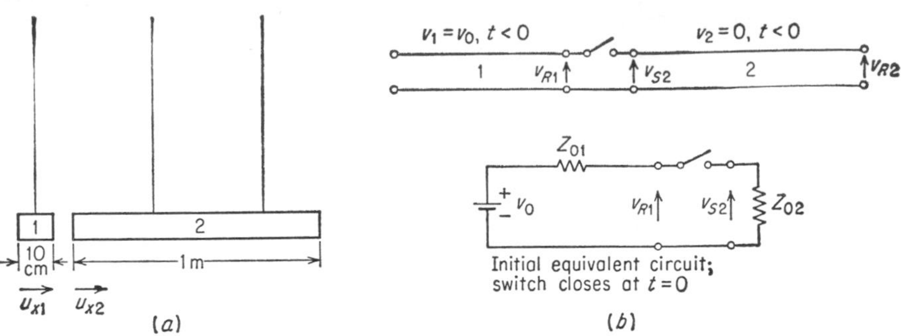

FIG. 4-12. Ballistic pendulum.

The electric-circuit equivalent has two transmission lines joined by a switch. Transmission line 1 is initially charged to voltage v_0, whereas there is no charge on line 2. The equivalent circuit for the moment of impact is also shown in the figure. At this instant, when the switch has been closed, the voltage at the terminals of line 1 (and also of line 2) has dropped from the open-circuit value v_0 to $v_0/2$ (provided that $Z_{01} = Z_{02}$). The terminal voltage is simply that which would appear across Z_{02} in the initial equivalent circuit.

The initial conditions on the bar itself may be described in terms of the velocities. The net velocity of the striker (u_{x1}) after impact is the sum of the incident particle velocity and the particle velocity of the reflected wave in the striker:

$$u_{x1} = u_{x0} + u_{x1}^{-}$$

where u_{x0} is the incident velocity (the velocity of the striker before impact) and u_{x1}^{-} is the particle velocity of the reflected wave in the striker. At the time of impact, the only velocity in bar 2 is

$$u_{x2} = u_{x2}^{+}$$

since the reflected wave in this bar cannot be established before the incident wave arrives at the far end. At the moment of impact, the surface joining the two bars must move at the same velocity for each bar; that is,

$$u_{x1} = u_{x2}$$

At the point of impact, the pressures in the two bars must be equal (action = reaction); that is,

$$p_{x1} = p_{x2}$$

but

$$p_{x1} = p_{x1}^{-} \qquad p_{x2} = p_{x2}^{+}$$

For the wave reflected in the striker, and hence traveling in the negative x direction, we have

$$u_{x1}^{-} = -p_{x1}^{-}Z_{01}$$

Furthermore, we have

$$u_{x2} = p_{x2}Z_{02}$$

Hence

$$u_{x1}^{-} = -u_{x2}^{+}$$

Substituting these results in the equation that states that the two velocities are equal at the point of impact, we have

$$u_{x0} + u_{x1}^{-} = u_{x2} = -u_{x1}^{-}$$

Solving this, we find

$$u_{x1}^{-} = -\frac{u_{x0}}{2} = -5 \text{ m/sec}$$

Conservation of momentum techniques would also have produced the same result, namely, that the struck particles move with half the initial velocity and that a backward motion is set up relative to the initial motion in the striking bar at a velocity half the initial velocity. Thus, the adjoining faces of both bars move at half the initial velocity. The velocity in bar 1 is the summation of the initial 10 m/sec and the −5 m/sec in the reflected wave, whereas that of the particles in bar 2 is just the velocity of the incident wave in that bar. The pressure in the incident and reflected waves is given by

$$\begin{aligned} p_x = \frac{5}{Z_0} = 5\sqrt{Y_0\rho_v} &= 5\sqrt{21 \times 7.8 \times 10^{13}} \\ &= 6.40 \times 10^7 \text{ newtons/m}^2 \\ &= 9{,}300 \text{ lb/in.}^2 \end{aligned}$$

We are now ready to consider the way the pulses bounce in the two bars. We need separate bounce diagrams for each, and these are indicated in Fig. 4-13. The velocity of propagation for the wave in steel in a thin bar is 5,190 m/sec (see Chap. 3), so the time of travel in bar 1 is 0.0193 msec and that in bar 2 is 0.193 msec. The bounce diagrams are shown in Fig. 4-13 for velocity.

Consider first the bounce diagram for bar 1. Both the initial 10 m/sec velocity and −5 m/sec velocity of the reflected wave appear initially, so that the right-hand side has a net velocity of 5 m/sec, as indicated. The reflection coefficient for the open end of the bar corresponds to an open circuit, so the pressure is reversed, but the velocity in the wave reflected from that end is the same as that in the wave incident upon the end. Thus, arrival of the wave at the right-hand side causes the bar to stop (the velocity to go to zero). The negative pressure wave (a tension wave) arriving at that end can have no effect on bar 2; hence the bars are free to separate at this point. This has the effect in the equivalent electric circuit of opening the switch, so that, when the voltage wave has been wiped off line 1 and the switch is opened at the right instant, line 1 is discharged, and the pulse continues on down line 2. On the bars, this separation causes the wave to disappear, just as happened in Example 4-2 when the thyratron opened, so bar 1

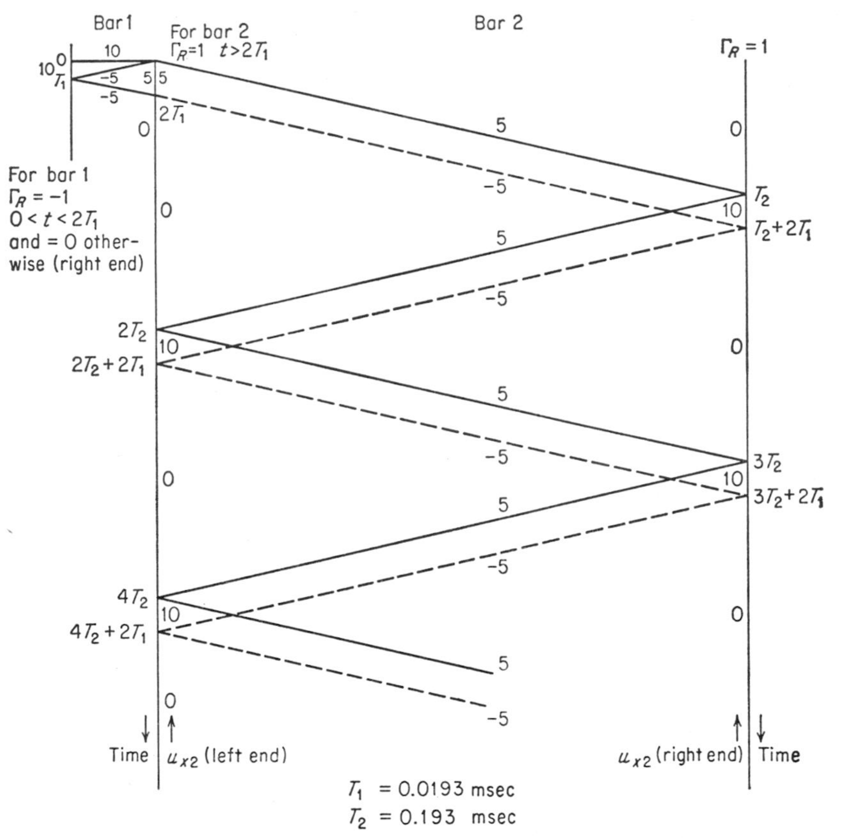

FIG. 4-13. Velocity bounce diagrams for ballistic pendulum.

remains stationary, with no more waves in it, and a *pulse* goes down bar 2.

Consider now what this has caused in bar 2. The initial pulse of 5 m/sec velocity travels down the bar and does not reach the far end until 0.193 msec has elapsed. At the time of the return of the reflection in bar 1 (when the bars separate), there is no more source of velocity or pressure for bar 2, so the velocity and pressure at its left-hand end go to zero. This can be represented on the bounce diagram by a negative step function, just as in Fig. 4-3.

The result is a negative step following closely behind the positive step on the bounce diagram. The corresponding pressure bounce diagram is shown in Fig. 4-14, and Figs. 4-15 to 4-17 show the pulses as a function of time at different positions on the bars. These time

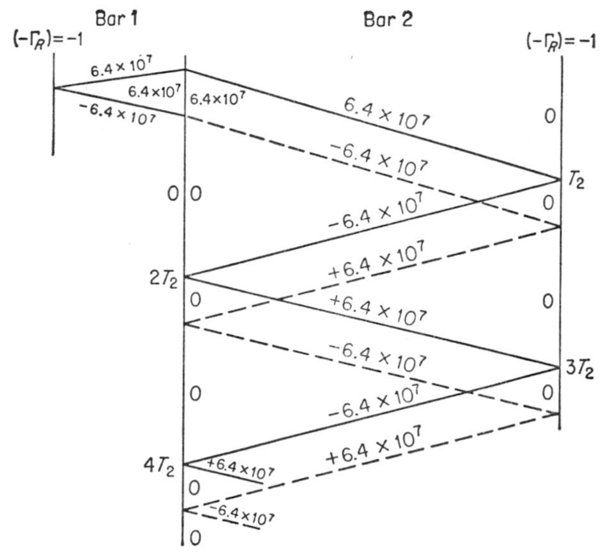

Fig. 4-14. Pressure bounce diagrams for ballistic pendulum.

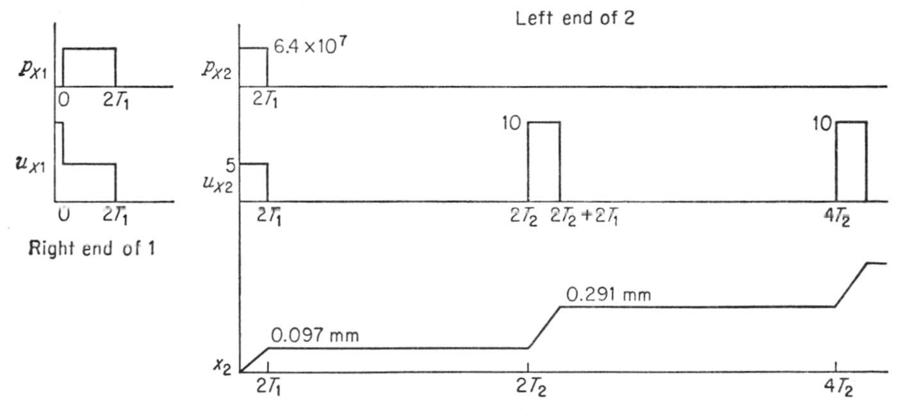

Fig. 4-15. Pulses at junction in ballistic pendulum.

diagrams may all be drawn by direct reference to Figs. 4-13 and 4-14. It is interesting to note, in Fig. 4-15, that the displacement of the left-hand end of bar 2 advances stepwise—not continuously, as we might normally expect.

In Fig. 4-16, the pulses are shown in the middle of the bar where the phase reversal of the pressure pulse is most important.

Of course, there is no pressure pulse at the end of bar 2 after contact has been broken with bar 1, because the open end offers no resistance. The composite pressure and velocity pulses which occur *near* the end of the bar, where there is some overlap, are shown in Fig. 4-17. The pulse shapes in this region are especially interesting.

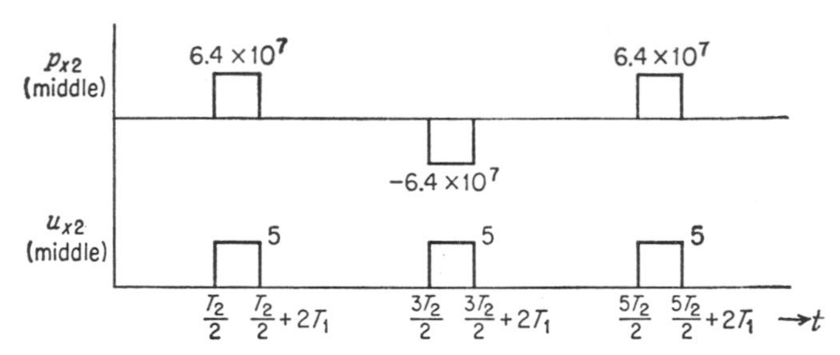

FIG. 4-16. Pulses at middle of ballistic pendulum.

These various pressure pulses may be observed on a bar by use of strain gauges. Equation 3-28 shows that the pressure is directly proportional to the strain, so the converse is also true. Hence, if one knows the modulus of elasticity (Young's modulus) for the bar, all he needs to do is measure strain to determine the stress (or pressure) pulse.

Because of the complexities of this example, a timetable has been prepared to show what happens both on the ballistic pendulum itself and on its electrical analog.

TABLE 4-2. TIMETABLE FOR BALLISTIC-PENDULUM EXAMPLE

Time	Pendulum	Line analogy
0	Two bars contact	Switch closed
$T_1 = 0.0193$ msec	Wave reflected from end of 1	Wave reflected from end of 1
$2T_1$	Reflection arrives at junction; bars lose contact; bar 1 stops; end of pulse starts down bar 2	Switch opened; currents and voltages stop in 1; end of pulse starts down line 2
$T_2 = 0.139$ msec	Leading edge of pulse arrives at end of 2 and is reflected	Leading edge of pulse arrives at end of 2 and is reflected
$T_2 + 2T_1$	End of pulse arrives at end of 2 and is reflected	End of pulse arrives at end of 2 and is reflected
$2T_2$	Reflected pulse leading edge arrives at left end of bar 2; left end starts to move again	Reflected pulse leading edge arrives at left end of line 2
$2T_2 + 2T_1$	End of reflected pulse arrives at left end of bar 2 and is reflected	End of reflected pulse arrives at left end of line 2 and is reflected

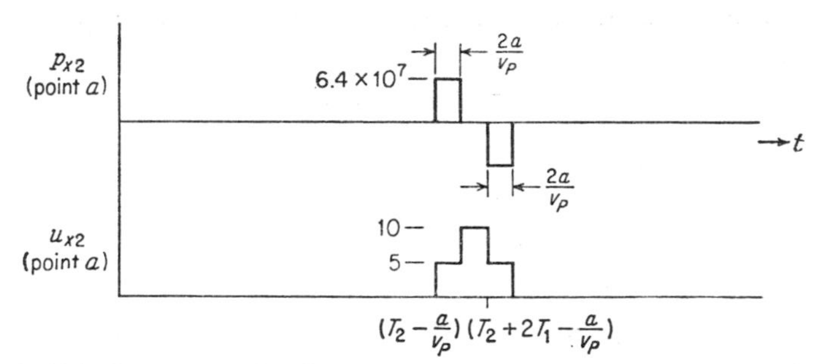

FIG. 4-17. Pulses *near* end of ballistic pendulum. (Distance from end = a.)

4-4. Transient Diffusion

Transient waves traveling in lossless media have been treated. The transient solution of a general lossy wave is a difficult problem, although it can be treated by Fourier and Laplace methods.

One special case of a transient solution with loss is of particular interest because it occurs so frequently. This is the solution of the diffusion equation (for which $\alpha = \beta$ in the steady-state solution) when the input is a step function at time zero. This situation is particularly important in heat-flow problems; it occurs, for example, when a solid heated object is dropped into a quenching bath. Diffusion applications occur in the case-hardening of steel, in the manufacture of transistors, and in related problems. Some telephone cables, particularly those for which a large number of wires are closely packed in a single cable, have very low inductance and conductance and therefore are very closely represented by the diffusion equation. In addition, the diffusion equation applies for the travel of electromagnetic-plane-wave pulses into conducting regions.

The general equation was given as

$$\frac{\partial^2 v}{\partial x^2} = \frac{1}{D}\frac{\partial v}{\partial t} \tag{3-69}$$

Particular cases of this equation include the heat-flow equation:

$$\nabla^2 \tau = \frac{S\rho_v}{k}\frac{\partial \tau}{\partial t} = \frac{1}{D_t}\frac{\partial \tau}{\partial t} \tag{3-42}$$

the equation for chemical concentration:

$$\frac{\partial^2 c}{\partial x^2} = \frac{1}{D}\frac{\partial c}{\partial t} \tag{3-54}$$

which may also be written in a three-dimensional form:

$$\nabla^2 c = \frac{1}{D}\frac{\partial c}{\partial t} \tag{3-54a}$$

and the equation for electromagnetic waves in a conducting medium:

$$\nabla^2 \mathbf{E} = \mu\sigma \frac{\partial \mathbf{E}}{\partial t} \tag{4-10}$$

This latter equation is simply the wave equation of (2-3) with the second-order time-derivative term omitted.

To determine the one-dimensional (plane-wave) solution of these equations, we shall treat (3-69). The solution will be assumed and shown to satisfy the equation. Let us assume a solution of the form

$$v = A \int_0^{x/2\sqrt{Dt}} e^{-Y^2}\, dY + B \tag{4-11}$$

To verify that this is indeed a solution of (3-69), note

$$\frac{\partial v}{\partial x} = Ae^{-x^2/4Dt} \frac{1}{2\sqrt{Dt}} \qquad \frac{\partial^2 v}{\partial x^2} = Ae^{-x^2/4Dt} \frac{-x}{4(Dt)^{\frac{3}{2}}}$$

$$\frac{\partial v}{\partial t} = Ae^{-x^2/4Dt} \frac{-x}{4\sqrt{D}\,t^{\frac{3}{2}}} \qquad \frac{\partial v}{\partial t} = Ae^{-x^2/4Dt}\, D \frac{-x}{4(Dt)^{\frac{3}{2}}}$$

Hence,

$$\frac{\partial^2 v}{\partial x^2} = \frac{1}{D} \frac{\partial v}{dt}$$

A common boundary condition for this type of problem states that

$$\begin{aligned} v &= v_0 \qquad \text{at } t = 0+,\ x = 0 \\ &= 0 \qquad \text{at } t = 0,\ \ x > 0 \end{aligned}$$

Substituting these conditions into (4-11), we have for the first,

$$v_0 = A \int_0^0 e^{-Y^2}\, dY + B = B$$

and for the second, with x infinite,

$$0 = A \int_0^\infty e^{-Y^2}\, dY + v_0 = A \frac{\sqrt{\pi}}{2} + v_0$$

Hence,

$$A = -v_0 \frac{2}{\sqrt{\pi}}$$

$$v = v_0 \left(1 - \frac{2}{\sqrt{\pi}} \int_0^{x/2\sqrt{Dt}} e^{-Y^2}\, dY\right) \tag{4-12}$$

The integral in this equation must be evaluated numerically, but it is a common one and is tabulated. The second term, involving the integral, is known as the *error function*, or *normal probability integral*. It is the value of the cumulative probability associated with the "normal," or Gaussian, probability distribution. It is found in one form or another in most mathematical tables. Here we define

$$\text{erf } z = \int_0^z \frac{2}{\sqrt{\pi}} e^{-Y^2} \, dY \tag{4-13}$$

so that (4-12) is expressible as

$$v = v_0 \left(1 - \text{erf} \frac{x}{2\sqrt{Dt}}\right) \tag{4-14}$$

In evaluating problems using the error function, it is important to note the form of its definition for the particular tables used. Sometimes the limits are from $-\infty$ to z or from $-z$ to $+z$. Also, sometimes the integrand is $\exp(-Y^2/2)$ instead of $\exp(-Y^2)$. The function of (4-13) is plotted in Fig. 4-18. It is interesting to consider what happens to a

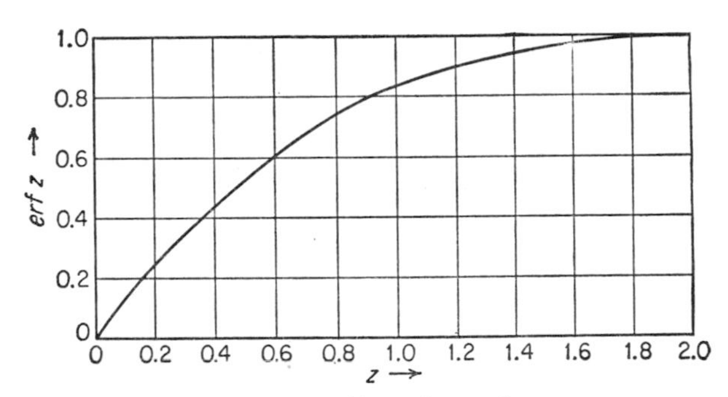

FIG. 4-18. Error integral.

pulse as it travels through a highly dissipative medium (or diffuses, as with a heat or gaseous diffusion pulse). It is possible to treat a square pulse at the sending end as the superposition of two steps such as those used to find the solution of (4-12), a positive one at the beginning of the pulse and a negative one at the end. Thus, the solution for the voltage, or other parameter, is in terms of the superposition of a positive solution like (4-12) followed by a negative one that is equal in initial amplitude but delayed in time. An example of such distortion at different distances is shown in Fig. 4-19.

In thermal problems it is sometimes desirable to consider instantaneous rates of cooling or heating $\partial\tau/\partial t$. If, in (4-12), we replace v by τ, we find

$$\tau = \tau_0 \left(1 - \text{erf} \frac{x}{2\sqrt{Dt}}\right)$$

$$\frac{d\tau}{\partial t} = \frac{\tau_0}{2t\sqrt{\pi Dt}} e^{-x^2/4Dt}$$

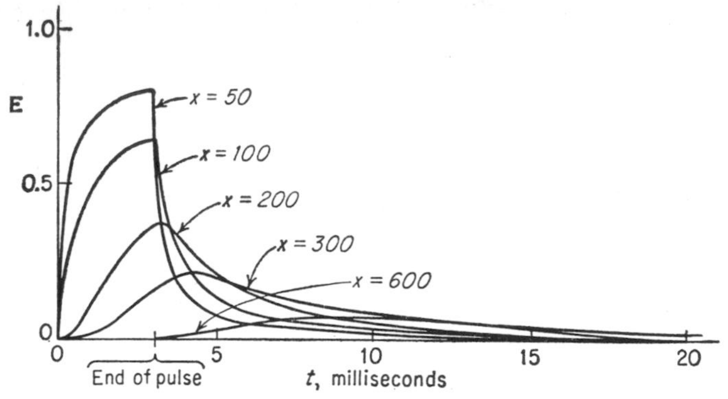

Fig. 4-19. Pulse distortion vs. distance. Pulse length, 3 msec. Electromagnetic wave in water, $\sigma = 0.1$ mho/m.

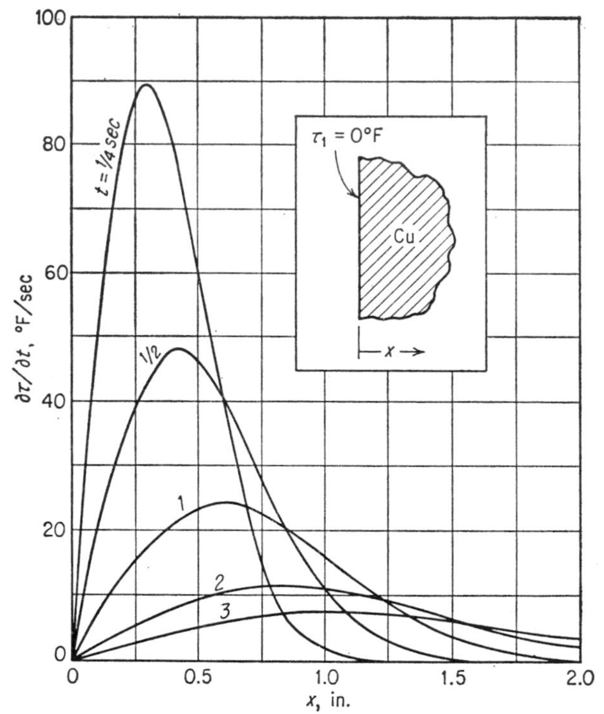

Fig. 4-20. Instantaneous cooling rates in a large mass of copper initially at 100°F. (*From P. J. Schneider, "Conduction Heat Transfer," Addison-Wesley Publishing Company, Reading, Mass.*, 1955; *by permission.*)

This is plotted for the example of a large mass of copper initially at 100°F and suddenly exposed at its surface to a cooling bath at 0°F in Fig. 4-20.

Example 4-5. An oven is insulated with 2 cm of asbestos sheet. Assuming that the oven is heated instantaneously from 20 to 300°C, how long does it take the halfway point in the insulation to rise to 100°C? (The halfway point was chosen rather than the outside, so that surface heat transfer did not have to be considered, since it is not a traveling-wave process. In a practical case, however, it would be important to consider the surface heat transfer in this example and calculate the temperature at the outside edge.)

Substituting temperature for voltage in Eq. (4-14), we note that the other conditions are the same as the boundary conditions listed, except that our scale of temperature must start at 20°C rather than at 0°C. The diffusion coefficient for asbestos is given by Table 3-8 as 2.48×10^{-7} m²/sec. The temperature rise which is called for is 80°C, and the temperature rise internally is 280°C. Thus, we may substitute in this equation and get

$$80 = 280\left(1 - \operatorname{erf} \frac{10^{-2}}{2 \times \sqrt{2.48 \times 10^{-7}t}}\right)$$

Thus

$$\operatorname{erf} \frac{10^{-2}}{2 \times \sqrt{2.48 \times 10^{-7}t}} = 1 - \tfrac{80}{280} = 1 - 0.286$$
$$= 0.714$$

From a table of error functions we find that this means

$$\frac{10^{-2}}{2 \times \sqrt{2.48 \times 10^{-7}t}} = 0.76$$

whence

$$t = \frac{10^{-4}}{4 \times 0.76^2 \times 2.48 \times 10^{-7}} = 174 \text{ sec}$$

Example 4-6. A cover is suddenly removed from a swimming pool at 8°C. Assuming that there is no wind, that the air is initially dry, and that the pool is at sea-level pressure, find the relative humidity 3 m above the pool after 10 hr have passed.

Immediately after the pool is uncovered, the partial pressure of water vapor rises to saturation at the surface of the water. This pressure is listed in psychometric tables as 8.05 mm of mercury at 8°C. Table 3-9 lists the diffusion coefficient for water vapor in air at this temperature as 0.329×10^{-4} m²/sec. Thus we may write for the partial pressure at time t, substituting from Eq. (4-14), the following equation:

$$p = p_0 \left(1 - \text{erf} \frac{x}{2\sqrt{Dt}}\right)$$
$$= 8.05 \left(1 - \text{erf} \frac{3}{2\sqrt{0.329 \times 10^{-4} \times 3.6 \times 10^4}}\right)$$

The equation is written this way since there are 3.6×10^4 sec in 10 hr.

$$\begin{aligned} p &= 8.05(1 - \text{erf } 1.38) \\ &= 8.05(1 - 0.949) \\ &= 8.05(0.051) \\ &= 0.4025 \text{ mm of mercury} \end{aligned}$$

Hence,

$$\text{Relative humidity} = \frac{0.4025}{8.05} = 5\%$$

At a height of 1 m above the pool the argument of the error function is only 0.46, the error function is 0.485, and the relative humidity is 51.5 per cent.

This indicates how slow the diffusion processes really are. Usually, of course, in a situation like this there is some mass transfer due to either wind or vertical currents of air, because water and air are not at the same temperature (as they were assumed to be in this case). Liquid diffusion is even slower, as can be seen from Table 3-9.

Example 4-7. *Transistor Manufacture.* An indium surface is placed next to a germanium surface in transistor manufacture. We assume that the indium concentration at the boundary is maintained at a value c_0 kg/m^3 and that the desired junction value of indium concentration is $10^{-6}c_0$. We wish to know at what depth the required concentration occurs after 10^3 sec of treatment at 700°C. We find

$$10^{-6}c_0 = c_0 \left(1 - \text{erf} \frac{x}{2\sqrt{Dt}}\right)$$
$$10^{-6} = 1 - \text{erf} \frac{x}{2\sqrt{Dt}}$$

The right-hand side of this equation, for large values of $x^2/4Dt$ may be approximated by[1]

$$1 - \text{erf} \frac{x}{2\sqrt{Dt}} = \frac{e^{-x^2/4Dt}}{\sqrt{\pi}\,(x/2\sqrt{Dt})}$$

[1] L. P. Hunter, "Handbook of Semiconductor Electronics," pp. 7–13, McGraw-Hill Book Company, Inc., New York, 1956.

The resulting transcendental equation has an approximate solution in this range

$$x \approx 8\sqrt{Dt}$$

Substituting from Table 3-10,

$$x \approx 8\sqrt{10^{-18} \times 10^{3}} = 8\sqrt{10} \times 10^{-8} = 2.58 \times 10^{-7}\ \text{m}$$

This is a very small distance, but it is realistic in terms of junction geometry.

4-5. The General Approach

In general, transient traveling-wave solutions may be obtained by Laplace-transforming the wave equation (2-3). This results in an ordinary differential equation in x which must still be solved. Sometimes this is solved by Laplace-transforming once more, and sometimes it is solved by other differential-equation techniques.

The general wave equation for the transmission line is

$$\frac{\partial^2 v}{\partial x^2} = RGv + (RC + LG)\frac{\partial v}{\partial t} + LC\frac{\partial^2 v}{\partial t^2} \tag{2-3}$$

Laplace-transforming this equation with respect to time, we have

$$\begin{aligned} \frac{\partial^2 V(s)}{\partial x^2} &= RGV(s) + (LG + RC)[sV(s) - v(0)] \\ &\qquad + LC[s^2V(s) - sv(0) - v'(0)] \\ &= V(s)[RG + s(LG + RC) + s^2LC] + \text{initial conditions} \end{aligned} \tag{4-15}$$

A general solution of this equation is exceedingly complex, but specific examples need not be.

$V(s)$ is the voltage as a function of position on the line and of the "complex decrement"

$$s = \sigma + j\omega$$

where ω is, as usual, the radian frequency associated with sinusoidal time variation and σ describes an exponential envelope on the sinusoid. Thus, at any point x, $V(s)$ has the same meaning as the transformed voltage has in any circuit problem. The difference here is that V is also a function of position, and solution of the differential equation of (4-15) is required to find its variation with position.

As with steady-state sinusoidal excitation, the propagation constant may be described by

$$\gamma^2 = RG + s(LG + RC) + s^2LC$$

although the separation into an attenuation coefficient and a phase factor is meaningless in this situation unless σ is zero. The initial conditions $v(0)$ and $v'(0)$ are functions of x and represent the voltage on the line at time zero and its *time* derivative.

It would also be possible to solve (2-3) by the method of separation of variables, a common technique of partial differential equations. Thus, we could assume

$$v = X(x)T(t)$$

where X and T are each a function of only one of the variables. Substituting into (2-3) and dividing through by v, we have

$$\frac{1}{X}\frac{d^2X}{dx^2} = RG + \frac{RC + LG}{T}\frac{dT}{dt} + \frac{LC}{T}\frac{d^2T}{dt^2}$$

The left-hand side of this equation is a function of x alone, and the right-hand side is a function of t alone, so each is equal to a constant. Thus

$$\frac{1}{X}\frac{d^2X}{dx^2} = a^2 \qquad RG + \frac{RC + LG}{T}\frac{dT}{dt} + \frac{LC}{T}\frac{d^2T}{dt^2} = b^2 \qquad a^2 = b^2$$

These two ordinary differential equations may be solved, the boundary and initial conditions substituted, and the results combined.

Both the above approaches yield the same results. In the general case, these results are rather complicated to express mathematically, because Fourier series result for both X and T. For this reason, the general case is not treated here, as a lengthy exercise in the solution of boundary problems is required to illustrate principles already illustrated by the two special cases of lossless waves and diffusion.

4-6. Summary

Special emphasis has been placed on transient solutions for two special cases, the wave in a lossless medium and the wave which satisfies the diffusion equation. The general approach for solving any transient problem has been indicated but not treated in detail.

Reflection is extremely important in nearly all traveling-wave problems. The reflection coefficient is defined as

$$\Gamma_R = \frac{v^-}{v^+} \tag{4-7}$$

This was shown to be

$$\Gamma_R = \frac{Z_R/Z_0 - 1}{Z_R/Z_0 + 1} = \frac{Z_R - Z_0}{Z_R + Z_0} \tag{4-8}$$

Reflections also occur from the generator end of the line, and the reflection coefficient for this purpose is obtained by simply shorting the emf and treating the resultant problem as if the internal impedance were a load impedance for the transmission line.

The wave initially starting down a transmission line is easily determined by letting the characteristic impedance for the transmission line represent the transmission line itself in a circuit diagram involving the generator and its internal impedance. This situation is somewhat more complicated when the wave is started by changing the conditions on a charged line (or a moving object). When this is the case, the voltage to which the line is charged must be considered as a source in series with the internal impedance of the line, or the circulating current in the line must be considered as a current source in shunt with the line. These problems were discussed in Examples 4-2 and 4-4.

The very handy bookkeeping technique for keeping track of reflections on a line, the bounce diagram, was introduced in Sec. 4-2 and applied later to both electrical and nonelectrical reflections.

The special case of transient diffusion with a step-function input results in a solution involving the error function. Some interesting properties of this solution are disclosed in Figs. 4-18 and 4-19. The complete distortion of a square pulse when the diffusion equation applies makes it very difficult to use such pulses for any type of communication or location system.

The general solution of the traveling-wave transient problem has only been indicated here, since it involves mathematical complexity so great that any general meaning is obscured and the results may only be interpreted in terms of the particular problem being treated.

PROBLEMS

4-1. A 100-volt battery is suddenly connected through its 20-ohm internal resistance to a transmission line with a 50-ohm characteristic impedance. The line is terminated in a short circuit. Use the bounce diagram to plot voltage and current at the sending end and center of the line as a function of time. Assume a traveltime of 1 msec.

4-2. A 5,000-ohm characteristic-impedance artificial line is charged initially to 50 volts, with both ends open-circuited. A 2,500-ohm resistor is suddenly connected across one end. The time for a wave to travel the length of the line is 10 msec. Plot voltage across and current through the resistor for 55 msec.

4-3. A commercial pulse generator uses a mercury switch to connect suddenly a charged external pulse cable to a resistor equal to its 50-ohm characteristic impedance. If a 10-ft length of cable having $v_p = 0.75$ (velocity of light) is connected to this generator, how long is the pulse? Assume that the cable is charged to 100 volts. What does the waveform look like if the load is 45 ohms instead of 50? Use bounce diagrams and sketch to scale.

4-4. A radar modulator uses the pulse-generating circuit of Fig. 4-9. The line has $Z_0 = 1{,}000$ ohms and a one-way transmission delay of 5 μsec. If it is initially charged to 10,000 volts, plot the current on the line at various appropriate times, the power in the load, and the load voltage.

4-5. A 100-volt battery with 10-ohm internal resistance is connected at time zero to a lossless 100-m length of 50-ohm transmission line. The end of this line is the junction between a 100-ohm line and a 200-ohm line, both also 100 m long. The former is terminated in 100 ohms and the latter in a short circuit. Assume that the velocity of propagation is 3×10^8 m/sec in all three lines. Sketch the voltage at the sending end and the junction.

4-6. Show that the reflection coefficient at the generator is as given by Eq. (4-9).

4-7. In this chapter, the effects of a purely resistive load have been considered. Suppose that a pure capacitance C_L is used as a load for a lossless line. Establish the boundary conditions which apply to determination of the reflected wave. Sketch the reflected voltage of a step-function incident wave, assuming that the load capacitor is initially uncharged.

4-8. Repeat Prob. 4-7, assuming an inductor (zero initial current) instead of a capacitor for the termination.

4-9. Show that the solution for a wave in the negative x direction satisfies the wave equation and that $v/i = -Z_0$ for this wave.

4-10. A lossless transmission line with a 100-ohm characteristic impedance, properly terminated and fed with an alternating voltage of 10,000 volts rms by a generator with 1-ohm internal resistance, is short-circuited at a point a half-wavelength from the generator. Assume that the voltage was at a maximum at the point of the short when it occurred. Plot the incident, reflected, and resultant voltages for 2 cycles after the short. Determine the steady-state current in the short circuit and the steady-state voltage at the sending end of the line. Actually, this is unrealistic, because such lines are invariably shorter.

4-11. A fault occurring on a power transmission line is in the form of a short circuit 100 miles from the generator. The line is carrying a 60-cps rms voltage of 110,000. Assume that the generator has an internal resistance of 100 ohms and that the transmission line has a characteristic impedance of 500 ohms, resistive. Neglecting losses and assuming a velocity of 175,000 miles/sec, sketch the voltage and current waveforms on the line for 0.25 cycle, assuming that the short occurred when the voltage at the short was zero.

4-12. A pile driver is driving steel piles into a sandstone of density 2.5×10^3 kg/m^3 with a longitudinal wave velocity of 2,130 m/sec. The hammer and the pile are the same cross section. The pile is 15 m long, and the hammer is 1 m long. The hammer is also made of steel. If the hammer is initially started from rest at a point 5 m above the top of the pile, calculate the pressure and velocity of the pulse which strikes the material at the bottom of the pile. What is its duration? Compare the pressure exerted by the pulse with that exerted by the weight of the hammer alone after it has come to rest.

4-13. Figure P4-13 shows an arrangement with two bars initially in contact. Bar 1 is held in compression with stress of 8×10^7 newtons/m^2. Bar 1 is made of steel and bar 2 of copper. At $t = 0$, bar 1 is suddenly released from its compression. Describe the motion of the two bars. Illustrate with an electrical analog.

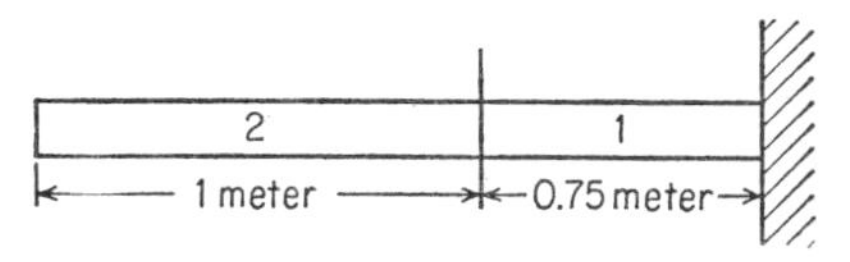

FIG. P4-13

4-14. Repeat Example 4-4 if the far end of the long bar is fixed.

4-15. You are asked to determine the feasibility of measuring the depth of a lake by determining the time delay for an electromagnetic pulse to travel from the surface to the bottom and back. The function is similar to that of a sound-ranging (sonar) fathometer, except that a transient electromagnetic wave replaces the sound pulse.

Assume that a plane wave is established instantaneously at the surface, with a steady electric field of 1 volt/m, at $t = 0$. This field is maintained at the surface for 3 msec, thus establishing a pulse of 3 msec duration. If the lake is 300 m deep and the wave is reflected as if from a short circuit at the bottom, plot what it looks like when it returns. Neglecting signal strength, does this scheme look feasible?

Assume that the conductivity of the water is 0.1 mho/m and that $\mu = \mu_0$. Neglect displacement currents. Treat the pulse as a positive step function followed by a negative step function. You will have to use a table of error functions.

4-16. Plot the pulse of Prob. 4-15 at a distance of 50 and 200 m from the surface (downward pulse only).

4-17. A large mass of copper having a plane face is dropped into a cooling bath whose temperature is maintained constant at 10°C. The copper was at 500°C prior to being dropped in the water. Find the temperature at a depth of 10 cm after 1 min and after 5 min.

4-18. Repeat Prob. 4-17 for aluminum.

4-19. Repeat Prob. 4-17 for glass (assuming, of course, that the thermal stress does not crack the glass—as it would be likely to do in this case). Replace 1 min by 1 hr.

4-20. In manufacturing a "solar battery," the use of boron as an acceptor impurity in silicon results in a P-type semiconductor. A vapor of a boron compound is exposed to a silicon bar at 1140°C for 10^3 sec. Plot the boron concentration vs. distance from the surface, as a function of the surface concentration c_0. Use a logarithmic concentration scale and plot down to $c = 10^{-7}c_0$, using the approximate expression for the error function from Example 4-7.

4-21. Case-hardening of steel is accomplished by diffusing carbon into the surface. Frequently the carbon is applied in the form of a gas (CH_3, methane, for example), with surface chemical reactions making the carbon atoms available for diffusion. Suppose that pure iron (great idealization) is heated to 1000°C in the presence of a gaseous carburizing agent. At this temperature the maximum concentration of carbon is 1.6 per cent, and this can be assumed at the surface. Calculate the concentration at a depth of 1 mm after 2, 5, and 10 hr.

BIBLIOGRAPHY

Churchill, R. V.: "Operational Mathematics," 2d ed., McGraw-Hill Book Company, Inc., New York, 1958.

5. Steady-state Lossless Traveling and Standing Waves

It might appear more logical, in a discussion of traveling waves, to treat the general solutions first and then specialize to lossless and very lossy lines and waves. Actually, however, for the steady state, as well as for the transients treated in Chap. 4, there are a number of concepts associated with waves traveling in two directions (incident and reflected waves) which are more easily understood when attenuation need not be considered. These concepts have to do with *reflected waves* and the *standing waves* which result after the steady state has been established.

As we noted in Chap. 3, it is customary to use the lossless form of the wave equation for acoustic waves, vibrating strings, electromagnetic waves in space, and some transmission lines operating at high frequencies. Hence, the steady-state solution for waves when reflections are present but attenuation is negligible is important in its own right.

5-1. General Considerations

Heretofore, in discussing steady-state solutions (in Chaps. 2 and 3), waves were assumed to be on an infinite or properly terminated line—or plane waves in a semi-infinite medium. In this chapter, lines treated for steady-state alternating current ordinarily have reflections present, as did those with transients treated in Chap. 4. The lines discussed in this chapter are of finite length and are not "properly terminated." Membranes have fixed edges, strings are tied down at the ends, and acoustic waves reflect from surfaces.

The treatment in this chapter deals largely with the transmission line, but it is equally applicable to the various types of lossless waves already discussed. Some of the problems deal with these, and references are made to them throughout.

The logical reference point on an infinite line is the sending end, where the generator is located. In fact, if the line is truly infinite, this is the only "end." For properly terminated finite lines, there is no reason to change the point of reference; but when reflections are present, it is more convenient to locate the reference point at the source of reflection, the receiving end of the line.

The coordinates used are indicated in Fig. 5-1. Here x measures the distance from the sending end as before. The distance from the receiving end is given by d. Phasor voltages are shown at sending and receiving ends, and phasor currents at these ends are also shown, along with

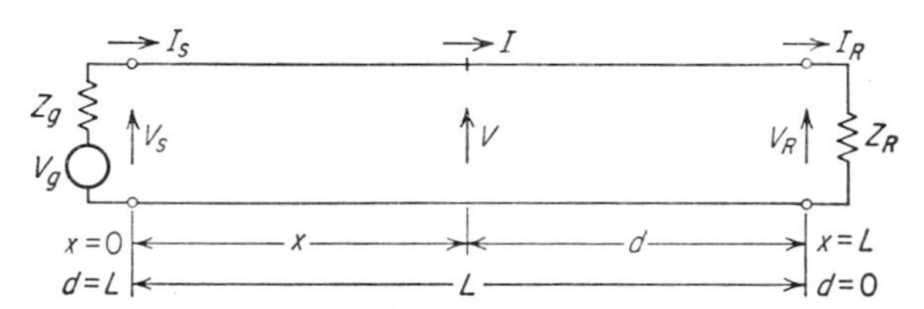

FIG. 5-1. Coordinates used in line analysis.

the generator impedance Z_g, the load impedance Z_R, and the generator voltage V_g. Note that the line length is L.

It can be seen from the sketch that

$$d = L - x \tag{5-1}$$

We originally solved the wave equations for a transmission line to obtain

$$V = Ae^{-\gamma x} + Be^{\gamma x} \tag{2-7}$$

$$I = \frac{A}{Z_0} e^{-\gamma x} - \frac{B}{Z_0} e^{\gamma x} \tag{2-10}$$

These equations could as well have been solved, with different constants, in terms of d instead of x. Such solutions are

$$V = V^+e^{\gamma d} + V^-e^{-\gamma d} \tag{5-2}$$

$$I = I^+e^{\gamma d} + I^-e^{-\gamma d} = \frac{V^+}{Z_0} e^{\gamma d} - \frac{V^-}{Z_0} e^{-\gamma d} \tag{5-3}$$

Using (5-1), it is apparent that

$$Ae^{-\gamma x} = V^+e^{\gamma d} = V^+e^{\gamma L}e^{-\gamma x}$$

$$A = V^+e^{\gamma L}$$

Similarly,

$$B = V^-e^{-\gamma L}$$

In discussing finite lines, it is frequently important to distinguish between the two wave components traveling in opposite directions; thus we write

Incident waves:

$$\mathcal{V}^+ = V^+ e^{\gamma d} \tag{5-4}$$

$$\mathcal{I}^+ = I^+ e^{\gamma d} = \frac{V^+}{Z_0} e^{\gamma d} \tag{5-5}$$

Reflected waves:

$$\mathcal{V}^- = V^- e^{-\gamma d} \tag{5-6}$$

$$\mathcal{I}^- = I^- e^{-\gamma d} = -\frac{V^-}{Z_0} e^{-\gamma d} \tag{5-7}$$

Here $\mathcal{V}^+$ and $\mathcal{I}^+$ represent the phasor incident voltage and current at different points on the line. Likewise, the phasor reflected voltage and current at various points on the line are given by $\mathcal{V}^-$ and $\mathcal{I}^-$. Thus, V^+, I^+, V^-, and I^- are reserved for the phasors corresponding to the incident and reflected voltage and current *at the load end of the line.* A similar technique was used in Chap. 4 in deriving the reflection coefficient, without the restriction to steady-state alternating current used here.

One more pair of definitions is required. The impedance at any point on the line, at distance d from the end, is

$$Z(d) = \frac{V(d)}{I(d)} \tag{5-8}$$

and the sending-end impedance is

$$Z_s = \frac{V_s}{I_s} \tag{5-9}$$

The Phasor Traveling Wave. As in ordinary a-c circuit theory, it is convenient to describe steady-state traveling waves in terms of phasor diagrams. In lumped-constant circuits, we are concerned with the relative phases of voltages and currents at particular points in a circuit, and it is customary to draw a separate phasor diagram for each significant part of the circuit or to show voltages and currents on a combined diagram for different points. In the case of traveling waves, we are concerned with a *continuum* of separate points, so that it is frequently desirable to give a *third dimension* to our phasor diagram.

Figure 5-2 shows such a three-dimensional phasor diagram. It can be seen that the third dimension is the distance down the line—this being added to the two dimensions necessary to indicate phase. This is a representation of a wave traveling in the positive x direction, given by

$$e^{j(\omega t - \beta x)}$$

It will be noted that the picture is such that phasors are shown only every quarter-wavelength (90° of space-phase shift). This is only for convenience, and they could be shown at any and all points. The locus of the ends of the phasors is a spiral around the x axis.

The solid phase vectors (phasors) represent time zero. They are what would be observed if all the rotating phasors were stopped for a "snapshot" at time zero. The long-dash phasors are stopped a quarter-cycle later in time, so each has rotated counterclockwise 90°. Note that, in each case, the phasor at $\beta x = \pi/2$ is 90° behind (clockwise) that at $x = 0$, as indicated by the above expression. Likewise, that at $\beta x = \pi$ is 180° behind that at 0. The short-dash phasors are for a time a quarter-cycle after the long-dash phasors and a half-cycle after the solid phasors.

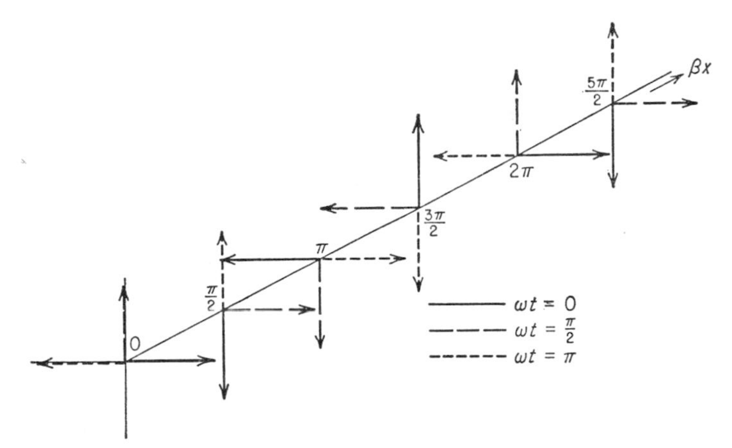

Fig. 5-2. Wave in positive x direction.

Note the progression of the zero phase (a vector horizontal to the right). Thus, it is at the origin at zero time; a quarter-cycle later, it is a quarter-wavelength down the line; and a half-cycle later, it is a half-wavelength down the line. Thus it can be seen that this phase is, indeed, traveling down the line at velocity ω/β. An observer traveling at this speed, and able to observe the phase at only his own point on the line, would always see the same phasor.

An observer traveling the other way would, on the other hand, go by successive phasors faster than would an observer staying at one point. For him to see the same phase, he must travel with a wave going in the negative x direction. Such a wave is shown in Fig. 5-3 and is described by

$$e^{j(\omega t+\beta x)}$$

In this wave, the phasors for the zero-time snapshot (solid) are progressively more advanced (lead by a greater amount) as one looks farther down the line. After a quarter-cycle, they have rotated counterclockwise a quarter-turn and each phase has advanced to a position a quarter-wavelength *closer* to the origin, and similarly after a half-cycle. Thus, this wave is traveling in the negative x direction. In our usual terminology it represents a reflected wave.

It is important that the student fully understand the ideas conveyed by Figs. 5-2 and 5-3. In what follows, phasor diagrams will be drawn

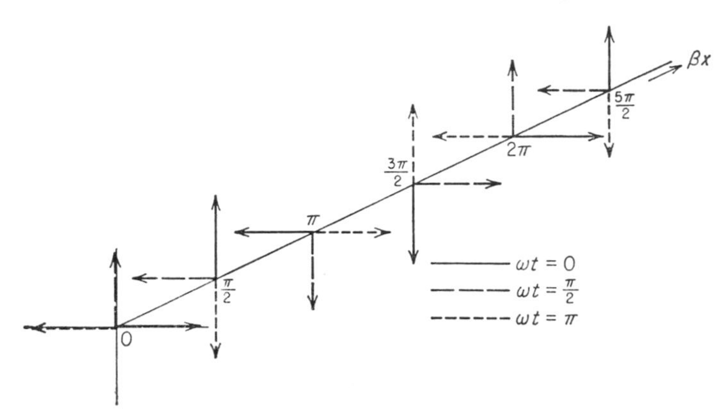

FIG. 5-3. Wave in negative x direction.

in two dimensions, with each representing one or more discrete points on the line. In each problem, however, it is possible, though not necessary, to construct diagrams similar to those shown in Figs. 5-2 and 5-3.

5-2. The Short-circuited Line—Standing Waves

The first line to be studied that has reflections is the short-circuited line. Not only is this simpler to analyze than a line with general termination, but it has many practical applications. Short-circuited lines are extensively used as UHF and microwave circuit elements. The analogous problem for plane electromagnetic waves involves reflections from a metal boundary, important in microwave transmission and antenna problems. In the field of transverse vibrations, this corresponds to a vibrating drumhead and to a vibrating string (or bridge cable) with at least one end held firm. For acoustic waves in liquids or solids it is the situation when a wave strikes a hard surface.

Consider a short-circuited transmission line, shown in Fig. 5-4. Because of the short circuit, the voltage at the end of the line must be zero—that is, the reflected voltage must be equal and opposite to the incident voltage. Thus, since d is zero at the short circuit, (5-2) becomes

$$V_R = V^+ + V^- = 0$$

so

$$V^- = -V^+ \tag{5-10}$$

Using the reflection coefficient defined in Eq. (4-7), we find for the phasor voltages (as for the transients of Chap. 4)

$$\Gamma_R = \frac{V^-}{V^+} = -1 \tag{5-11}$$

To find the current, we use (5-3), obtaining

$$I_R = \frac{1}{Z_0}(V^+ - V^-) = 2I^+ = \frac{2V^+}{Z_0} \tag{5-12}$$

It can be seen that, in general, (5-2) and (5-3) may be rewritten in terms of the reflection coefficient:

$$V = V^+(e^{j\beta d} + \Gamma_R e^{-j\beta d}) \tag{5-13}$$

$$I = \frac{V^+}{Z_0}(e^{j\beta d} - \Gamma_R e^{-j\beta d}) \tag{5-14}$$

In the case of short-circuited lines, this becomes

$$V = V^+(e^{j\beta d} - e^{-j\beta d}) = \mathcal{V}^+ + \mathcal{V}^- \tag{5-15}$$

$$I = \frac{V^+}{Z_0}(e^{j\beta d} + e^{-j\beta d}) = \mathcal{I}^+ + \mathcal{I}^- \tag{5-16}$$

Phasor diagrams for different points on the line are shown in Fig. 5-4. At the receiving end of the line, the incident and reflected voltages are

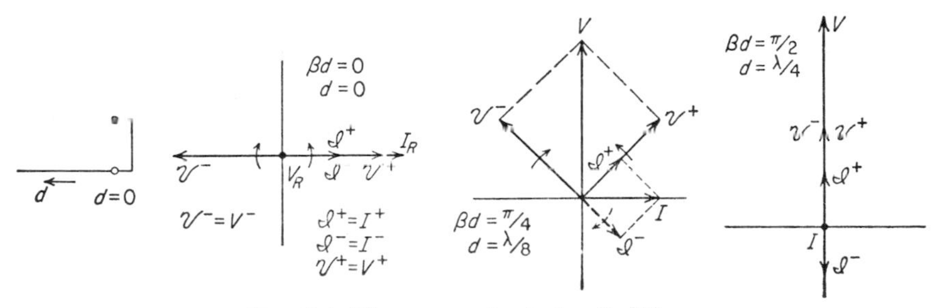

FIG. 5-4. Phasors on short-circuited line.

out of phase, and the incident and reflected currents are in phase. As one considers for this same instant a point on the line closer to the generator, the incident-wave phasors appear rotated counterclockwise and the reflected-wave phasors rotated clockwise. Thus, at $\beta d = \pi/4$, an eighth-wavelength from the short, the voltage phasors are both in the upper half-plane, making equal angles with the imaginary axis. Hence, since they are the same length, the resultant is along that axis. At this point, rotation is 45° from the original position, so the resultant voltage is

$$V^+ \sqrt{2}\, e^{j\pi/2} \qquad \text{or} \qquad jV^+ \sqrt{2}$$

The current is the resultant of two phasors balanced about the axis of reals, so it is

$$I^+ \sqrt{2} \qquad \text{or} \qquad \frac{V^+ \sqrt{2}}{Z_0}$$

When the total distance from the load is doubled to a quarter-wavelength ($\beta d = \pi/2$), both incident- and reflected-voltage phasors lie along

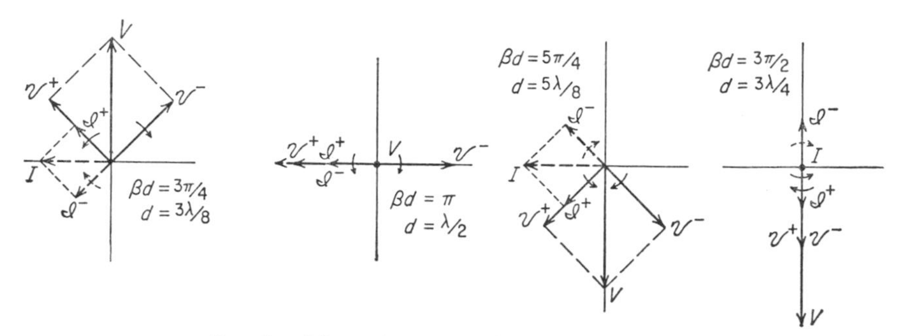

FIG. 5-5. More phasors on short-circuited line.

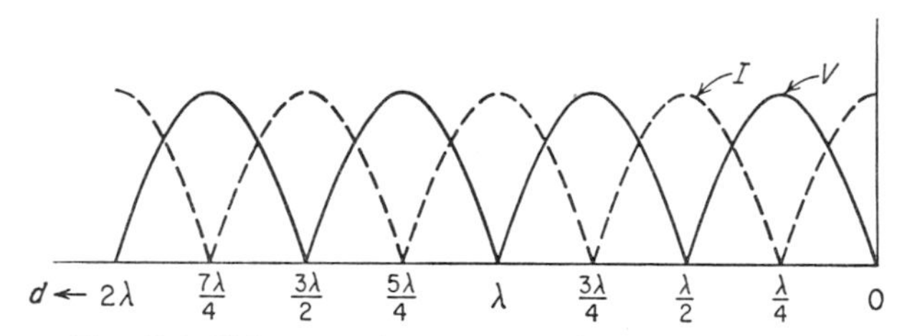

FIG. 5-6. Voltage and current on short-circuited line.

the imaginary axis, so the resultant voltage is $j2V^+$. Both current phasors also lie along this axis, but since the incident-current phasor is up and the reflected-current phasor down, they add up to zero. Hence, no current whatever flows at a point a quarter-wavelength from the load. In Fig. 5-5, the phasors are shown for several more positions, farther from the short circuit.

We may plot the length of the resultant phasors as a function of position on the line, obtaining the result shown in Fig. 5-6. Here it can be seen that the voltage and current are 90° out of phase in *space;* that is, the maximum of current and minimum of voltage occur at the same point, and the maximum of current is a quarter-wavelength from the maximum of voltage. From the phasor diagrams and the expressions listed above, it can be seen that the voltage and current are *also* in *time* quadrature.

Figure 5-6 shows a plot of the *amplitude* of the wave at each point on the line. This is a measure of the excursion of voltage or current from zero at that point, and it is a picture of instantaneous values only at the times when the sinusoidal time variation passes through its maximum. Because of the quadrature relation, this happens for voltage and current at different times. Thus, at the instant when the voltage is as plotted in Fig. 5-6, the current is zero everywhere on the line, and vice versa.

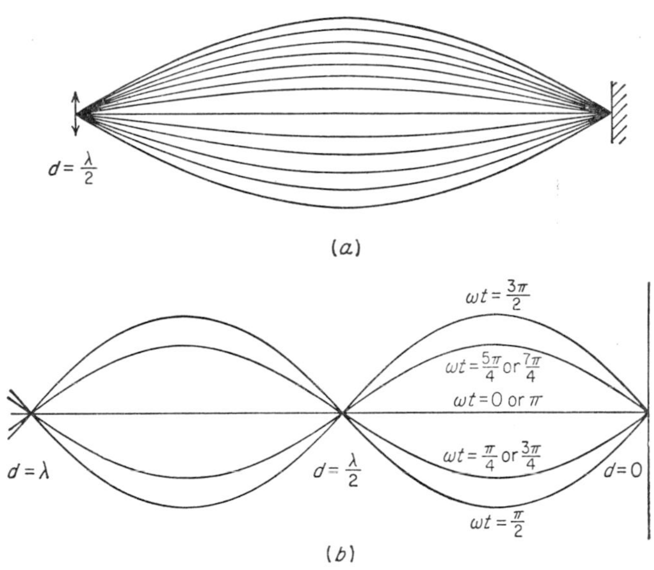

FIG. 5-7. Standing waves on a short-circuited line and string: (*a*) vibrating string, with standing wave; (*b*) voltage standing wave.

To aid in comprehension of this situation, it is instructive to consider such a wave on a string. Figure 5-7 shows what is observed on a string tied solidly at one end (equivalent to a short circuit) and vibrated at a point a half-wavelength from the tie-down. The various lines represent the positions of the string at different times—or of the voltage on a line at different times. Figure 5-7*b* shows the voltage from Fig. 5-6, but at different specific instants of time. Thus, Fig. 5-6 gives the voltage amplitudes, but Fig. 5-7*b* gives instantaneous values.

The waves plotted above are known as *standing waves*, rather than as *traveling waves*. They "stand" on the line but do not carry power down the line, since the voltage and current are in quadrature (see Fig. 5-5). When the load is a short or an open circuit, or a pure reactance, only standing waves are present. When the load has a resistive component, different from Z_0, both standing and traveling waves are present. If the load resistance is equal to Z_0, of course, only the incident wave is present. *Standing waves result from the superposition of traveling waves going in opposite directions.*

In practical problems, it is frequently not possible to measure V^+ separately. In fact, until only a few years ago, it was never measured separately. Today, directional couplers may be used to separate incident and reflected waves, so they may be measured separately. Nevertheless, it is often easier to measure the sending- or receiving-end voltage and current than to measure components traveling different ways.

To obtain the voltage and current at any point on a line in terms of those at the sending end, it is merely necessary to draw a phasor diagram

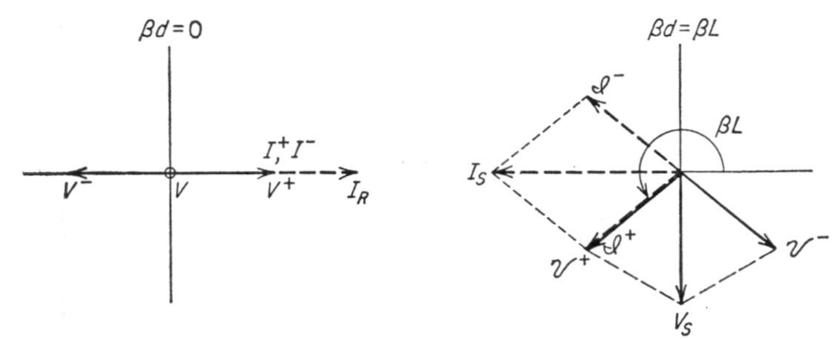

FIG. 5-8. Phasor diagrams used in computation.

for the sending end and determine V^+ in terms of V_s geometrically on this diagram. Figure 5-8 is an example of this method for a line whose length is slightly more than a half-wavelength. Note that

$$V_s = 2V^+ \sin \beta L \, e^{j90^\circ}$$

so

$$V^+ = \frac{V_s}{2j \sin \beta L}$$

From this we see that

$$V = V_s \frac{e^{j\beta d} - e^{-j\beta d}}{2j \sin \beta L}$$

but

$$\frac{e^{jx} - e^{-jx}}{2j} = \sin x$$

so

$$V = V_s \frac{\sin \beta d}{\sin \beta L} \tag{5-17}$$

Similar expressions may be derived for I.

The instantaneous value of voltage on the short-circuited line may be determined analytically as

$$
\begin{aligned}
v &= \text{Re } V^+[e^{j(\omega t+\beta d)} - e^{j(\omega t-\beta d)}] \\
&= \text{Re } V^+e^{j\omega t}(e^{j\beta d} - e^{-j\beta d}) = \text{Re } 2jV^+e^{j\omega t} \sin \beta d \\
&= \text{Re } 2V^+ \sin \beta d e^{j(\omega t+\pi/2)} = 2V^+ \sin \beta d \cos \left(\omega t + \frac{\pi}{2}\right)
\end{aligned}
$$

This last expression should be compared with the phasor expression

$$V = 2jV^+ \sin \beta d$$

Note that this means that the amplitude is

$$2V^+ \sin \beta d$$

and that the phase is 90° ahead of the reference (V^+).

A similar development for the current is

$$
\begin{aligned}
I &= \frac{V^+}{Z_0} e^{j\beta d} - \frac{V^-}{Z_0} e^{-j\beta d} \\
&= \frac{V^+}{Z_0} (e^{j\beta d} + e^{-j\beta d}) \\
&= \frac{2V^+}{Z_0} \cos \beta d \\
i &= \text{Re } Ie^{j\omega t} = \frac{2V^+}{Z_0} \cos \beta d \cos \omega t
\end{aligned}
$$

Note, as mentioned before, that space and time phases are *both* quadrature.

The impedance may be obtained by taking the ratio of voltage to current at any point on a line. Thus, for the short-circuited line,

$$Z = \frac{2jV^+ \sin \beta d}{(2V^+/Z_0) \cos \beta d}$$

$$Z = jZ_0 \tan \beta d \tag{5-18}$$

The most important impedance on a line, other than that at the load, is usually the sending-end impedance—where $d = L$. Thus,

$$Z_s = jZ_0 \tan \beta L = jZ_0 \tan \frac{\omega L}{v_p} = jZ_0 \tan \frac{2\pi}{\lambda} L \tag{5-19}$$

This impedance is plotted in Fig. 5-9. Note here that the *electrical length* of the line may be varied by changing either the actual length or the frequency.

This effect is used in a number of ways. For example, it is possible to determine the distance to a short circuit by varying the frequency and observing the frequencies for which the impedance is zero (or infinite).

This method is used for locating faults in power and telephone lines. With acoustic waves, it is used for fault location in solids.

It can be seen that, for lengths an odd-integer multiple of a quarter-wavelength, the impedance behaves like that of a parallel resonant circuit—and it is used as such in many tuning applications at UHF and microwaves. In radar sets, the arrangement of Fig. 5-10 is used to isolate the receiver and transmitter, so that none of the transmitter

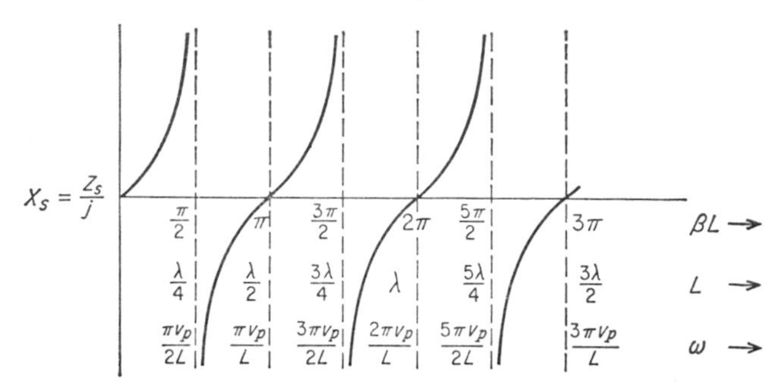

FIG. 5-9. Input impedance of short-circuited line.

power is wasted on the receiver and, even more important, so that the transmitter does not burn out the receiver. The box marked TR is the so-called TR tube, a gas-discharge tube in a cavity. When the transmitter turns on, it causes the TR tube to break down, providing essentially a short circuit at that point in the line. This means that the receiver appears to the transmitter as a shunt open circuit (or as a parallel resonant circuit), which does not affect transmission from source to antenna. The quarter-wavelength line is also used as a "metallic

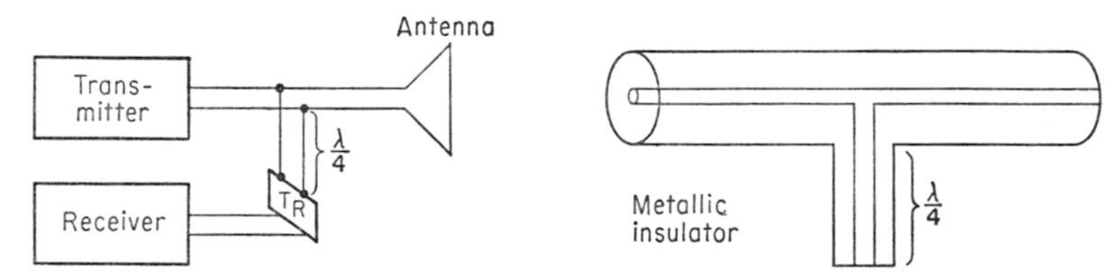

FIG. 5-10. Use of quarter-wavelength line in radar.

insulator" to support the center conductor of coaxial lines. This is shown in Fig. 5-10.

It should also be noted that, near the half-wavelength points, the impedance behaves like a series resonant circuit. This effect is also used in tuning devices of various kinds.

To give examples, acoustic resonators with roughly a short-circuit termination are found in organ pipes and horns. Stringed instruments, from the violin to the piano, depend on short-circuit terminations at both ends of the string, so that the vibration is as shown in Fig. 5-7*a*. Similar terminations give standing waves, which cause trouble in turbine-blade vibrations and in helicopter rotor motion.

5-3. The Open-circuited Line

Another instructive special case is the open-circuited line. It does not have as many practical transmission-line applications as the short-circuited line, for a true open circuit is harder to approximate closely than a short circuit, because of the capacitance between the ends of a line. Nevertheless, where its properties are needed, a near open circuit may be achieved by using a parallel resonant termination. For some other types of wave, with appropriate electrical analogies, open circuits are easier to achieve. For example, the pressure of particles in an acoustic wave is usually considered analogous to current on a transmission line—and the boundary condition for a wave in a solid at an air boundary is (essentially) zero pressure.

The boundary condition for an open-circuited line is that the current at the receiving end of the line, I_R, must be zero. Thus,

$$I_R = 0 = \frac{V^+}{Z_0} - \frac{V^-}{Z_0}$$

hence,

$$V^- = V^+$$

so

$$\Gamma_R = 1$$

Thus, at the load, the incident and reflected voltages are in phase and equal, whereas the currents are out of phase and equal. At any point on the open-circuited line, the voltage and current are, therefore, given by

$$V = V^+(e^{j\beta d} + e^{-j\beta d}) \tag{5-20}$$

$$I = \frac{V^+}{Z_0}(e^{j\beta d} - e^{-j\beta d}) \tag{5-21}$$

Figure 5-11 presents the phasor diagrams required to show the build-up of the standing wave on the open-circuited line. This wave itself is plotted in Fig. 5-12. The latter figure should be compared with Fig. 5-6, the comparable figure for the short-circuited line. Note that the roles of current and voltage are interchanged between short- and open-circuited lines. In both cases, however, both voltage and current go to zero at some point on the line, and these points are 90° out of phase. At any point where either the voltage or current is always zero, there can be no transfer of power down the line; therefore, these are strictly *standing waves.* This can also be seen from the phasor diagrams, which show that the voltage and current are everywhere in quadrature.

Since voltage and current are always in quadrature, open- and short-circuited lines are energy-storage elements, just as are capacitors and inductors, or combinations thereof. Of course the transmission line has both electric-field energy storage (corresponding to a capacitor) and

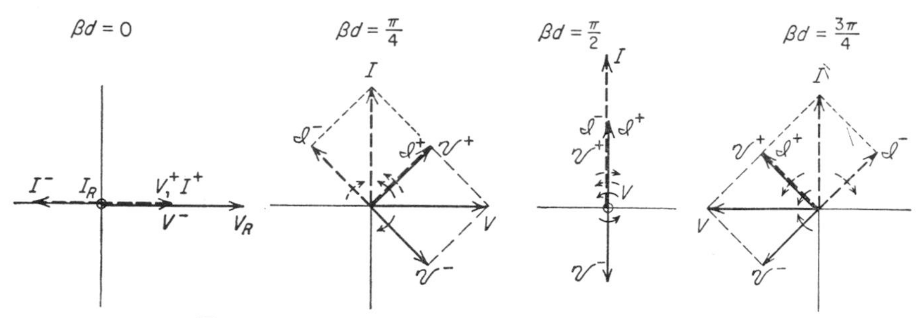

FIG. 5-11. Phasor diagrams on open-circuited line.

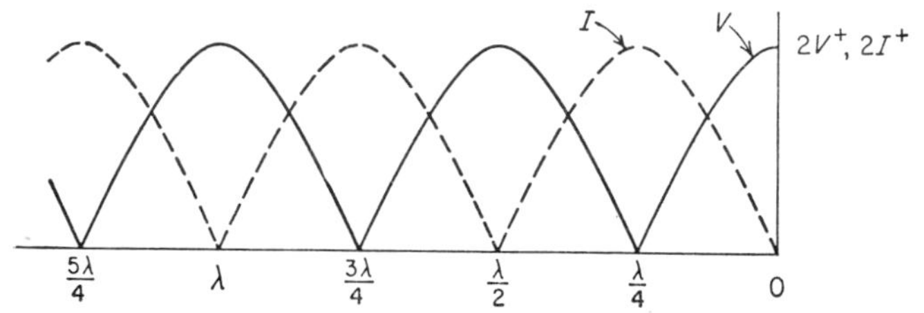

FIG. 5-12. Standing waves on open-circuited line.

magnetic-field energy storage (corresponding to an inductor), with oscillatory transfer from one to the other, as in a resonant circuit.

Analytically, the instantaneous voltage on the open-circuited line is

$$\begin{aligned} v &= \text{Re}\, V^+[e^{j(\omega t+\beta d)} + e^{j(\omega t-\beta d)}] \\ &= V^+ \cos \omega t [e^{j\beta d} + e^{-j\beta d}] \\ &= [2V^+ \cos \beta d] \cos \omega t \end{aligned}$$

and the current is

$$\begin{aligned} i &= \text{Re}\, \frac{V^+}{Z_0} [e^{j(\omega t+\beta d)} - e^{j(\omega t-\beta d)}] \\ &= \left[2 \frac{V^+}{Z_0} \sin \beta d\right] \cos\left(\omega t + \frac{\pi}{2}\right) \\ &= \left[2 \frac{V^+}{Z_0} \cos\left(\beta d - \frac{\pi}{2}\right)\right] \cos\left(\omega t + \frac{\pi}{2}\right) \end{aligned}$$

From these, the time and space quadrature relations are readily seen. Note that the quantities in brackets are the *magnitudes*, as a function of position, of the time-varying sine waves. In phasor notation, the voltage and current on the line are

$$V = 2V^+ \cos \beta d$$
$$I = j \frac{2V^+}{Z_0} \sin \beta d$$

From these it can be seen that the impedance on an open-circuited line is, at any point,

$$Z = -jZ_0 \cot \beta d \tag{5-22}$$

This is plotted in Fig. 5-13. Note, as with the short-circuit case, that varying either length or frequency causes this variation of impedance. For points less than a quarter-wavelength from the open circuit, the

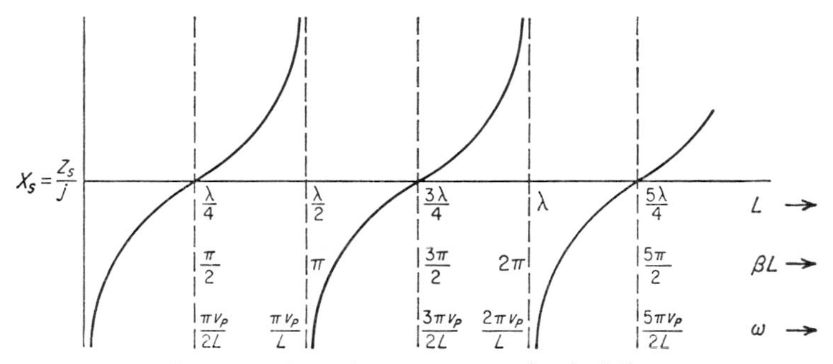

FIG. 5-13. Impedance on open-circuited line.

reactance is capacitive (it is inductive for these lengths in the short-circuited case). This is to be expected, for a very short length of line would even look like a "capacitor." At a point a quarter-wavelength from the open circuit, on the other hand, the impedance is zero—the open circuit has been transformed into a short circuit! This fact is made use of in the antenna switching system of radar sets.

An interesting effect occurs on open-circuited lines whose length is in the vicinity of a quarter-wavelength. This *Ferranti effect* is that the voltage at the receiving end of the line greatly exceeds that at the sending end. This is of considerable importance in lines carrying large amounts of power (either low-frequency or radio-frequency), for the large voltages at the receiving end can cause breakdown of insulation. Figure 5-14 shows the phasor diagrams for this situation. Note that the sending-end voltage is the resultant of incident and reflected phasors nearly out of phase, so it is small compared with the incident voltage—and the receiving-end voltage is twice the incident voltage! Analytically, this is easy to see, for

$$V_s = 2V^+ \cos \beta L$$
$$V_R = 2V^+$$
$$= \frac{V_s}{\cos \beta L}$$

Note that this becomes infinite for finite V_s and βL of 90°. Of course, this does not happen in practice (even neglecting insulation breakdowns), for the input impedance for a quarter-wavelength open-circuited line is zero, and finite V_s cannot be achieved.

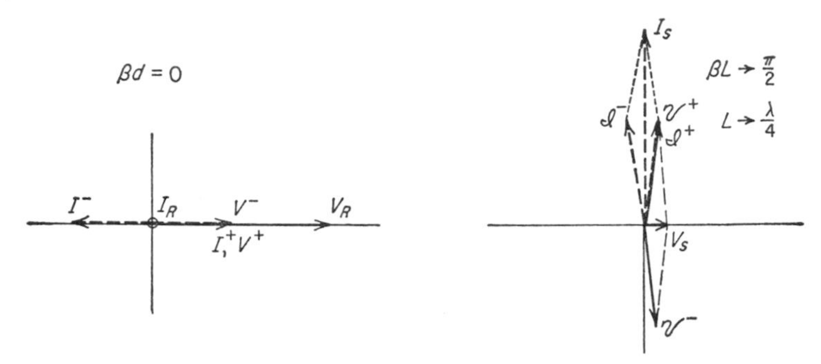

FIG. 5-14. Phasor diagrams illustrating Ferranti effect.

An example of this effect may be seen by considering a line with a fixed sending-end voltage of 100. When its electrical length (βL) is 70°, V_R is 292 volts. When βL becomes 80°, V_R goes up to 575 volts. For 85° (if the 100 volts can be maintained with this low an impedance), V_R is 1,140 volts.

It is interesting to consider the equivalent lumped-constant circuit of a quarter-wavelength short- or open-circuited line. In the latter case,

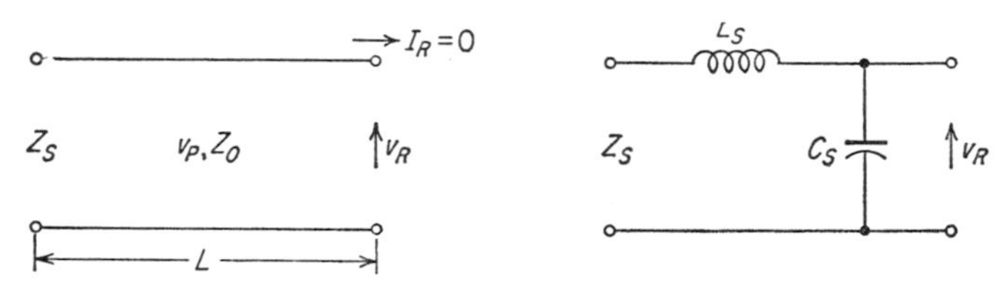

FIG. 5-15. Lumped-circuit equivalent of quarter-wavelength open-circuited line.

we have an input impedance of zero and a significant output voltage. This can be achieved by making the input to the line appear as the terminals of a series resonant circuit. The question then arises, Is the output across the capacitor or the inductor? Since V_R lags I_s, the current through the series resonant circuit, the output must be across the capacitor.

Figure 5-15 shows the open-circuited transmission line and its equivalent LC circuit. To determine the approximate equivalent circuit for the quarter-wavelength line, it is necessary first to write the input impedance to the line in a form appropriate for the input impedance to the series resonant circuit. Then the impedance to the series resonant circuit is written in the same form, and appropriate substitutions are made.

The input impedance of the quarter-wavelength open-circuited line may be written as

$$Z_s = -jZ_0 \cot \beta L = -jZ_0 \tan \left(\frac{\pi}{2} - \beta L\right)$$

Where the length of the line is very near a quarter-wavelength, we have

$$\frac{\pi}{2} - \beta L \ll \frac{\pi}{2}$$

Hence, we may write the approximation

$$\tan\left(\frac{\pi}{2} - \beta L\right) = \left(\frac{\pi}{2} - \beta L\right) + \frac{(\pi/2 - \beta L)^3}{3} + \cdots$$
$$\approx \left(\frac{\pi}{2} - \beta L\right)$$

Thus, we may write

$$Z_s \approx -jZ_0\left(\frac{\pi}{2} - \beta L\right)$$

Since

$$\beta = \frac{\omega}{v_p}$$

this is

$$Z_s = jZ_0 \frac{\omega L}{v_p}\left(1 - \frac{\pi v_p}{2\omega L}\right)$$

If the angular frequency at which the line is a quarter-wavelength is called ω_0, then

$$\frac{\omega_0 L}{v_p} = \frac{\pi}{2}$$

so

$$\omega_0 = \frac{\pi v_p}{2L}$$

and

$$Z_s = \frac{jZ_0\omega L}{v_p}\left(1 - \frac{\omega_0}{\omega}\right)$$

The comparable equation may be developed for the circuit as follows:

$$Z_s = j\left(\omega L_s - \frac{1}{\omega C_s}\right)$$
$$= j\omega L_s\left(1 - \frac{1}{\omega^2 L_s C_s}\right)$$

but the resonant frequency for the circuit ω_0 is given by

$$\omega_0^2 = \frac{1}{L_s C_s}$$

Hence,

$$Z_s = j\omega L_s\left(1 - \frac{\omega_0^2}{\omega^2}\right)$$
$$= j\omega L_s\left(1 - \frac{\omega_0}{\omega}\right)\left(1 + \frac{\omega_0}{\omega}\right)$$

When ω is close to ω_0, this becomes approximately

$$Z_s \approx 2j\omega L_s\left(1 - \frac{\omega_0}{\omega}\right)$$

Comparing this equation with the one for the quarter-wavelength transmission line, we see that the same impedance may be obtained at the input for the circuit and a line if

$$2L_s = \frac{Z_0 L}{v_p}$$

The capacitance may be obtained by noting that

$$\omega_0 = \frac{\pi v_p}{2L} = \frac{1}{\sqrt{L_s C_s}}$$

Hence,

$$C_s = \frac{4L^2}{\pi^2 L_s v_p^2} = \frac{8v_p L}{\pi^2 Z_0 v_p^2} = \frac{8L}{\pi^2 Z_0 v_p}$$

This technique for obtaining an equivalent circuit may be applied to a half-wavelength open- or short-circuited line or to a quarter-wavelength short-circuited line, and the results for the quarter- or half-wavelength line may be extended to comparable situations at an odd or even number of quarter-wavelengths. Note that this equivalence is effective only at the input terminals.

5-4. General Resistive Termination

The concept of standing waves has been introduced in this chapter by discussion of open- and short-circuited lines, on which *only* standing waves are present. Obviously, in a practical wave problem, neither a perfect short circuit nor a perfect open circuit is possible. The next more complicated situation is that in which the load impedance is finite, but resistive. This happens in many practical problems, at least to a first approximation. Where the actual load contains some reactance, this may be tuned out, if desired, resulting in, effectively, a resistive load—at least at a single frequency.

With short- and open-circuited lines, only standing waves are present; that is, there is no net transfer of power from generator to load. Short and open circuits cannot absorb power. When the load is resistive, on the other hand, power is absorbed, so there is a net traveling-wave component from generator to load. Unless the resistance of the resistor is equal to the characteristic impedance, however, there is also a reflected wave—and therefore a standing wave. Thus, the situation in this case is a mixture of the traveling wave of Chap. 2 with the standing wave of this chapter.

Here we must consider the reflection coefficient, which has been defined previously, and the *standing-wave ratio.* The reflection coefficient is just the ratio of reflected to incident voltage at the load. The standing-wave ratio is the ratio of maximum voltage on the line to minimum voltage on the line and is directly related to the reflection coefficient. Figs. 5-6 and 5-12 show that this ratio is infinite when only a standing wave is present, because there is a finite voltage at some points on the line whereas at other points on the line the voltage is always zero. If there is no reflected wave (proper termination), the steady-state magnitude (not instantaneous value) is the same everywhere, so the ratio of maximum to minimum is 1.

The derivation in Chap. 4 for the reflection coefficient was based simply upon the boundary condition and was not restricted to a lossless line or a particular type of load impedance. Hence, the reflection coefficient is given *in general* by

$$\Gamma_R = \frac{Z_R/Z_0 - 1}{Z_R/Z_0 + 1} = \frac{Z_R - Z_0}{Z_R + Z_0} \tag{4-8}$$

The reflection coefficient was plotted in Fig. 4-4 for a resistive load. The reflection coefficient is negative for resistances less than Z_0 and positive for those greater than Z_0. Since it is -1 for a short-circuit load, it is convenient to think of negative reflection coefficients as being on the "short-circuit side" of Z_0. Likewise, it is convenient to think of positive reflection coefficients as corresponding to loads on the "open-circuit side" of Z_0.

A positive reflection coefficient indicates that the reflected voltage is in phase with the incident voltage, so that the voltage at the load lies between the incident voltage (proper termination) and twice the incident voltage (open circuit). Likewise, a negative reflection coefficient means that the reflected voltage is 180° out of phase with the incident voltage, so that the voltage at the load lies between zero (short circuit) and the incident voltage (proper termination).

As with short- and open-circuited lines, phasors may be used for the solution of problems with resistive loads. In fact, whereas the phasor solution may have seemed somewhat clumsy, compared with the analytical forms, for simple terminations, it is much less so for more complicated terminations, where the analytical expressions are also more complicated.

Example 5-1. The easiest way to see what happens to voltage, current, and impedance on a line with resistive termination is to plot

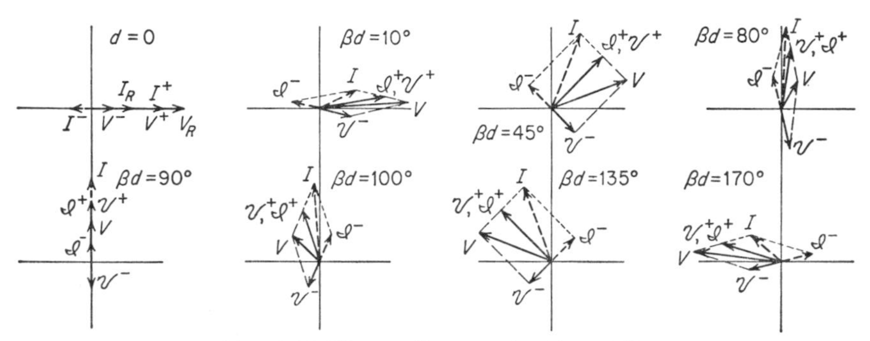

FIG. 5-16. Phasor diagrams for $Z_R = 2Z_0$.

a set of phasor diagrams and the resulting voltage, current, and impedance curves. The example chosen is one for which the load impedance is twice the characteristic impedance. Hence,

$$\Gamma_R = \frac{2Z_0 - Z_0}{2Z_0 + Z_0} = \frac{1}{3}$$

Figure 5-16 shows the phasor diagrams for this example at selected points on the line. The points have been crowded near the quarter-wavelength points on the line to illustrate the behavior of the voltage and current phasors in those regions. Note that, in this case, where the reflection coefficient is positive, the maximum of voltage occurs at the load, as for the open circuit. The voltage does not go to zero a quarter-wavelength from the load, however, nor does the current go to zero at the load.

Note that the phase of the current shifts rapidly near the load (its minimum) and that the phase of the voltage shifts rapidly a quarter-wavelength down the line (near its minimum). In just 10°, there is a phase shift of about 45° near the minimum but of only about 7° near the maximum. There is also a rapid shift in amplitude near the minimum, and the amplitude change near the maximum is slow. Thus, the curves of Fig. 5-17 definitely are *not* sine curves. They have narrow minima, but the slope is continuous at the minima, unlike that for the short and open circuits (Figs. 5-6 and 5-12). These are typical standing-wave curves.

Since this line is, to some extent, like an open-circuited line, we should expect the current to lead the voltage in the first quarter-wavelength back from the load. This is indeed the case, but the lead is less than the 90° lead for the open-circuited case. Thus, the impedance of the line in this part has not only capacitive reactance but also resistance. Between a quarter- and a half-wavelength, as with the open-circuited line, the current lags the voltage—but not by 90°. The resistive component of the impedance in each case is

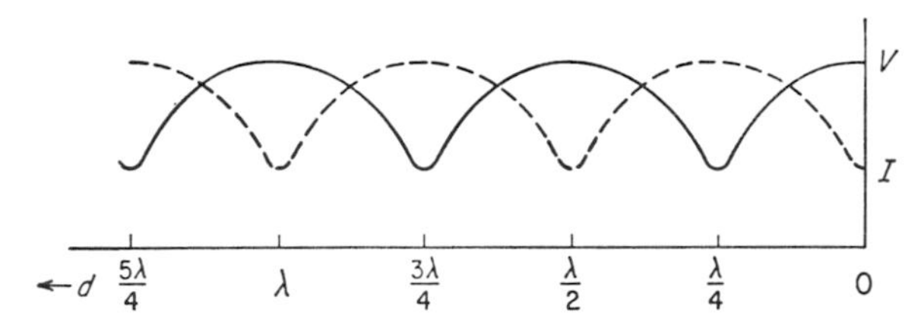

FIG. 5-17. Voltage and current for $Z_R = 2Z_0$.

associated with net power flow toward the load, whereas the reactive component, as with the short- and open-circuit load, is associated with energy storage in the standing wave.

Had the load impedance in Example 5-1 been less than the characteristic impedance, the reflection coefficient would have been negative, and the characteristics of the line would have been similar to those of a short-circuited line; that is, the voltage minimum and current maximum would have occurred at the load. The impedance in the first quarter-wavelength would have had inductive reactance.

The standing-wave ratio for the voltage on the line is defined as the ratio of maximum to minimum voltage on the line. It can be seen that the maximum is

$$V_{max} = V^+ + |\Gamma_R| V^+ = V^+(1 + |\Gamma_R|)$$

and the minimum is

$$V_{min} = V^+ - |\Gamma_R| V^+ = V^+(1 - |\Gamma_R|)$$

Hence,

$$\text{VSWR} = \frac{1 + |\Gamma_R|}{1 - |\Gamma_R|} \tag{5-23}$$

For the line of Example 5-1 it is $\frac{4}{3}/\frac{2}{3}$, or 2.

It can be shown for resistive load (left as a problem) that

$$\text{VSWR} = \frac{Z_R}{Z_0} \text{ or } \frac{Z_0}{Z_R}$$

whichever is greater than unity. This dual relation results because the standing-wave ratio is defined so that it always equals or exceeds unity. Sometimes, with resistive loads, it is convenient to consider that the standing-wave ratio can be less than unity, in which case VSWR $= Z_R/Z_0$ always.

The standing-wave ratio for current, of course, is the same as that for voltage.

Although we have talked only in terms of phasor-diagram solutions for voltage and current on a line, it is frequently possible to save work in problem solutions by using analytical expressions. Because of their value in understanding what is happening on a line, the phasor diagrams should be sketched, even when analytical methods are used.

We have given the analytical expression for voltage at any point on a line as

$$V = V^{+}(e^{j\beta d} + \Gamma_R e^{-j\beta d}) \tag{5-13}$$

To find the voltage in terms of that at the load, note that d is zero at the load; thus,

$$V_R = V^{+}(1 + \Gamma_R) \qquad I_R = \frac{V^{+}}{Z_0}(1 - \Gamma_R) \tag{5-24}$$

Hence,

$$V = V_R \frac{e^{j\beta d} + \Gamma_R e^{-j\beta d}}{1 + \Gamma_R} \qquad I = I_R \frac{e^{j\beta d} - \Gamma_R e^{-j\beta d}}{1 - \Gamma_R} \tag{5-25}$$

Since V_s occurs where d is L, we find

$$V_s = V^{+}(e^{j\beta L} + \Gamma_R e^{-j\beta L}) \tag{5-26}$$

so

$$V = V_s \frac{e^{j\beta d} + \Gamma_R e^{-j\beta d}}{e^{j\beta L} + \Gamma_R e^{-j\beta L}} \tag{5-27}$$

and a similar expression may be developed for current.

One of the most important quantities to determine for transmission lines and other wave problems is the impedance. Probably the most important impedance to determine is the input impedance, provided that the load is specified.

The impedance at any point on a line is defined as the ratio of voltage to current at that point on the line. For various types of waves in space, there are other impedances of importance—for example, the ratio of particle velocity to pressure for an acoustic wave. The input impedance is simply the voltage-to-current ratio at the input end of the line, and similar relations may be defined for other kinds of waves.

As an example of what happens to the impedance on a line terminated with a resistance, consider the situation of Example 5-1, shown in Fig. 5-16. Here a line is terminated by a resistance greater than Z_0, so the reflected wave is in phase with the incident voltage wave, at the load end of the line. Note that, in the quadrant between the end of the line and a quarter-wavelength from the end, the current leads the voltage. Here the reactive component of the impedance is capacitive, as it would be in the first quarter-wavelength for an open-circuited line. Near the load, the current is small compared with the voltage, so the impedance is high. Near the quarter-wavelength point, the voltage is small and the current is large, so the impedance is low.

The impedance may be determined for any point on this line by taking the ratio of the voltage phasor to the current phasor. On the diagram of Fig. 5-16, note that

$$\begin{aligned} V &= V^+(\cos \beta d + j \sin \beta d) + \Gamma_R V^+(\cos \beta d - j \sin \beta d) \\ &= V^+[(1 + \Gamma_R) \cos \beta d + j(1 - \Gamma_R) \sin \beta d] \end{aligned}$$

Similarly,

$$\begin{aligned} I &= I^+(\cos \beta d + j \sin \beta d) + \Gamma_R I^+(- \cos \beta d + j \sin \beta d) \\ &= I^+[(1 - \Gamma_R) \cos \beta d + j(1 + \Gamma_R) \sin \beta d] \end{aligned}$$

Since

$$1 + \Gamma_R = \frac{2Z_R}{Z_R + Z_0}$$

and

$$1 - \Gamma_R = \frac{2Z_0}{Z_R + Z_0}$$

the ratio of voltage to current, after taking out the common factor $1/(Z_R + Z_0)$, is given by

$$Z = Z_0 \frac{Z_R \cos \beta d + jZ_0 \sin \beta d}{Z_0 \cos \beta d + jZ_R \sin \beta d} \tag{5-28}$$

A form of this expression which is easier to remember is obtained directly from the usual expression for voltage and current on the line. Thus,

$$Z = \frac{V^+(e^{j\beta d} + \Gamma_R e^{-j\beta d})}{I^+(e^{j\beta d} - \Gamma_R e^{-j\beta d})}$$

This requires less computational effort if exp $(j\beta d)$ is removed from the parentheses in both numerator and denominator and canceled out. The resulting expression, *which should be remembered,* is

$$Z = Z_0 \frac{1 + \Gamma_R e^{-j2\beta d}}{1 - \Gamma_R e^{-j2\beta d}} \tag{5-29}$$

5-5. Complex Termination

When the load impedance contains storage elements (is complex), the reflection coefficient is complex. Otherwise, the treatment of a line so terminated is the same as that for resistive termination. A complex reflection coefficient means that the reflected voltage and current at the load are out of phase with the incident voltage and current.

To see what can happen with different load impedances, it is convenient to consider some examples.

Example 5-2. *Reflection Coefficient in First Quadrant.* Let

$$\frac{Z_R}{Z_0} = 1 + j1$$

then

$$\Gamma_R = \frac{1 + j1 - 1}{1 + j1 + 1} = \frac{j1}{2 + j1} = \frac{1e^{j90°}}{2.23e^{j26.6°}} = 0.448e^{j63.4°}$$
$$= 0.2 + j0.4$$

Note that, for $|Z_R/Z_0|$ the same size but Z_R real,

$$\Gamma_R = \frac{1.41 - 1}{1.41 + 1} = \frac{0.41}{2.41} = 0.171$$

so the angle of the load impedance has a large effect on the *magnitude*, as well as on the angle, of Γ_R. This example and those which follow are illustrated in Fig. 5-18.

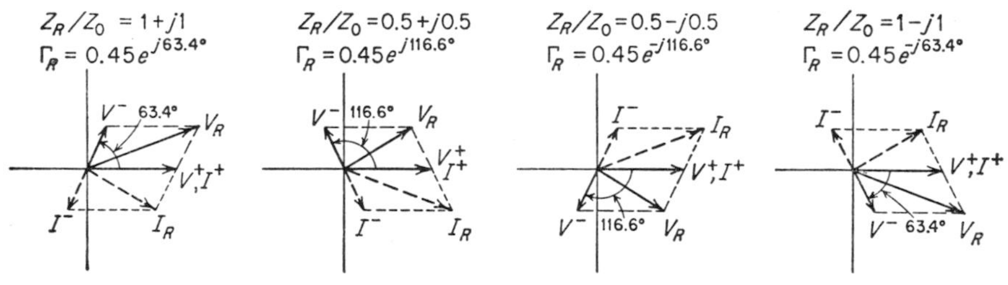

FIG. 5-18. Reflection coefficients.

Example 5-3. *Reflection Coefficient in Second Quadrant.* Let

$$\frac{Z_R}{Z_0} = 0.5 + j0.5$$

then

$$\Gamma_R = \frac{0.5 + j0.5 - 1}{0.5 + j0.5 + 1} = \frac{-0.5 + j0.5}{1.5 + j0.5} = \frac{\frac{1}{2}\sqrt{2}\,e^{j135°}}{\frac{1}{2}\sqrt{10}\,e^{j18.4°}}$$
$$= \frac{1}{\sqrt{5}}e^{j116.6°} = 0.45e^{j116.6°} = -0.202 + j0.404$$

For $|Z_R/Z_0|$ the same size but Z_R real,

$$\Gamma_R = -0.171$$

but for Z_R the same size but purely reactive,

$$\Gamma_R = \frac{j0.707 - 1}{j0.707 + 1} = \frac{1.22\underline{/180^\circ - 35.1^\circ}}{1.22\underline{/35.1^\circ}} = e^{j109.8^\circ}$$

This is a *general result*—that $|\Gamma_R| = 1$ for pure reactance load.

Example 5-4. *Reflection Coefficient in Third Quadrant.* Let

$$\frac{Z_R}{Z_0} = 0.5 - j0.5$$

then

$$\Gamma_R = \frac{0.5 - j0.5 - 1}{0.5 - j0.5 - 1} = \frac{-0.5 - j0.5}{1.5 - j0.5} = \frac{0.707e^{-j135^\circ}}{1.58e^{-j18.4^\circ}}$$
$$= 0.45e^{-j116.6^\circ}$$
$$= -0.202 - j0.404$$

Example 5-5. *Reflection Coefficient in Fourth Quadrant.* Let

$$\frac{Z_R}{Z_0} = 1 - j1$$

then

$$\Gamma_R = \frac{1 - j1 - 1}{1 - j1 + 1} = \frac{-j1}{2 - j1} = \frac{e^{-j90^\circ}}{2.23e^{-j26.2^\circ}} = 0.45e^{-j63.4^\circ}$$
$$= 0.2 - j0.4$$

Note that Γ_R is in the lower half-plane for capacitive loads and in the upper half-plane for inductive loads. It is in the right half-plane for large $|Z_R/Z_0|$ and in the left half-plane for small $|Z_R/Z_0|$.

The method of calculation on a line terminated with a complex impedance is also best illustrated by example. It is important to realize that termination of a line by other than a resistance means that the voltage maximum and minimum are *not* at the load or some multiple of a quarter-wavelength from the load. Thus, in our examples, we shall draw a phasor diagram at the end of the line and then one at the voltage maximum or minimum point (whichever comes first). From there on, we can keep track of what happens on the line by drawing phasor diagrams every quarter-wavelength.

Example 5-6. Consider Example 5-2 above. Figure 5-18 shows the situation at the load. When the phasors are rotated in their usual directions through an angle of 31.7°, $\mathcal{V}^+$ and $\mathcal{V}^-$ are coincident, so they add magnitudes; therefore, this is the point of maximum voltage. At the same point, the current phasors are rotated to the position where they are in opposition, so this is the point of minimum current. Figure 5-19 shows the phasor diagrams for this and for points 90° farther along. Figure 5-20 shows the corresponding voltage and current on the line.

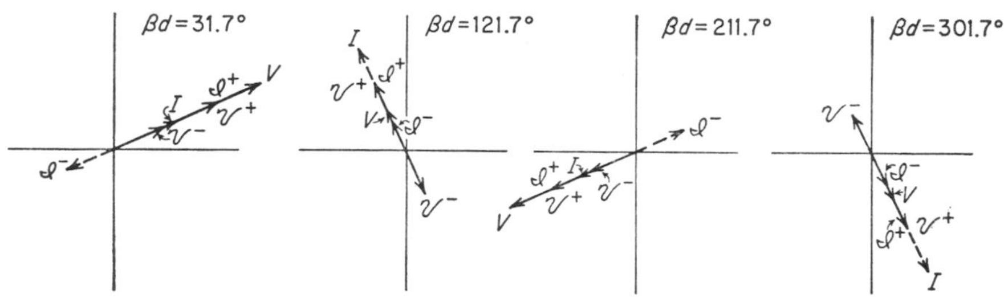

FIG. 5-19. Phasor diagrams for $Z_R/Z_0 = 1 + j1$.

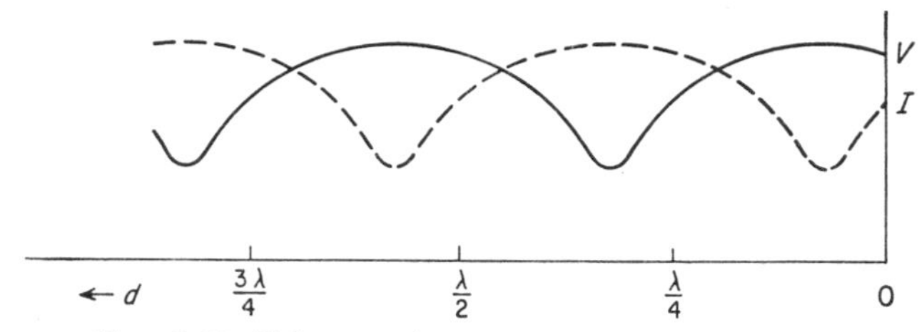

FIG. 5-20. Voltage and current for $Z_R/Z_0 = 1 + j1$.

The same methods may be used for finding input voltage, output voltage, and impedance from the phasor diagrams as for resistive loads. A typical example follows.

Example 5-7. Consider Example 5-2 above. The input voltage is measured to be 100. Assume that the line is 0.6 wavelength, or 216°, in length. Find the output voltage and current.

The process by which this is done is as follows:

1. Draw the phasor diagram for the receiving end of the line, in terms of V^+.
2. Rotate the phasors to the sending end of the line.
3. Sketch the sending-end voltage phasor, in terms of V^+.
4. Determine V_s trigonometrically on the phasor diagram, in terms of V^+.
5. Solve for the magnitude of V^+.
6. Use this magnitude to determine V_R and I_R.

For the example, we already have the receiving-end phasor diagram in Fig. 5-18. Figure 5-21 shows the sending-end phasor diagram.

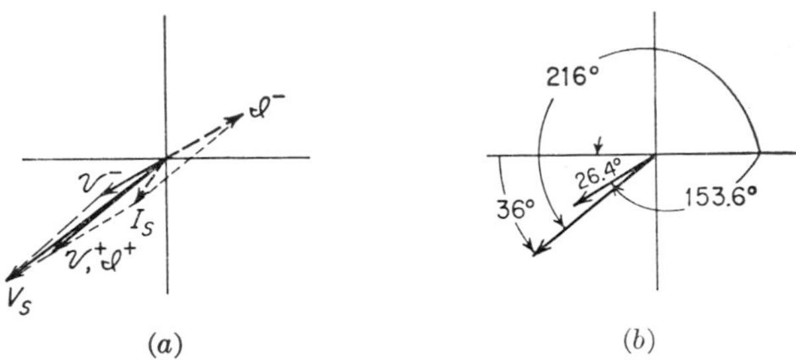

FIG. 5-21. Sending-end phasor diagram for Example 5-7: (a) complete diagram; (b) some pertinent angles.

From this, we have, for V_s,

$$\begin{aligned} V_s &= V^+(-\cos 36° - j \sin 36°) + 0.45V^+(-\cos 26.4° - j \sin 26.4°) \\ &= -V^+[(\cos 36° + 0.45 \cos 26.4°) + j(\sin 36° + 0.45 \sin 26.4°)] \\ &= -V^+[(0.81 + 0.404) + j(0.589 + 0.200)] \\ &= -V^+(1.214 + j0.789) \\ &= V^+e^{j180°}(1.44)e^{j33.0°} \end{aligned}$$

So

$$|V^+| = \frac{100}{1.44} = 69.1$$

This may now be used to find V_R (Fig. 5-18) for

$$\begin{aligned} V_R &= V^+(1 + 0.45 \cos 63.4° + j0.45 \sin 63.4°) \\ &= 69.1(1.202 + j0.404) = 83 + j27.9 \\ &= 87.2\underline{/18.6°} \text{ volts} \end{aligned}$$

$$I_R = \frac{V_R}{Z_R} = \frac{87.2\underline{/18.6°}}{Z_0(1 + j1)} = \frac{87.2}{Z_0\sqrt{2}}\underline{/-26.4°} = \frac{61.9}{Z_0}\underline{/-26.4°}$$

In this example it is important to note that the *magnitude* of V^+ was found from V_s and that the angle found in that calculation was discarded. The reason for discarding it is that, in all calculations other than that for finding V^+ from V_s, V^+ (at the receiving end) is considered the reference. In this special calculation, V_s is assumed to have zero angle. It is obvious from Fig. 5-21 that this is inconsistent with using V^+ as a reference. It would, of course, be possible to use any other reference throughout the calculations—including V_s—but it would not be so convenient.

In this example it was not necessary to find Z_s, because the sending-end voltage was given. In many cases the generator emf is known, along with the source impedance. In such a situation the input impedance must be known before V_s and I_s and must therefore be computed. The input impedance can be calculated easily from the geometry of Fig. 5-21*a*, since it is simply the ratio of V_s to I_s. Of course, the analytical solution given by (5-29) is also applicable here, and it is somewhat easier to calculate. Note, in using this form, that the *complex* reflection coefficient applies, not just the magnitude.

Before proceding further, let us consider another example.

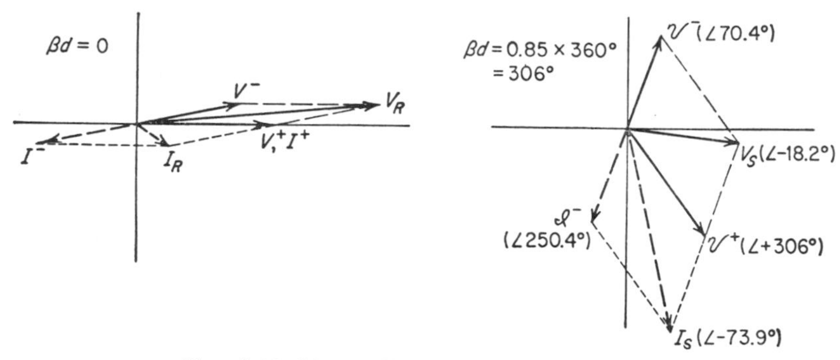

FIG. 5-22. Phasor diagrams for Example 5-8.

Example 5-8. Let

$Z_R = 150 + j75$ ohms
$Z_0 = 50$ ohms
Length $= 1.85$ wavelengths
$V_g = 100$ volts
$Z_g = 10 + j5$ ohms

Find V_s, I_s, V_R, I_R.

$$\Gamma_R = \frac{Z_R - Z_0}{Z_R + Z_0} = \frac{100 + j75}{200 + j75} = \frac{125\underline{/36.9^\circ}}{214\underline{/20.5^\circ}} = 0.585\underline{/16.4^\circ}$$

$$Z_s = \frac{V^+[(\cos 54^\circ + 0.585 \cos 70.4^\circ) + j(-\sin 54^\circ + 0.585 \sin 70.4^\circ)]}{I^+[(\cos 54^\circ - 0.585 \cos 70.4^\circ) + j(-\sin 54^\circ - 0.585 \sin 70.4^\circ)]}$$

$$= 50\,\frac{[(0.59 + 0.196) + j(-0.81 + 0.551)]}{[(0.59 - 0.196) + j(-0.81 - 0.551)]} = 50\,\frac{0.786 - j0.259}{0.394 - j1.361}$$

$$= 50\,\frac{0.829\underline{/-18.2^\circ}}{1.42\underline{/-73.9^\circ}} = 29.1\underline{/55.7^\circ} = 16.4 + j24.0$$

$$I_s = \frac{V_g}{Z_g + Z_s} = \frac{100}{10 + 16.4 + j29} = \frac{100}{39.1\underline{/47.7^\circ}} = 2.56\underline{/-47.7^\circ}$$

This is referred to V_g. We wish to refer I_s to V^+. Thus,

$$I_s = \frac{V^+}{50}[(\cos 54^\circ - 0.585 \cos 70.4^\circ) + j(-\sin 54^\circ - 0.584 \sin 70.4^\circ)]$$

$$= \frac{V^+}{50}(1.42)\underline{/-73.9^\circ}$$

So

$$V^+ = \frac{|I_s|50}{1.42} = \frac{2.56 \times 50}{1.42} = 90.2$$

$$V_s = V^+(0.829)\underline{/-18.2°} = 74.9\underline{/-18.2°}$$

$$V_R = 90.2(1 + 0.585 \cos 16.4° + j0.585 \sin 16.4°)$$

$$= 141 + j14.9 = 141\underline{/5.9°}$$

$$I_R = \frac{90.2}{50}(1 - 0.585 \cos 16.4° - j0.585 \sin 16.9°)$$

$$= 0.795 - j0.298 = 0.85\underline{/-20.5°}$$

5-6. Lines in Tandem

In some transmission-line problems, lines are used in tandem. By using lines of different characteristic impedance in tandem, it is possible to obtain a match between two different impedances. This method is

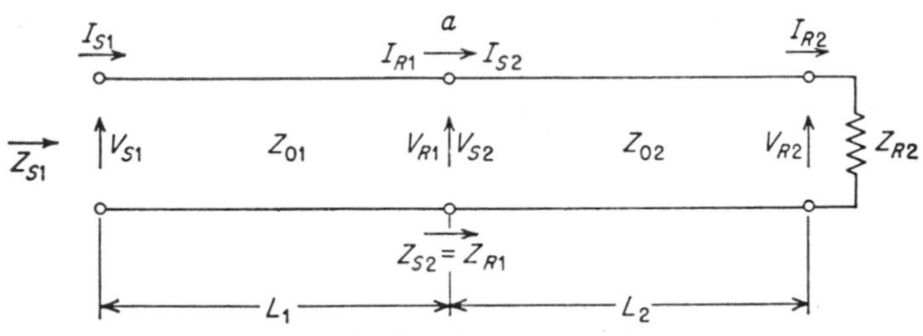

FIG. 5-23. Two lines in tandem.

used in some antennas to match the transmission-line impedance to the antenna impedance, but it is more common in problems involving waves of various types in space. For example, an acoustic wave traveling through a windowpane or a wall is first in air; then in the window or wall, where the acoustic intrinsic impedance is different, and then in air again. An electromagnetic wave from a radar antenna must usually travel through a "radome," which protects the antenna from the elements. The intrinsic impedance in the radome is different from that in air, so this, like the acoustic problem, is an air-dielectric-air one. A vibration wave on an airplane wing may go through a region where the skin is thicker and another where it is thinner; here again two different impedances are represented. A heat wave going through a building wall is first in air, then in the wall, and then in air again.

All these problems may be treated in a straightforward manner as tandem-line problems. A problem of this type is illustrated in Fig. 5-23. Here the subscript 1 refers to the line to the left, and the subscript 2 to the line to the right. They have different characteristic impedances, different velocities of propagation (and therefore different wavelengths), and different lengths. The point a is common to both lines. Thus,

$$V_a = V_{s2} = V_{R1} \qquad I_a = I_{s2} = I_{R1}$$

The load impedance for line 1 is the sending-end impedance for line 2. Thus,

$$Z_{s2} = Z_{02} \frac{1 + \Gamma_{R2} e^{-j2\beta_2 L_2}}{1 - \Gamma_{R2} e^{-j2\beta_2 L_2}} = Z_{R1}$$

At point a, then, we can calculate a reflection coefficient for line 1 in the usual manner, so that

$$\Gamma_{R1} = \frac{V_1^-}{V_1^+} = \frac{Z_{R1} - Z_{01}}{Z_{R1} + Z_{01}} = \frac{Z_{s2} - Z_{01}}{Z_{s2} + Z_{01}}$$

We may also calculate the ratio of V_{s2} to V_1^+. This is known as the *transmission coefficient*, and it is commonly used in plane-wave problems. It is given by

$$\text{(5-30)} \quad \Gamma_t = \frac{V_{s2}}{V_1^+} = \frac{V_1^+ + V_1^-}{V_1^+} = 1 + \frac{V_1^-}{V_1^+} = 1 + \Gamma_R = \frac{2Z_{s2}}{Z_{s2} + Z_{01}}$$

Since V_{s2} may be found in terms of V_2^+, it is possible to find all quantities in terms of V_2^+, although it is usually more convenient to consider each line separately.

Example 5-9. Consider two transmission lines in tandem as shown in Fig. 5-23, with the following parameters:
$Z_{01} = 50$ ohms
$Z_{02} = 100$ ohms
Length of 1 = 0.6 wavelength
Length of 2 = 0.2 wavelength
$Z_{R2} = 50 + j0$ ohms
We wish to find the sending-end impedance Z_{s1}.

$$\Gamma_{R2} = \frac{50 - 100}{50 + 100} = -\frac{1}{3}$$

$$\begin{aligned} Z_{s2} = Z_{R1} &= 100 \frac{1 - \frac{1}{3} e^{-j2(0.2 \times 2\pi)}}{1 + \frac{1}{3} e^{-j2(0.2 \times 2\pi)}} \\ &= 100 \frac{1 - \frac{1}{3}(\cos 144^\circ - j \sin 144^\circ)}{1 + \frac{1}{3}(\cos 144^\circ - j \sin 144^\circ)} \\ &= 100 \frac{1 - 0.27 + j0.196}{1 + 0.27 - j0.196} \\ &= 100 \frac{0.73 + j0.196}{1.27 - j0.196} = 100 \frac{0.758\underline{/15.0^\circ}}{1.285\underline{/-8.75^\circ}} \\ &= 58.9\underline{/23.75^\circ} = 54.0 + j23.75 \end{aligned}$$

$$\mathbf{T}_{R1} = \frac{54.0 + j23.75 - 50}{54.0 + j23.75 + 50} = \frac{4.0 + j23.75}{104.0 + j23.75}$$

$$= \frac{24.01\underline{/80.4^\circ}}{106.5\underline{/12.8^\circ}} = 0.225\underline{/67.6^\circ}$$

$$Z_{s1} = 50\,\frac{1 + 0.225e^{j67.6^\circ}e^{-j2(0.6\times 360)^\circ}}{1 - 0.225e^{j67.6^\circ}e^{-j2(0.6\times 360)^\circ}}$$

$$= 50\,\frac{1 + 0.225[\cos\,(432^\circ - 67.6^\circ) - j\sin 364.4^\circ]}{1 - 0.225(\cos 364.4^\circ - j\sin 364.4^\circ)}$$

$$= 50\,\frac{1 + 0.225 - j0.0173}{1 - 0.225 + j0.0173}$$

$$= 50\,\frac{1.225 - j0.0173}{0.775 + j0.0173} = 50\,\frac{1.225\underline{/-0.81^\circ}}{0.775\underline{/1.28^\circ}}$$

$$= 79.0\underline{/-2.09^\circ}$$

It would be possible from this to calculate the voltages at various points along the line using the analytic expressions given previously or using phasor diagrams. Since this procedure is straightforward but involved numerically, it is not repeated here.

5-7. Summary

When reflection is introduced into the steady-state solution for transmission lines or other waves, the combination of the incident and reflected waves results in a *standing wave*. Such a wave represents energy storage but not transmission. If the termination of the line is a short circuit, an open circuit, or a reactance, the reflected wave is of the same size as the incident wave, and no net power is transmitted. If, on the other hand, the terminal or load impedance has a resistive component, there is a combination of a standing and a traveling wave, due to the reflected wave being smaller than the incident wave, and some power is transmitted down the line, although significant energy is stored in the standing wave as well.

The concept of the standing-wave ratio, introduced in this chapter, is of great importance in microwave work, where it is usually easy to measure the standing-wave ratio but difficult to measure other quantities on the line. This is discussed in more detail in Chap. 9. The standing-wave ratio is the ratio of the maximum voltage amplitude on the line to the minimum voltage amplitude on the line. This ratio is infinite for open- or short-circuit and reactive terminations and is finite for terminations containing a resistive component.

To understand what happens on a transmission line involving standing waves, it is necessary to use the phasor diagram, either in its three-dimensional form, as in Figs. 5-2 and 5-3, or in the two-dimensional forms of the later figures. The phasor-diagram technique makes it possible to determine voltage or current on the line in terms of the incident voltage at the load, and this method should be used when there is any question about the meaning of a particular problem.

The short-circuited line less than a quarter-wavelength long has an inductive reactance; between a quarter- and a half-wavelength long its reactance is capacitive; and between a half-wavelength and three quarter-wavelengths it becomes inductive again. This pattern continues, with the reactance changing every quarter-wavelength from then on. The expression for the input impedance for such a short-circuited line is given by

$$Z_s = jZ_0 \tan \beta L \tag{5-18}$$

When the line is open-circuited, the input impedance looks like a capacitive reactance for a line less than a quarter-wavelength long and like an inductive reactance for a line between a quarter- and a half-wavelength long. The expression for the input impedance for the open-circuited line is given by

$$Z_s = -jZ_0 \cot \beta L \tag{5-22}$$

Near resonance, a series or parallel resonant circuit with an inductor and a capacitor may be used to represent the transmission line, or conversely. Exact equivalents may be developed.

The voltage on a line may be expressed in terms of the incident voltage at the load by

$$V = V^+(e^{j\beta d} + \Gamma_R e^{-j\beta d}) \tag{5-13}$$

Similarly, the current may be expressed by

$$I = I^+(e^{j\beta d} - \Gamma_R e^{-j\beta d}) \tag{5-31}$$

The impedance at any point on a line is therefore expressed by the ratio of these two equations, whence we get, after some simplification,

$$Z = Z_0 \frac{1 + \Gamma_R e^{-j2\beta d}}{1 - \Gamma_R e^{-j2\beta d}} \tag{5-29}$$

Expressions may also be developed from this for voltage and current in terms of sending- and receiving-end values.

The voltage and current standing waves are always 90° apart in space on the transmission line. For reactive or short- and open-circuit terminations, they are also 90° apart in time. For other terminations, although the 90° space-phase relationship is maintained, the voltage and current are less than 90° apart in time; this is necessary in order to have some power transmitted down the line to be absorbed in the resistive part of the termination.

Lines may be connected in tandem, in which case the sending-end impedance for one line becomes the receiving-end impedance for the other. Straightforward, but lengthy, calculations are necessary to determine the performance of lines in tandem.

The lossless line has been treated first in order to familiarize the student with the traveling- and standing-wave concepts in phasor notation without the complications of loss. In Chap. 6, losses are introduced, and it is shown that their effect is easily handled by the methods of this chapter, with a few additional steps. Chapter 9 shows some graphical techniques which greatly simplify steady-state a-c calculations.

PROBLEMS

5-1. Consider a quarter-wavelength open-circuited line fed from a generator of internal impedance Z_g and emf V_g. $v_p = 3 \times 10^8$ m/sec. (*a*) Find the voltage and current at both ends of the line. (*b*) Find the equivalent *LC* circuit if Z_0 is 100 ohms and the wavelength is 100 m.

5-2. Using counterrotating phasors, determine the output voltage of a 2.35-wavelength open-circuited line, having a Z_0 of 600 ohms, for an input a-c voltage of 100. Determine the current at the input to the line.

5-3. Plot the current standing waves, using counterrotating phasors, for lines with the following loads: (*a*) $R_R = 3Z_0$; (*b*) $R_R = 2Z_0/3$. Determine the standing-wave ratio for the lines of (*a*) and (*b*).

5-4. Use counterrotating phasors to determine load voltage and current for each of the lines of Prob. 5-3 when

$V_s = 100$ volts
$Z_0 = 600$ ohms
$\beta L = 215°$

5-5. A line having a 200-ohm characteristic impedance is terminated in $200 + j300$ ohms. This might be a television receiving-antenna line terminated in a somewhat detuned receiver. (*a*) Plot for 2 wavelengths the voltage and current standing waves, using counterrotating phasors. (*b*) For 1.0 volt input, find the load voltage and current. Assume the line is 2.81 wavelengths long. (*c*) Find the standing-wave ratio. (*d*) Plot the impedance on the line for 2 wavelengths, showing magnitude and phase.

5-6. This is an illustration of a practical measurement technique used with UHF and microwaves. A line having a Z_0 of 100 ohms is found to have a standing-wave ratio of 3. The first voltage minimum occurs 1 m from the load, and the next occurs 2.5 m from the load. Find the wavelength and load impedance.

5-7. A 50-ohm (Z_0) transmission line is found to have a standing-wave ratio of 1.39. The first minimum is 5 cm from the end of the line, and the next is 30 cm from the end of the line. Find the load impedance, using phasor diagrams.

5-8. A 2.6-wavelength 50-ohm transmission line is terminated by a 0.45-wavelength 120-ohm line, which itself is terminated by an impedance of 40 ohms (resistive). Find the output voltage when the input voltage is 100.

5-9. An electromagnetic plane wave in air passes through a dielectric sheet (normal to the direction of travel). The sheet is 0.58 wavelength thick. The plane wave continues on to infinity after passing through the sheet. If the relative dielectric constant of the sheet is 2, find the reflected electric field and the intensity of the wave which goes on, for an incident intensity of 1 watt/m². This problem is a simpler version of practical problems having to do with radomes.

5-10. An acoustic plane wave is traveling in air when it encounters a windowpane normal to the direction of travel. If the sound pressure was 10 newtons/m² outside the window, what is it inside? The following parameters apply to the media, and the window is 0.10 wavelength thick.

Air density	*Glass density*
1.21 kg/m³	2,400 kg/m³
v_p	v_p
344 m/sec	6,000 m/sec

From these data and your knowledge of the passage of sound through windows from practical experience, do you think the sound we normally hear comes through windows in this manner or in some other manner? If in some other manner, what is it?

5-11. Show that the standing-wave ratio for a line terminated in resistance is Z_R/Z_0 if $Z_R > Z_0$ and is Z_0/Z_R if $Z_R < Z_0$.

5-12. Find the acoustic-wave impedance presented at the end of a pipe 1.20 m long, at a frequency of 1,000 cps, if the other end of the pipe is a solid plug and if the pipe contains air at standard temperature and pressure. Use phasors.

5-13. A section of lossless open-circuited transmission line is fed from a generator with an internal emf of 100 volts and an internal resistance of 10 ohms. If the line has a characteristic impedance of 100 ohms and is 0.70 wavelength long, find the output voltage, using counterrotating phasors.

5-14. A 50-cm section of a lossless 50-ohm (Z_0) line is operated at 1,000 Mc from a generator having an internal impedance of $10 + j20$ ohms and is terminated with an impedance of 80 ohms. If the generator emf is 1 mv, find the voltage and current in the load, using counterrotating phasors.

5-15. From instantaneous power calculations, describe the flow of power in a short-circuited line a half-wavelength long. Express your results in terms of energy storage.

5-16. Determine an approximate lumped-circuit equivalent for a half-wavelength open-circuited line.

5-17. Determine an approximate lumped-circuit equivalent for a quarter-wavelength short-circuited line.

5-18. A 75-ohm transmission line is terminated in a capacitive reactance of 100 ohms. Sketch voltage and current on the line for 1 wavelength from the termination.

5-19. A 100-ohm line is terminated in a 20-ohm inductive reactance. Sketch voltage and current on the line for 1 wavelength.

6. Traveling Waves with Losses

Loss on a transmission line modifies the standing-wave patterns produced by reflections. As the observer moves toward the generator from the load on a lossless line, the observed voltage and current pattern is the same within each wavelength as in the preceding one. On a lossy line, as he moves from load toward generator, he finds that the attenuation of the reflected wave causes it to become less important. At the same time, the incident wave is becoming larger. Thus, the relative size of the reflected wave is doubly decreased in going toward the source. The net result is that the line begins to appear more and more like an infinite line, regardless of its actual termination, as the observer gets farther and farther from the load.

6-1. Qualitative and Graphical Interpretation

The effect of attenuation may be understood best by reference to phasor diagrams such as those used in Chap. 5 to describe the lossless line. Consider the expressions for steady-state a-c voltage and current on a line:

$$V = V^+e^{\gamma d} + V^-e^{-\gamma d} \tag{5-2}$$

$$I = I^+e^{\gamma d} + I^-e^{-\gamma d} = \frac{V^+}{Z_0}\,e^{\gamma d} - \frac{V^-}{Z_0}\,e^{-\gamma d} \tag{5-3}$$

Similar equations could be set up for any of the various nonelectrical quantities or for electric and magnetic fields in a plane wave. To show the effect of attenuation, we rewrite these equations as

$$V = V^+e^{\alpha d}e^{j\beta d} + V^-e^{-\alpha d}e^{-j\beta d}$$

$$I = \frac{V^+}{Z_0}\,e^{\alpha d}e^{j\beta d} - \frac{V^-}{Z_0}\,e^{-\alpha d}e^{-j\beta d}$$

Here it can be seen that the exponentials involving the attenuation constant α increase the incident voltage as one goes from the load toward the source and at the same time decrease the reflected voltage. This, of course, is to be expected. Since the incident wave is attenuated in going from source to load, it would naturally be expected to appear larger as one goes from load toward source. Likewise, the reflected wave is attenuated in going toward the generator, because its source is the load.

Following the technique used in Chap. 5 for a lossless line, we may refer all voltages and currents to the incident-voltage phasor at the load end of the line. When we do this, Eqs. (5-2) and (5-3) become

$$V = V^+(e^{\alpha d}e^{j\beta d} + \Gamma_R e^{-\alpha d}e^{-j\beta d})$$

$$I = \frac{V^+}{Z_0}(e^{\alpha d}e^{j\beta d} - \Gamma_R e^{-\alpha d}e^{-j\beta d})$$

The magnitudes of the phasors associated with the incident voltage and current are thus

$$|\mathcal{V}^+| = V^+ e^{\alpha d} \qquad |\mathcal{I}^+| = \frac{V^+}{|Z_0|} e^{\alpha d} \tag{6-1}$$

These quantities may be used in the phasor diagrams in the same way that the comparable quantities were used for the lossless line in Chap. 5. However, it should be noted that the length of the phasors involved here is changing, whereas it was fixed in Chap. 5. Likewise, the magnitudes of the reflected voltage and current are given by

$$|\mathcal{V}^-| = |\Gamma_R| V^+ e^{-\alpha d} \qquad |\mathcal{I}^-| = \left|\frac{\Gamma_R}{Z_0}\right| V^+ e^{-\alpha d} \tag{6-2}$$

Here the absolute magnitude signs have been placed around both the reflection coefficient and the characteristic impedance, since both of these may be complex and, in fact, are likely to be so on lossy lines.

To illustrate the effect of losses on the phasor traveling and standing waves, three examples are treated. Example 6-1 deals with a low-loss line ($\alpha\lambda = 0.1$); Example 6-2 deals with a medium-loss line ($\alpha\lambda = 1.0$); and Example 6-3 deals with a high-loss line having the same loss as a thermal or diffusion wave ($\alpha\lambda = 2\pi$). The definitions of "low-," "medium-," and "high-loss" are based upon the effect on the standing waves. The "low-loss line" chosen might actually have a great deal of total loss, if the wavelengths were short and the distance required for transmission was large. In such a case, the so-called medium-loss line of Example 6-2 would be considered very lossy indeed. The choice of terminology means simply that, for the low-loss line, the magnitude of the phasor voltages and currents does not change significantly during a wavelength; for the medium-loss line, the magnitude changes significantly, but the effect of reflection is still very much present after a wavelength; and for the high-loss line, the effect of reflection is essentially negligible beyond a half-wavelength or so.

Example 6-1. *Low-loss Line.* For this example we consider an attenuation of 0.1 neper/wavelength and a line with an open-circuit

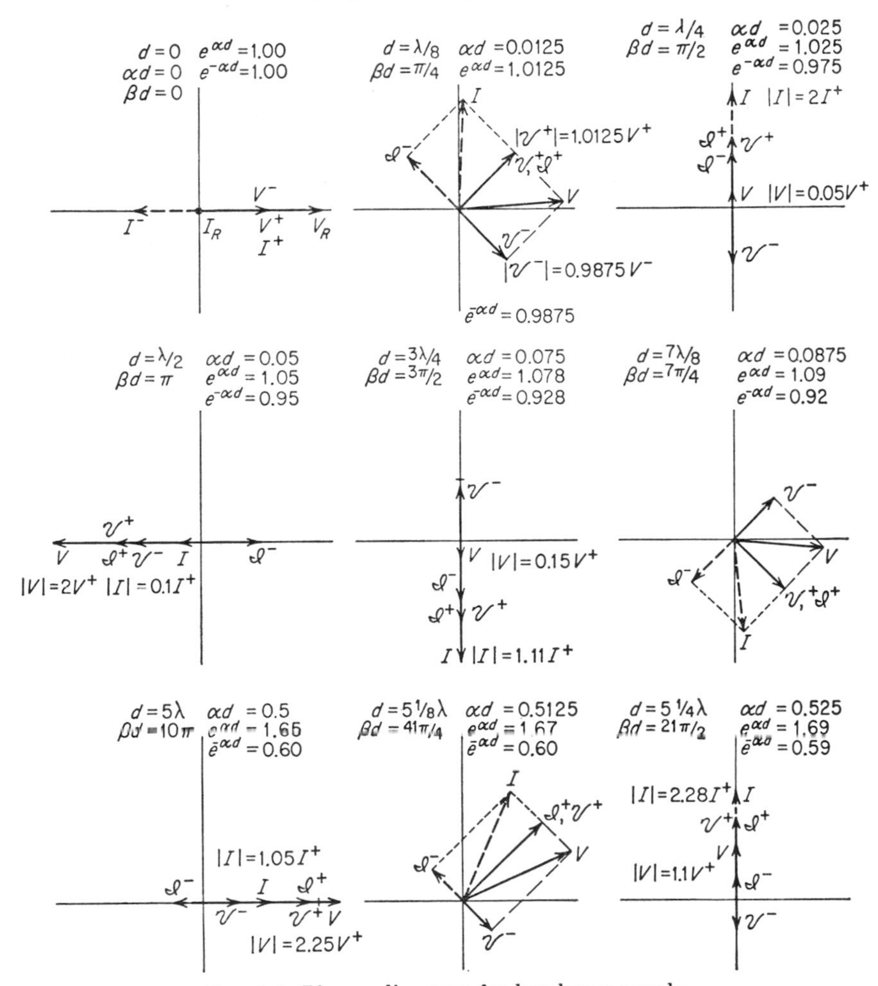

FIG. 6-1. Phasor diagrams for low-loss example.

termination. Thus, the reflection coefficient is $+1$ for voltage and -1 for current.

Figure 6-1 illustrates the phasor diagrams for this type of attenuation in the vicinity of the load and at a point 5 wavelengths from the load.

Figure 6-2 shows the amplitude of the voltage and current as a function of distance along the line. It can be seen that there is little difference from the lossless line in the current a quarter-wavelength from the open circuit (about $2I^+$) and in the voltage a half-wavelength from the open circuit (about $2V^+$). On the other hand, there is a big difference between the voltages at a quarter-wavelength and the currents at a half-wavelength—these would be zero for a lossless line, and here the voltage at a quarter-wavelength is 5 per cent of V^+ and the current at a half-wavelength is 10 per cent of I^+.

In the 5-wavelength region, the incident phasor is larger, by a factor of about two-thirds, than that at the load, and the reflected

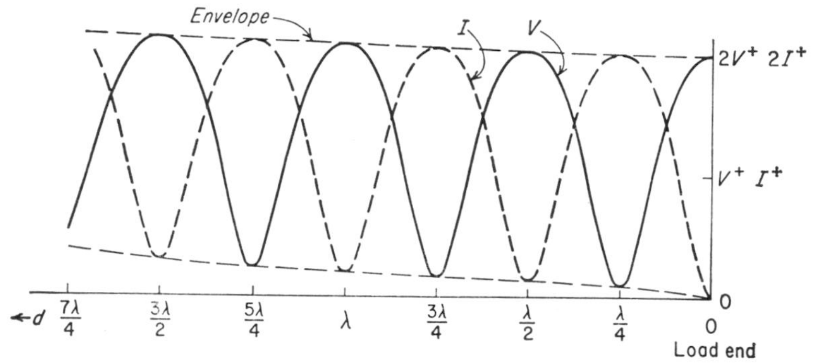

FIG. 6-2. Voltage and current for low-loss example.

phasor is less by about 40 per cent. Even at this distance the maxima are not changed greatly (only $12\frac{1}{2}$ per cent at 5 wavelengths for V and 14 per cent at $5\frac{1}{4}$ wavelengths for I). However, the minima have shown a really large change, for the *minimum* current at 5 wavelengths is larger than the *incident* current at the load, whereas the minimum current for a lossless line would still be zero. Similarly, the *minimum* voltage at $5\frac{1}{4}$ wavelengths is 10 per cent *greater than* V^+ (at the load). Another interesting result is that the voltage and current are no longer 90° out of phase, as indicated by the diagram for $5\frac{1}{8}$ wavelengths. This is necessary since power must flow down the line to be dissipated in the loss between the 5-wavelength point and the load.

By taking the ratio of the voltage to the current shown in the phasor diagram, it is possible to plot the magnitude of the impedance as a function of position on the line, and this is shown in Fig. 6-3.

The large, almost-periodic fluctuations of voltage, current, and impedance in Figs. 6-1 to 6-3 indicate that it should be quite easy to plot the envelopes for this variation. The maximum voltage

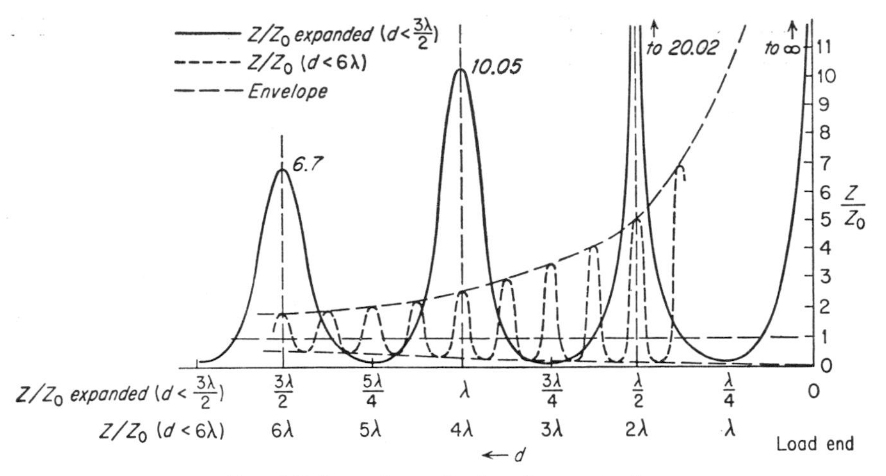

FIG. 6-3. Impedance for low-loss example.

possible at any particular point is that which would be obtained if the incident and reflected voltages were in phase at that point. Thus, the envelope for the maximum of the voltage curve is given by

$$V_{\text{max}} = V^+(e^{\alpha d} + e^{-\alpha d}) = 2V^+ \cosh \alpha d \tag{6-3}$$

Likewise, the minimum occurs when the incident and reflected voltages are 180° out of phase, so that the envelope for the minimum is

$$V_{\text{min}} = V^+(e^{\alpha d} - e^{-\alpha d}) = 2V^+ \sinh \alpha d \tag{6-4}$$

These envelopes are shown on the diagram of Fig. 6-2. The maximum value of impedance occurs where the voltages add in phase and the currents subtract. Thus,

$$Z_{\text{max}} = \frac{V_{\text{max}}}{I_{\text{min}}} = \frac{V^+(e^{\alpha d} + e^{-\alpha d})}{(V^+/Z_0)(e^{\alpha d} - e^{-\alpha d})}$$

Simplifying, this becomes

$$Z_{\text{max}} = \frac{Z_0(e^{\alpha d} + e^{-\alpha d})}{e^{\alpha d} - e^{-\alpha d}} \tag{6-5}$$

These limits are also shown on the diagram of Fig. 6-3. It should be noted that the impedance fluctuations are considerably greater than the voltage and current fluctuations, because they are the ratios of maximum voltage to minimum current and minimum voltage to maximum current, as given in

$$Z_{\text{min}} = \frac{Z_0(e^{\alpha d} - e^{-\alpha d})}{e^{\alpha d} + e^{-\alpha d}} \tag{6-6}$$

Close to the load, the maximum value for the impedance (which would be infinite if there were no loss) is given by

(6-5a) $$Z_{max} \approx \frac{1}{\alpha d} Z_0 \qquad \alpha d \ll 1$$

Likewise, the minimum impedance is given by

(6-6a) $$Z_{min} \approx \alpha d Z_0 \qquad \alpha d \ll 1$$

On the other hand, it can be seen that the impedances both tend toward Z_0 for large values of αd, as the $e^{-\alpha d}$ terms in (6-5) and (6-6) tend to zero. Thus,

(6-5b) $$Z_{max} \approx Z_0 \qquad \alpha d \gg 1$$
(6-6b) $$Z_{min} \approx Z_0 \qquad \alpha d \gg 1$$

The voltage standing-wave ratio was defined for the lossless line as the ratio of maximum to minimum voltage on the line. Since, as indicated by Fig. 6-2, this ratio changes and approaches unity as one gets farther away from the load, the standing-wave ratio is not so easy to define for a lossy line. One may, of course, define an average standing-wave ratio by defining the ratio of the average of two successive maxima to the minimum in between or by defining it as a ratio of a maximum to the average of the two minima on either side of it. This, in fact, is done in practice for standing-wave measurements on low-loss lines.

Here, however, we shall define a fictitious quantity, the *standing-wave ratio at a point*. Obviously, this cannot be measured directly, since the standing-wave ratio is defined in terms of quantities which occur a quarter-wavelength apart. However, it is a convenient parameter in describing lossy lines, and so will be used.

The standing-wave ratio at a point is described as

$$\text{VSWR} = \frac{|V_{inc}| + |V_{refl}|}{|V_{inc}| - |V_{refl}|} = \frac{e^{\alpha d} + |\Gamma_R|e^{-\alpha d}}{e^{\alpha d} - |\Gamma_R|e^{-\alpha d}}$$

This can be simplified to

(6-7) $$\text{VSWR} = \frac{1 + |\Gamma_R|e^{-2\alpha d}}{1 - |\Gamma_R|e^{-2\alpha d}}$$

It can be seen that this approaches unity as d approaches infinity; that is, the standing-wave ratio is the same as that for a properly terminated line if the observer is a long way from the point of reflection.

For a relatively low-loss line, the standing-wave ratio is a common

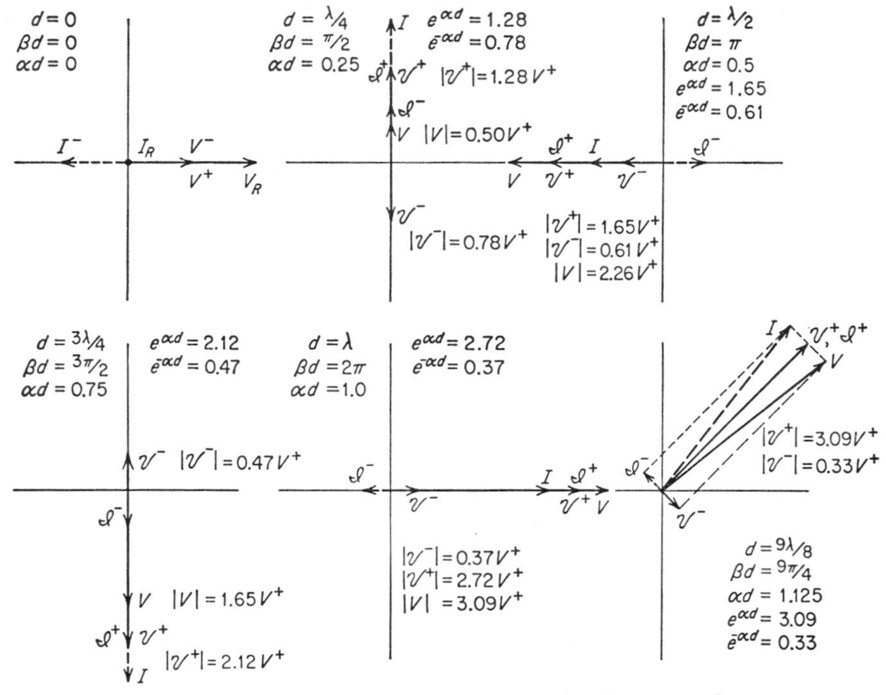

Fig. 6-4. Phasor diagrams for intermediate-loss example.

measured quantity. For intermediate- and high-loss lines, however, it is difficult to measure.

Example 6-2. *Intermediate Loss.* Here we take $\alpha\lambda = 1.0$ and again consider an open-circuit termination. The appropriate phasor diagrams are shown in Fig. 6-4. The voltage and current are shown in Fig. 6-5, and the impedance is shown in Fig. 6-6. With this intermediate loss, it can be seen that the effect is significant even

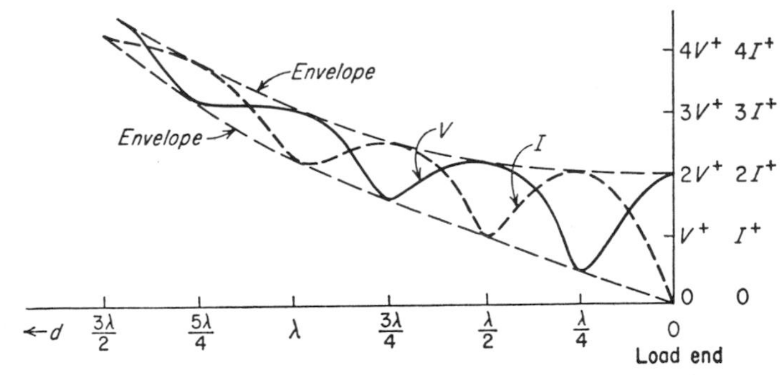

Fig. 6-5. Voltage and current for medium-loss example.

at a quarter-wavelength. Thus, whereas the current at a quarter-wavelength is only $2.06I^+$, as compared with $2I^+$ for a lossless line, the voltage (which is at a minimum a quarter-wavelength from the open circuit) is up to $0.5V^+$, as compared with 0 for the lossless case. Only a half-wavelength from the end of the line the minimum current is greater than the incident current at the end of the line (as compared with zero for a lossless line). The last phasor diagram shown indicates that voltage and current are almost in phase at a distance just shortly over a wavelength from the load, even though they are of necessity 90° apart in phase at the load and would be 90° apart at the 1-wavelength point on a lossless line.

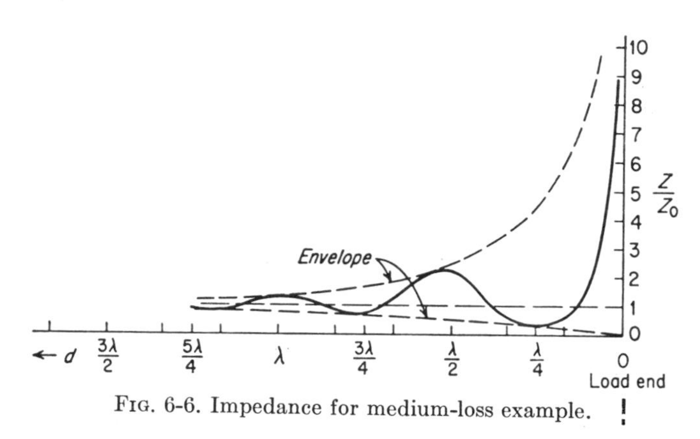

FIG. 6-6. Impedance for medium-loss example.

The way in which the voltage and current approach the infinite-line values is shown in Fig. 6-5, and the quick reduction in the oscillation of the impedance is shown in Fig. 6-6.

Example 6-3. *High Loss.* This high-loss example deals with a wave such as an electromagnetic wave in a conducting medium, a wave on a telephone cable, or a thermal or other diffusion wave. That is, $\alpha = \beta$, so that $\alpha\lambda = 2\pi$. The termination is again an open circuit. Figure 6-7 shows the phasor diagrams for this case. Figure 6-8 shows the voltage and current, and Fig. 6-9 shows the impedance. Because of the great attenuation, it is difficult to draw a phasor diagram to scale even at a distance of only 1 wavelength from the open circuit. It can be seen from examination of these phasor diagrams that the effect of even the open-circuit reflection is essentially negligible a half-wavelength from the termination.

The effect of different amounts of attenuation is illustrated in Fig. 6-10, where the impedances plotted from Figs. 6-3, 6-6, and 6-9 are all shown on the same axes.

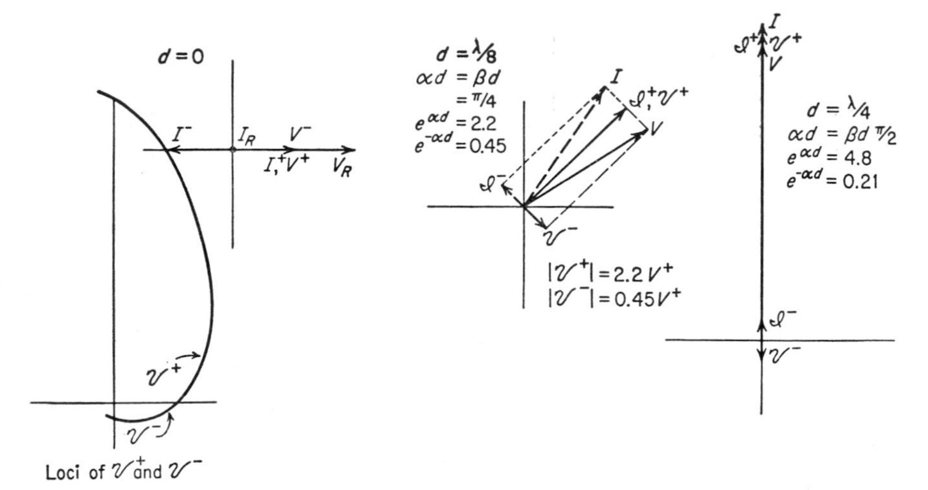

FIG. 6-7. Phasor diagrams for high-loss example ($\alpha = \beta$).

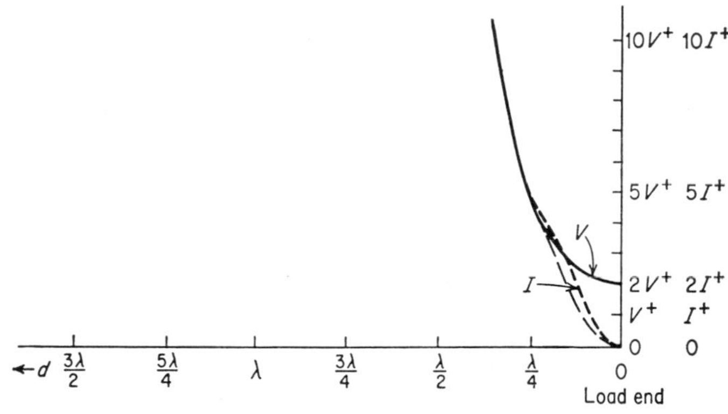

FIG. 6-8. Voltage and current for high-loss example ($\alpha = \beta$).

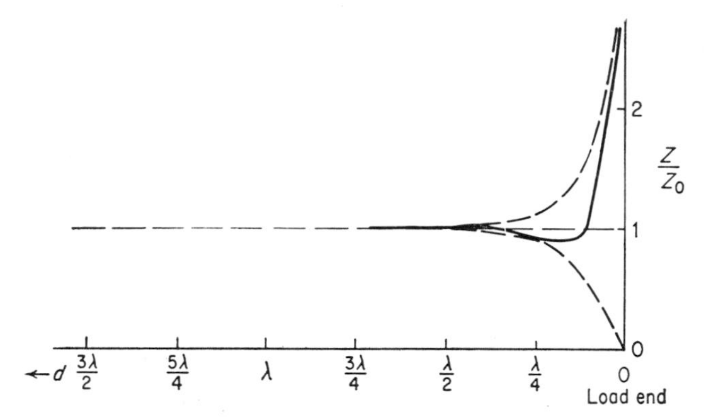

FIG. 6-9. Impedance for high-loss example ($\alpha = \beta$).

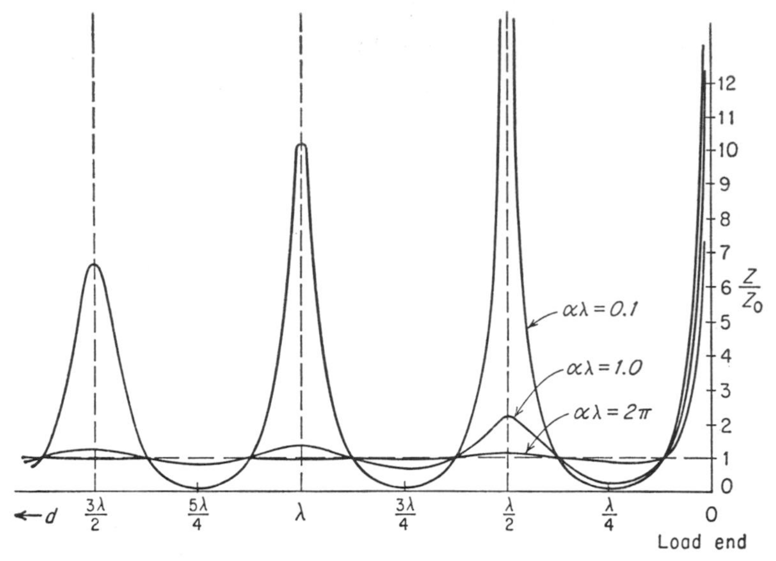

FIG. 6-10. Comparison of impedance variation for different losses.

6-2. Analysis of Lossy Lines

In practice, it is usually simpler to use analytical expressions than phasor diagrams, once the principles illustrated by the phasor diagrams are understood. Hence, a few of the analytical expressions for voltage, current, and impedance are developed in this section.

Recall that

$$V = V^+(e^{\alpha d}e^{j\beta d} + \Gamma_R e^{-\alpha d}e^{-j\beta d})$$

and

$$I = I^+(e^{\alpha d}e^{j\beta d} - \Gamma_R e^{-\alpha d}e^{-j\beta d})$$

The impedance at any point d on the line can therefore be determined by taking the ratio of these two, which is

$$Z = Z_0 \frac{e^{\alpha d}e^{j\beta d} + \Gamma_R e^{-\alpha d}e^{-j\beta d}}{e^{\alpha d}e^{j\beta d} - \Gamma_R e^{-\alpha d}e^{-j\beta d}}$$

Although the origin of the impedance expression is most readily illustrated by the above form, for practical computations it is more convenient to divide numerator and denominator by the common first term, so that the first terms cancel out. The result is

$$Z = Z_0 \frac{1 + \Gamma_R e^{-2\alpha d}e^{-2j\beta d}}{1 - \Gamma_R e^{-2\alpha d}e^{-2j\beta d}} \tag{6-8}$$

This equation should be compared with Eq. (5-29) for the lossless line. It can be seen that Γ_R in Eq. (5-29) is replaced by $\Gamma_R e^{-2\alpha d}$ here. The reflection coefficient is effectively reduced by distance between the point where the impedance is measured and the end of the line, so the effective reflection coefficient is

$$\Gamma_{R,\text{eff}} = \Gamma_R e^{-2\alpha d}$$

Using this expression, we have, instead of (6-8),

$$Z = Z_0 \frac{1 + \Gamma_{R,\text{eff}} e^{-j2\beta d}}{1 - \Gamma_{R,\text{eff}} e^{-j2\beta d}}$$

This, of course, is almost the form of Eq. (5-29).

It can be seen that, as the distance from the load is increased, that is, as $d \to \infty$, the effective reflection coefficient $\Gamma_{R,\text{eff}} \to 0$. That is, the effect of reflection becomes negligible a large distance from the termination.

For the short-circuited line treated in lossless form in Sec. 5-2, the voltage is given by

$$V = V^+(e^{\gamma d} - e^{-\gamma d})$$

which should be compared with the lossless form in Eq. (5-15). The expressions which relate the hyperbolic sine and cosine to the exponentials are

$$\sinh x = \frac{e^x - e^{-x}}{2}$$

$$\cosh x = \frac{e^x + e^{-x}}{2}$$

Utilizing the first of these relations, we have

$$V = 2V^+ \sinh \gamma d \tag{6-9}$$

for the short-circuited line. Although this is a compact form for the expression for voltage, actual evaluation is not so simple as it looks, for a numerical evaluation of $\sinh \gamma d$ where γ is complex must proceed according to the following rules:

$$\begin{aligned}\sinh \gamma d &= \sinh (\alpha + j\beta)d \\ &= \sinh \alpha d \cosh j\beta d + \cosh \alpha d \sinh j\beta d \\ &= \cos \beta d \sinh \alpha d + j \cosh \alpha d \sin \beta d\end{aligned}$$

The expression for the voltage on a lossless line is easily obtained from these relations by noting that, as $\alpha \to 0$, $\sinh \alpha d \to 0$ and $\cosh \alpha d \to 1$. Hence,

$$V \to j2V^+ \sin \beta d$$

Similarly, for the short-circuited line,

$$I = I^+(e^{\gamma d} + e^{-\gamma d})$$

which may be represented by

$$I = 2I^+ \cosh \gamma d \tag{6-10}$$

As $\alpha \to 0$, this becomes

$$I \to 2I^+ \cos \beta d$$

A concise form for the impedance of the short-circuited line is also possible and is obtained by taking the ratio of (6-9) to (6-10):

$$Z = Z_0 \tanh \gamma d \tag{6-11}$$

Although this impedance looks compact, it is actually rather complicated to compute, because both numerator and denominator are of the form shown above for complex γ. In calculating the impedance for the lossy line, therefore, it is generally easier to use (6-8) than (6-11).

Similar relations may be developed for the open-circuited line. Thus,

$$V = 2V^+ \cosh \gamma d \tag{6-12}$$
$$I = 2I^+ \sinh \gamma d \tag{6-13}$$
$$Z = Z_0 \coth \gamma d \tag{6-14}$$

Again, however, for computational purposes it is better to use Eq. (6-8) than Eq. (6-14).

So far in this chapter, the voltage has always been referred to the incident value at the receiving end of the line, using such relations as

$$V = V^+(e^{\gamma d} + \Gamma_R e^{-\gamma d}) \tag{5-2}$$

It is sometimes convenient to express the voltage at any point in terms of the sending-end voltage (and also at times to express it in terms of the receiving-end voltage). To express it in terms of the sending-end voltage, we note that

$$V_s = V^+(e^{\gamma L} + \Gamma_R e^{-\gamma L})$$

Taking the ratio of Eq. (5-2) to this equation, we have

$$V = V_s \frac{e^{\gamma d} + \Gamma_R e^{-\gamma d}}{e^{\gamma L} + \Gamma_R e^{-\gamma L}} \tag{6-15}$$

A commonly used form for the voltage, current, and impedance expressions may be obtained by noting that

$$\begin{aligned} e^{\gamma d} &= \cosh \gamma d + \sinh \gamma d \\ e^{-\gamma d} &= \cosh \gamma d - \sinh \gamma d \end{aligned} \tag{6-16}$$

Utilizing these relations, Eq. (5-2) becomes

$$V = V^+[(1 + \Gamma_R) \cosh \gamma d + (1 - \Gamma_R) \sinh \gamma d]$$

but

$$1 + \Gamma_R = 1 + \frac{Z_R - Z_0}{Z_R + Z_0}$$

Hence,

$$1 + \Gamma_R = \frac{2Z_R}{Z_R + Z_0} \tag{6-17}$$

and

$$1 - \Gamma_R = \frac{2Z_0}{Z_R + Z_0} \tag{6-18}$$

In the voltage equation above, the factor $2/(Z_R + Z_0)$ is common. Hence, we may write for the voltage

$$V = \frac{2V^+}{Z_R + Z_0}(Z_R \cosh \gamma d + Z_0 \sinh \gamma d)$$

In terms of the sending-end voltage then,

$$V = V_s \frac{Z_R \cosh \gamma d + Z_0 \sinh \gamma d}{Z_R \cosh \gamma L + Z_0 \sinh \gamma L} \tag{6-19}$$

Similar expressions may be developed for the current and the impedance.

The expressions developed in this section, and others like them, are sometimes quite useful in numerical calculations for lossy lines. However, it usually turns out to be more convenient to use the exponential expressions than to use those involving hyperbolic functions. The principal reason for presenting the hyperbolic-function expressions here is that they are found commonly in the literature, and one should be familiar with their general form and the way in which they relate to the exponential expressions.

6-3. Effect of Reflections on Loss

Improperly terminating a line results in an increase in the loss. In this section, the magnitude of this increased loss is calculated under certain simplifying assumptions which give an indication of its size. The results are exact when the line length is an integral multiple of a half-wavelength and when a resistive load is used.

The approach used here is to calculate the ratio of power sent to power received for a properly terminated line and the comparable ratio for an improperly terminated line (one having reflections). It is then assumed that the same power is delivered to the load in both cases, and the required amount of power at the input to the line is calculated for the matched and unmatched cases. From this, it is easy to calculate both the ratio of these required powers and the extra amount of power required because of mismatch (improper termination).

For simplicity, we make four assumptions: (1) that the load is resistive; (2) that Z_0 is the same in both cases; (3) that line length is an integral multiple of $\lambda/2$; and (4) that Γ_R is real. The special terminology used here is

W_R = power received unmatched
W_{RM} = power received matched
W_S = power sent unmatched
W_{SM} = power sent matched

When the line is properly terminated, since we are assuming a resistive load, the received power is given by

$$W_{RM} = V_{RM}I_{RM} = V^+I^+$$

The power at the input when the line is matched is

$$W_{SM} = V_{SM}I_{SM} = (V^+e^{\alpha L})(I^+e^{\alpha L})$$

Hence,

$$W_{SM} = W_{RM}e^{2\alpha L} \tag{6-20}$$

When the line is not properly terminated, we have

$$\begin{aligned} W_R = V_RI_R &= (V^+ + V^-)(I^+ + I^-) \\ &= V^+(1 + \Gamma_R)I^+(1 - \Gamma_R) \end{aligned}$$

Hence,

$$W_R = V^+I^+(1 - \Gamma_R^2) \tag{6-21}$$

The power sent is given by

$$W_S = V_SI_S$$

When the line length is $n\lambda/2$,

$$\begin{aligned} V_S &= V^+(e^{\alpha L}e^{jn\pi} + \Gamma_Re^{-\alpha L}e^{-jn\pi}) \\ &= V^+[e^{\alpha L}(-1)^n + \Gamma_Re^{-\alpha L}(-1)^n] \end{aligned}$$

Hence,

$$V_S = (-1)^nV^+(e^{\alpha L} + \Gamma_Re^{-\alpha L})$$

Similarly, we may show that

$$I_S = (-1)^nI^+(e^{\alpha L} - \Gamma_Re^{-\alpha L})$$

Combining these, we have

$$\begin{aligned} W_S = V_SI_S &= V^+I^+(e^{\alpha L} + \Gamma_Re^{-\alpha L})(e^{\alpha L} - \Gamma_Re^{-\alpha L}) \\ &= V^+I^+e^{2\alpha L}(1 + \Gamma_Re^{-2\alpha L})(1 - \Gamma_Re^{-2\alpha L}) \end{aligned}$$

which simplifies to

$$W_S = V^+I^+e^{2\alpha L}(1 - \Gamma_R^2e^{-4\alpha L}) \tag{6-22}$$

But we know from Eq. (6-21) that

$$V^+I^+ = \frac{W_R}{1 - \Gamma_R^2}$$

so we may write (6-22) as

$$W_S = W_R e^{2\alpha L} \frac{1 - \Gamma_R^2 e^{-4\alpha L}}{1 - \Gamma_R^2} \tag{6-23}$$

If the same power is received in the matched and unmatched cases, we set

$$W_R = W_{RM}$$

In this case the ratio of transmitted power unmatched to transmitted power matched is given by

$$\frac{W_S}{W_{SM}} = \frac{1 - \Gamma_R^2 e^{-4\alpha L}}{1 - \Gamma_R^2} \tag{6-24}$$

To find the extra amount of power required in terms of the power which must be sent in the matched condition, we write

$$\frac{W_S - W_{SM}}{W_{SM}} = \frac{W_S}{W_{SM}} - 1 = \frac{\Gamma_R^2(1 - e^{-4\alpha L})}{1 - \Gamma_R^2} \tag{6-25}$$

The magnitude of the extra power required because of reflection is shown in Fig. 6-11 for lines having various attenuations and as a function

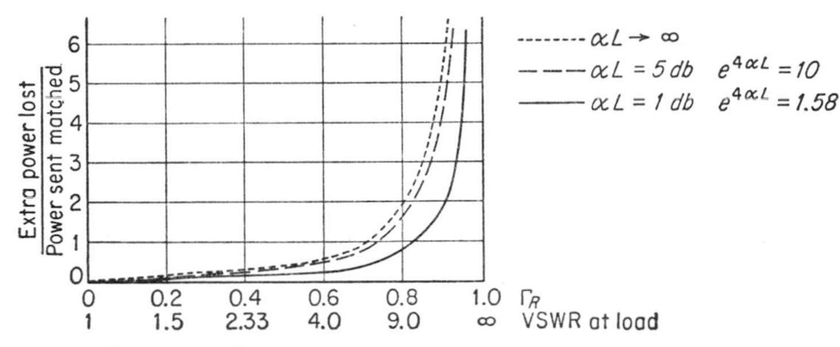

FIG. 6-11. Extra power required because of mismatch.

of the reflection coefficient. The figure shows that, for a given value of Γ_R, the percentage of extra power required is essentially the same for lines with any amount of total attenuation exceeding 5 db. Whereas it is quite small for reflection coefficients of $\frac{1}{4}$ or less, it becomes extremely large when the reflection coefficient approaches 1.

A scale of voltage standing-wave ratios at the load is also indicated. Thus, a standing-wave ratio of about 2 can be tolerated without a significant increase in power loss. However, when the standing-wave ratio exceeds 3, the power loss is quite significant.

From these curves it is apparent that, for situations in which power loss is a consideration, it is desirable to terminate the line with a resistance as close to its characteristic impedance as possible. This is desirable not only from the power-loss point of view but also because of the fact that the increased voltage due to the presence of standing waves may cause breakdown of the line or may require larger spacing, in order to prevent breakdown. Also, losses due to radiation of energy into space are more likely with the increased voltage and current associated with standing waves.

6-4. Skin Effect

An electromagnetic wave in a conductor is a very important type of lossy wave. As indicated previously, these fields satisfy the diffusion equation; this means that, for steady-state sinusoidal excitation, $\alpha = \beta$. That is, the attenuation is 2π nepers/wavelength, or about 55 db/wavelength. Such waves were treated in Example 6-3.

These waves are present at the surface of any conductor carrying current, and their effect is to concentrate the current near the surface and to move it away from those parts of the conductor (or wire) that are far from the surface. This tends to increase the effective resistance of the conductor at high frequencies by reducing the effective cross section through which current flows. The concentration of the current near the edge is known as the *skin effect*.

The wave equation for the electric field in general is

$$\nabla^2 \mathbf{E} = \mu\sigma \frac{\partial \mathbf{E}}{\partial t} + \mu\epsilon \frac{\partial^2 \mathbf{E}}{\partial t^2} \tag{2-3}$$

In steady-state form, it becomes

$$\nabla^2 \mathbf{E} = [j\omega\mu\sigma + (j\omega)^2\mu\epsilon]\mathbf{E}$$

This may be written as

$$\nabla^2 \mathbf{E} = j\omega\mu\sigma \left(1 + \frac{j\omega\epsilon}{\sigma}\right) \mathbf{E}$$

We say that a substance is a good conductor for a particular frequency when the second term in parentheses is negligible compared with the first term. That is, a material is a good conductor when

$$\omega\epsilon \ll \sigma \tag{6-26}$$

This is the same as saying that a material is a *good conductor when displacement current is negligible compared with conduction current.*

Likewise, we consider that a medium is a good dielectric, or a poor conductor, when

$$\omega\epsilon \gg \sigma \tag{6-27}$$

That is, a *good dielectric is a medium in which displacement current greatly exceeds conduction current.*

Copper is a good conductor. To see how high a frequency would be required before copper would cease to be a good conductor, let us consider the frequency for which $\sigma = \omega\epsilon$ for copper. To do so, we shall have to assume that $\epsilon = \epsilon_0$, for copper is such a good conductor that ϵ cannot actually be measured. If we do this, we find that the frequency associated with this condition is

$$f = \frac{5.8 \times 10^7}{2\pi\epsilon_0} = \frac{5.8 \times 10^7}{2\pi(1/36\pi)10^{-9}}$$
$$= 1.04 \times 10^{18} \text{ cps}$$

This is such a high frequency that it is well outside what we consider as the radio-frequency region. In fact, it is so high that classical electrodynamics cannot be used and quantum electrodynamics is required. Hence, we may assume, for all practical purposes, that copper is always a conductor.

This is true of all metals. However, sea water, soil, and some other materials change from conductor to dielectric within frequency ranges that the engineer ordinarily deals with. This situation is indicated in one of the problems.

For simplicity in developing the expression for skin effect, we shall assume initially that current is flowing in a solid sheet many wavelengths thick and, for all practical purposes, infinitely wide. The fact that it is many wavelengths thick makes the effect of any reflections from the back side negligible, because of the 55 db/wavelength attenuation in the conductor. Such a conductor is shown in Fig. 6-12.

Here it is assumed that power is flowing parallel to the surface of the conductor, that is, that the current is flowing along the conductor, and that we may calculate power flow either from the integration of the Poynting vector[1] in the space outside the conductor or from the product of in-phase components of voltage and current within the conductor. If the conductor were perfect, the electric field **E** would be normal to it, for a tangential component would result in infinite current flow. The electric field is set up by the charge associated with the flowing current. A magnetic field is set up parallel to the surface and normal to the direction of current flow. This is indicated in Fig. 6-12*a* for the perfect conductor, along with the direction of the Poynting vector and the current

[1] Recall that the Poynting vector

$$\mathbf{P} = \mathbf{E} \times \mathbf{H}$$

is a measure of power flow per unit area. The time average power flow per unit area is

$$\mathbf{P}_{av} = \tfrac{1}{2} \operatorname{Re} \mathbf{E} \times \mathbf{H}^*$$

where $\mathbf{H}^*$ is the complex conjugate of $\mathbf{H}$.

flow. The total power flow is, of course,

$$\int_{\substack{\text{area} \\ \text{normal to} \\ \text{current flow}}} (\mathbf{E} \times \mathbf{H}) \cdot d\mathbf{A}$$

where $d\mathbf{A}$ is a differential area element.

The imperfect conductor is shown in Fig. 6-12*b*. Here the Poynting vector **P** has both tangential and inward normal components, as does the electric field **E**. The normal component of **E** is associated with power flow along the surface. The tangential component of **E** is associated with losses in the surface and corresponds to the voltage drop due

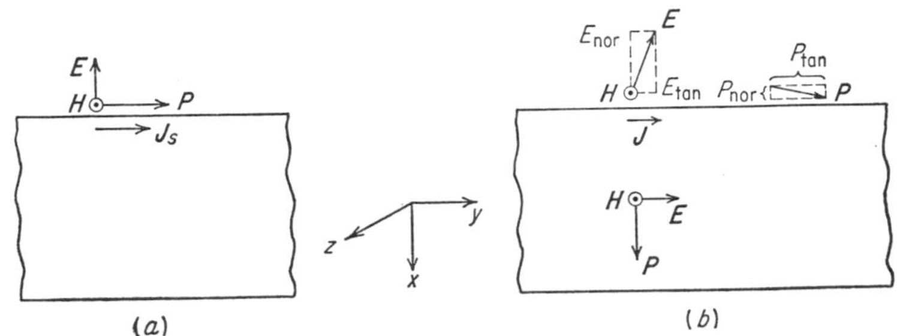

FIG. 6-12. Fields for skin effect: (*a*) perfect conductor; (*b*) imperfect conductor.

to resistance in the conductor. The normal component of the Poynting vector is, of course, given by

$$P_{\text{nor}} = E_{\text{tan}} H$$

A plane wave in the conductor satisfies the equation

$$\frac{\partial^2 E_y}{\partial x^2} = j\omega\mu\sigma E_y$$

Neglecting any reflections (because of the thickness of the conductor) and noting that the inward wave is in the positive x direction, we see that the expression for the electric field in this case is

$$E_{\text{tan}} = E_y = E_{y0} e^{-\alpha x} e^{-j\beta x} = E_{y0} e^{-\alpha x} e^{-j\alpha x}$$

where E_{y0} is the surface value of E_y. Likewise, the magnetic field for this wave is

$$H = H_0 e^{-\alpha x} e^{-j\alpha x} = H_z$$

where H_0 is the surface value of H. α in this case is

$$\alpha = \text{Re}\sqrt{j\omega\mu\sigma} = \sqrt{\frac{\omega\mu\sigma}{2}}$$

For copper this is

$$\alpha = \beta = \sqrt{\frac{2\pi f \times 4\pi \times 10^{-7} \times 5.8 \times 10^7}{2}}$$
$$= 15.1 \sqrt{f} \qquad \text{nepers/m}$$

with f in cycles per second.

The relation in the conductor between the tangential component of **E** and the magnetic field **H** is given by

$$E_y = \eta H_z = \sqrt{\frac{j\omega\mu}{\sigma}}\, H_z$$

To obtain the value for the current density, we apply Ampere's law, neglecting the displacement current because the fields are in a conductor. In this form, Ampere's law is

$$\nabla \times \mathbf{H} = \mathbf{J}$$

In our coordinate system, this is, for the inward-bound plane wave,

$$\frac{\partial H_z}{\partial x} = -J_y$$

Hence,

$$J_y = +\alpha(1 + j)H_0 e^{-\alpha x} e^{-j\beta x} = J_0 e^{-\alpha x} e^{-j\beta x} \tag{6-28}$$

where J_0 is the surface value of J. Thus, the current density decreases exponentially as one leaves the surface of the conductor. For copper, as indicated above, if the frequency is 1 Mc, $\alpha = 1.51 \times 10^4$, so the current density decreases to $1/e$ (37 per cent) in about 6.6×10^{-5} m, or 6.6×10^{-2} mm, a very short distance indeed. Hence, at 1 Mc the current is concentrated very close to the surface of a copper conductor. Even at 10 kc it is down to 37 per cent in less than 1 mm, and at 100 cps, in less than 1 cm.

One result of skin effect is that the resistance for a conductor is increased and an inductive reactance is created. To see this, consider a strip of the conductor of Fig. 6-12 whose width is w and whose length is h. The total current flowing in the strip of width w is obtained by integrating the current density for this width and the full depth of the strip. Thus,

$$I_w = w \int_0^\infty J_0 e^{-\alpha(1+j)x}\, dx$$

This integrates to

$$I_w = w \frac{J_0}{\alpha(1 + j)} = H_0 w \tag{6-29}$$

It is interesting that the total current is 45° out of phase with the current density at the surface but is in phase with the magnetic field at the surface. If we divide through by w in Eq. (6-29), we have the familiar expression showing that the magnetic field is equal to the surface current density, although in this case the "surface current density" extends through the full width of the body rather than being concentrated strictly at the surface.

To calculate the impedance for the length h on the assumption that

$$h \ll \lambda_{\text{air}}$$

we note that the voltage drop is given by

$$V = E_{y0}h$$

The expression would be more complicated, of course, if h were comparable with λ_{air}. Taking the ratio of this voltage to the current given by Eq. (6-29), we find that the impedance for the strip of width w and length h is given by

$$Z_i = \frac{V}{I_w} = \frac{hE_{y0}}{wH_0} = \frac{h\eta}{w}$$

Here the subscript i on the impedance indicates that this is the *internal* impedance and that it does not take into account external magnetic fields. The effect of the external magnetic fields is to increase the inductive reactance, but they are not considered here. Utilizing the expression for η in a conductor, we have

$$Z_i = \frac{h}{w}\sqrt{\frac{j\omega\mu}{\sigma}} = \frac{h}{w}\sqrt{\frac{\omega\mu}{2\sigma}}\,(1 + j) \tag{6-30}$$

It is very interesting that the impedance for a surface where skin effect must be considered always has a 45° phase angle (when external magnetic fields are neglected). That is, there is always as much inductive reactance as there is resistance. The internal resistance itself is given by

$$R_i = \frac{h}{w\sigma\delta}$$

where δ is a quantity known as the *skin depth* and is given by

$$\delta = \sqrt{\frac{2}{\omega\mu\sigma}} = \frac{1}{\alpha} \tag{6-31}$$

It can be seen that δ is just the depth corresponding to a reduction of the current density to $1/e$. Examination of the expression for resistance shows that the over-all resistance is just that for the d-c resistance of a strip of length h, width w, and thickness δ. That is, the resistance to alternating current, due to the effect of skin depth, is the same as the resistance to direct current that would be observed if the current were concentrated uniformly in the region between the surface and a depth δ and were zero everywhere else. Thus δ is the *effective* depth of penetration of the current into the conductor; hence, the name "skin depth."

Since the reactance is equal to the resistance, we may calculate the internal inductance by

$$L_i = \frac{R_i}{\omega} = \frac{h}{w}\sqrt{\frac{\mu}{2\omega\sigma}} \tag{6-32}$$

This could also have been calculated by application of Ampere's law to the current as calculated by Eq. (6-28).

It is instructive to consider the magnitude of the skin depth at various frequencies. Thus, for copper we find that it is given as follows:

Frequency	60 cps	600 kc	6,000 Mc
Skin depth	8.5 mm	0.085 mm	0.00085 mm

Since $\alpha\delta = 1$, $\alpha(10\delta) = 10$. From Table 2-1 it can be seen that this corresponds to a reduction of the current at a depth of 0.0085 mm at 6,000 Mc to 0.000045 times its value at the surface!

It is not customary to deal with current in flat conductors, although they do occur occasionally. Much more frequently we have the problem of a circular wire or outer conductor for a coaxial cable. Where the frequency is sufficiently high for the skin depth to be much less than the radius (see Fig. 6-13), that is, where

$$\delta \ll a$$

the calculations for the sheet carrying current may be applied to the wire. In Fig. 6-13, for example, the depth of penetration, or skin depth, is much less than the radius, and therefore the curvature of the thin layer carrying current on the outside of the wire may be considered negligible. Hence the total width transverse to the direction of current flow is just the circumference of the wire. That is,

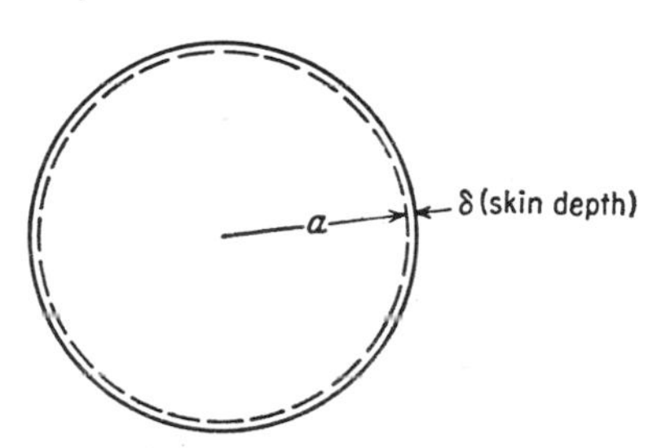

FIG. 6-13. Skin depth in a circular wire at high frequency ($w = 2\pi a$).

$$w = 2\pi a$$

The resistance of a length h of a wire is given by

$$R = \frac{h}{2\pi a\sigma\delta} \tag{6-33}$$

To see what this means, compare this expression with the d-c resistance of the wire. Thus,

$$\frac{R}{R_{dc}} = \frac{h}{2\pi a\sigma\delta}\frac{\sigma\pi a^2}{h} = \frac{a}{2\delta} \tag{6-34}$$

The resistance to alternating current, for a conductor, is a great deal more than it is to direct current, provided, of course, that the skin depth is small compared with the radius of the wire. This is discussed further in Example 6-4.

When the skin depth is not a small fraction of the radius, it is necessary to solve a boundary-value problem in cylindrical coordinates to determine the fields in the wire. The expression for the impedance (both resistive and reactive components) resulting from the exact solution is given by[1]

$$Z = \frac{-j\eta J_0(j\gamma a)}{2\pi a J_0'(j\gamma a)} \tag{6-35}$$

Here $\gamma = \sqrt{j\omega\mu\sigma}$ as usual, and J_0 is the Bessel function of the first kind, zero order. The prime in the denominator denotes differentiation of the Bessel function with respect to its argument, so the impedance is proportional to the ratio of the Bessel function of this complex argument to its derivative.

Example 6-4. *Skin Effect on a Circular Wire.* Consider a circular copper wire 1 cm in diameter. At 60 cps the skin depth δ is 8.5 mm, and the approximate formulas cannot be used. At 600 kc, where $\delta = 0.085$ mm, the approximate formulas can be used, as they can at 6,000 Mc, where $\delta = 0.00085$ mm.

The resistance per meter of length under the various conditions is given by

$$\begin{aligned} R_{dc} &= \frac{1}{\sigma\pi a^2} \\ &= \frac{1}{5.8 \times 10^7 \times (\pi/4)10^{-4}} \\ &= \frac{400 \times 10^{-5}}{5.8\pi} \\ &= 2.19 \times 10^{-4} \text{ ohm} \end{aligned}$$

At 60 cps we cannot use the approximate formula, but the use of (6-35) shows that the resistance is within a few per cent of the

[1] Simon Ramo and John R. Whinnery, "Fields and Waves in Modern Radio," 2d ed., pp. 230–253, John Wiley & Sons, Inc., New York, 1953.

d-c value. At 600 kc, using (6-34), we have

$$R = \frac{2.19 \times 10^{-4} \times 5 \times 10^{-3}}{2 \times 8.5 \times 10^{-5}}$$
$$= 29.5 \times 2.19 \times 10^{-4}$$
$$= 6.45 \times 10^{-3} \text{ ohm}$$

At 6,000 Mc,

$$R = 2.19 \times 10^{-4} \times 2{,}950$$
$$= 0.645 \text{ ohm}$$

Of course, at 600 kc and 6,000 Mc, the internal reactances are equal to the resistances, but at the latter frequency the external reactance is much larger.

For a wire of lesser diameter the approximation cannot be used at so low a frequency as for the 1-cm wire. However, it can be used reasonably well for a wire of about 1 mm diameter at 600 kc.

Because of corona, it is necessary on long-distance, high-voltage transmission lines to have conductors of large external diameter. Otherwise, the electric fields would be so large that the air would break down readily at the voltages used on these lines. With these large-diameter conductors, skin effect is important even at 60 cps. As a result, the large-diameter conductors are normally made hollow, since the current would be concentrated near the outer edge anyway, and the extra weight of copper is not justified by increased carrying capacity.

6-5. Wave Velocities in Dispersive Media

Several velocities may be defined for waves. The concept of phase velocity was introduced in Chap. 2 and has been used in the succeeding chapters. It is obvious from Fig. 4-18 that the phase velocity does not adequately describe the velocity of travel of the pulse indicated there. Neither does it adequately describe the propagation of a *signal* superimposed on a *carrier* in a lossy medium.

The term *dispersive* medium is frequently applied to a medium in which velocity is a function of frequency, so that phase relations at the source of a wave are not preserved at a distant point of observation. Among media with losses, only the *distortionless transmission line* (discussed in Chap. 8) has a phase velocity independent of frequency. All others are, in fact, dispersive. It is customary to treat low-loss situations as if they were lossless, as was done in Chaps. 3 and 4 for mechanical waves and in Chap. 5 for transmission lines. In fact, there is some dispersion in any medium, and we are extremely fortunate that we may often neglect it. Although almost all lossy media are dispersive, it is *not* true that all lossless media are nondispersive.

When phase velocity is a function of frequency, it often develops that several other velocities must be considered. One can always define some sort of *signal velocity* which measures the rate at which some characteristic of the signal, such as the peak value, travels. In the situation represented in Figs. 4-18 and 4-19, not only is this velocity different from the phase velocity, but it is a function of the distance from the source to the point of observation, whereas the phase velocity is determined completely by the medium properties and the frequency (where a frequency can be assigned).

A *velocity of energy transport* can also be defined. Thus, if the energy stored in a region of a line or medium flows from one place to another, a velocity can be defined for this energy, just as for particles flowing from one place to another.

Both the velocity of energy transport and the signal velocity are difficult to define in a general way. In many cases of interest some carrier frequency is *modulated* by a signal whose component frequencies are much less than the carrier frequency. The *group velocity* may be defined for such a medium, and it is frequently a very good approximation to both signal velocity and velocity of energy transport.[1]

Consider two frequencies, f_1 and f_2, that are close together compared with their magnitudes; that is,

$$f_2 - f_1 \ll f_1 < f_2 \tag{6-36}$$

The group velocity for these frequencies is defined as the phase velocity for the beat note at frequency $(f_2 - f_1)$.

Now consider that the *carrier* angular frequency ω_c lies midway between ω_1 and ω_2, so that

$$\omega_1 = \omega_c - \omega_s$$
$$\omega_2 = \omega_c + \omega_s$$

Then expand the phase constants in a Taylor series about the value for ω_c. Thus,

$$\beta_1 = \beta_c - \left.\frac{d\beta}{d\omega}\right|_{\omega_c} \omega_s + \frac{1}{2!}\left.\frac{d^2\beta}{d\omega^2}\right|_{\omega_c} \omega_s^2 - \cdots$$
$$\beta_2 = \beta_c + \left.\frac{d\beta}{d\omega}\right|_{\omega_c} \omega_s + \frac{1}{2!}\left.\frac{d^2\beta}{d\omega^2}\right|_{\omega_c} \omega_s^2 + \cdots$$

[1] For an excellent discussion of this whole problem, see Leon Brillouin, "Wave Propagation and Group Velocity," Academic Press, Inc., New York, 1960. Some of the mathematics in this reference is considerably beyond the level of the present work; however, a reader not prepared for the mathematics may benefit from the lucid physical discussions.

If the two frequencies are represented by equal cosinusoidal voltages, the sum is

$$v = \cos(\omega_1 t - \beta_1 x) + \cos(\omega_2 t - \beta_2 x)$$
$$= 2\cos\left(\frac{\omega_1 + \omega_2}{2}t - \frac{\beta_1 + \beta_2}{2}x\right)\cos\left(\frac{\omega_2 - \omega_1}{2}t - \frac{\beta_2 - \beta_1}{2}x\right)$$

In terms of ω_c and ω_s this becomes, when ω_s is small,

(6-37) $$v = 2\cos(\omega_c t - \beta_c x)\cos\left(\omega_s t - \left.\frac{d\beta}{d\omega}\right|_{\omega_c}\omega_s x\right)$$

This can be interpreted as a carrier frequency cosinusoid whose amplitude is varied at the rate ω_s. It is customary to refer to this relatively low ω_s as the *signal* frequency, although of course no information is transferred by the signal unless its amplitude or frequency is changing.

The phase velocity for the carrier is just the value of v_p at ω_c, and for the assumption of small ω_s it is very close to the phase velocities for ω_1 and ω_2. The phase velocity for the signal frequency is given by

(6-38) $$v_g = \left.\frac{d\omega}{d\beta}\right|_{\omega = \omega_c}$$

and is called the *group velocity*. Note that the assumption that higher-order terms may be neglected in the Taylor series for β_1 and β_2 means that the group velocity may also be defined as the limit as $\omega_s \to 0$ of the phase velocity for the beat between two sinusoids separated by angular frequency $2\omega_s$.

For many purposes the group velocity may be assumed to represent the velocity of travel of disturbances, or intelligence, superimposed on a carrier. Under the same conditions the group velocity is ordinarily very close to the signal velocity and the velocity of energy transport.

In the case of a *step* function in a diffusing medium (electromagnetic wave in a conductor, heat conduction, molecular diffusion, etc.), the restriction of a narrow range of frequencies is not applicable, and the group velocity is not particularly significant.

In a nondispersive medium (ordinarily this is a lossless medium) the phase and group velocities are the same. This is readily seen, since in such media v_p is independent of frequency and

$$\beta = \frac{\omega}{v_p}$$

Under these conditions,

(6-39) $$\frac{d\omega}{d\beta} = v_p = v_g$$

For the special case of a medium satisfying the diffusion equation, it is possible to use the group velocity, provided that the percentage bandwidth occupied by the signal is not too great. Here, from Eq. (3-44),

$$\beta = \sqrt{\frac{\omega}{2D}}$$

so

$$v_g = \frac{d\omega}{d\beta} = 4D\beta = 2v_p \tag{6-40}$$

It is interesting to note that the group velocity may be either greater or less than the phase velocity. An example of the latter occurs in waveguides.

Example 6-5. *Group Velocity in Waveguide.* As will be shown in Chap. 9, the phase velocity in a waveguide is given by

$$v_p = \frac{v_{p0}}{\sqrt{1 - \omega_0^2/\omega^2}}$$

with

$$\beta = \omega\sqrt{\mu\epsilon}\sqrt{1 - \omega_0^2/\omega^2}$$

Now,

$$\begin{aligned}\frac{d\beta}{d\omega} &= \sqrt{\mu\epsilon}\sqrt{1 - \omega_0^2/\omega^2} + \frac{\omega\sqrt{\mu\epsilon}}{2}\frac{2\omega_0^2}{\omega^3}\frac{1}{\sqrt{1 - \omega_0^2/\omega^2}} \\ &= \sqrt{\mu\epsilon}\,\frac{1 - \omega_0^2/\omega^2 + \omega_0^2/\omega^2}{\sqrt{1 - \omega_0^2/\omega^2}} \\ &= \frac{\sqrt{\mu\epsilon}}{\sqrt{1 - \omega_0^2/\omega^2}} = \frac{1}{v_{p0}\sqrt{1 - \omega_0^2/\omega^2}}\end{aligned}$$

so

$$v_g = v_{p0}\sqrt{1 - \omega_0^2/\omega^2}$$

Thus, when $\omega > \omega_0$, the phase velocity is greater than the velocity of light (v_{p0}) and the group velocity is less. These same effects are observed for propagation of electromagnetic waves in the ionosphere, where ω_0^2 is directly proportional to the density of free electrons.

6-6. Summary

As a result of attenuation, the phasors associated with incident and reflected traveling waves change in magnitude as they travel along the line. The phasor associated with the incident wave becomes larger as the generator is approached, and that associated with the reflected wave becomes smaller as the generator is approached and the load which is its source is left behind. This can be seen from

(6-1) $$|\mathcal{V}^+| = V^+e^{\alpha d}$$

(6-2) $$|\mathcal{V}^-| = V^+|\Gamma_R|e^{-\alpha d}$$

Because of the changing relative size of the incident and reflected waves, the effect of reflection is reduced close to the generator relative to its effect close to the load. Hence, the standing-wave ratio approaches unity as d approaches infinity. It is impossible to measure exactly the standing-wave ratio at a point defined by

(6-7) $$\text{VSWR} = \frac{1 + |\Gamma_R|e^{-2\alpha d}}{1 - |\Gamma_R|e^{-2\alpha d}}$$

When the attenuation per wavelength is small, this quantity may be readily approximated by measurement; but when the attenuation is large, measurement is difficult.

The most useful expression for the impedance at any point on a transmission line is

(6-8) $$Z = Z_0 \frac{1 + \Gamma_R e^{-2\alpha d} e^{-2j\beta d}}{1 - \Gamma_R e^{-2\alpha d} e^{-2j\beta d}}$$

Numerous expressions may be developed for voltage, current, and impedance in terms of hyperbolic functions of complex arguments. However, the resulting computational complexities are so great that it is usually more convenient to use the exponential forms or phasor diagrams.

The effect of reflection is to increase the amount of loss on a transmission line or other traveling-wave device. Thus, if the line is an integral number of half-wavelengths long, and if the other restrictions detailed in Sec. 6-3 are met, the amount of increased loss is given by

(6-25) $$\frac{W_S}{W_{SM}} - 1 = \frac{\Gamma_R^2(1 - e^{-4\alpha L})}{1 - \Gamma_R^2}$$

Whenever a current flows in an imperfect conductor, there is a lossy traveling wave representing the heat loss in the conductor. The resistance which results is the same as it would be if the current were concentrated uniformly within a depth δ below the surface, where $\alpha\delta = 1$. That is, the resistance is the same as if all the current were distributed uniformly and concentrated within a belt corresponding to the depth for attenuation of 1 neper.

There is an inductive reactance associated with the resistance when skin effect is important, and it is such that the reactance due to the magnetic field inside the conductor is essentially equal to the resistance. The expression for the over-all internal impedance is

(6-30a) $$Z_i = \frac{h}{w\sigma\delta}(1 + j)$$

For a circular wire whose skin depth is much less than its radius, the ratio of a-c to d-c resistance is given by

$$\frac{R}{R_{dc}} = \frac{a}{2\delta} \tag{6-34}$$

The same principles may be applied in determining the performance of lines with loss as are applied when the lines do not have loss. However, the effects are somewhat more complicated because of the changing sizes of incident and reflected voltages.

A wave traveling in a lossy medium is associated with several velocities. The phase velocity is not very useful in describing propagation of *signals* in such a medium. The most easily defined velocity for this purpose is the group velocity, which is the phase velocity for the beat between adjacent frequencies and, therefore, for the modulation on a carrier. In many cases, where the percentage bandwidth of a system is small, the group velocity is very close to the values of the signal velocity and the velocity of energy transport. The more complicated situations for which this is not true require special treatment for each problem, and they have not been covered here.

PROBLEMS

6-1. A 100-ohm transmission line has a loss of 0.05 neper/wavelength. It is terminated in 200 ohms resistance. Sketch voltage, current, and impedance as a function of distance over 1-wavelength intervals in the following regions: (*a*) 0 to 1 wavelength from the load; (*b*) 5 to 6 wavelengths from the load; (*c*) 10 to 11 wavelengths from the load; (*d*) 20 to 21 wavelengths from the load; and (*e*) 100 to 101 wavelengths from the load. Phasor diagrams should be sketched for each region, although one per region will be satisfactory after the first region, where at least five will be required.

6-2. A 50-ohm transmission line has a loss of 0.02 neper/wavelength. It is terminated in 33 ohms resistance. Sketch voltage, current, and impedance as a function of distance over the same 1-wavelength intervals listed in Prob. 6-1. At least one phasor diagram per interval is required.

6-3. A 50-ohm transmission line is terminated in $75 - j25$ ohms. It has a loss of 0.15 neper/wavelength. Sketch voltage, current, and impedance as a function of distance over 1-wavelength intervals in the following regions: (*a*) 0 to 1 wavelength from the load; (*b*) 5 to 6 wavelengths from the load; (*c*) 10 to 11 wavelengths from the load; (*d*) 30 to 31 wavelengths from the load. Phasor diagrams are required as in Prob. 6-1.

6-4. A 72-ohm transmission line is terminated in $25 + j10$ ohms. It has a loss of 0.20 neper/wavelength. Sketch voltage, current, and impedance as a function of distance over the 1-wavelength intervals of Prob. 6-3.

6-5. A 100-ohm transmission line has a loss of 1.5 nepers/wavelength. It is terminated in 200 ohms resistance. Sketch voltage, current, and impedance for this line over a 2-wavelength range, starting at the load. Phasor diagrams are required at least every quarter-wavelength for the first wavelength.

6-6. A plane ultrasonic wave in air is reflected from a completely hard surface. The wave is at a frequency of 1 Mc and has an incident peak particle velocity of

1 m/sec. Attenuation of sound waves in air is given approximately by

$$\alpha = 2 \times 10^{-13} f^2 \qquad \text{nepers/m}$$

where f is the frequency in cycles per second. Describe the standing wave in the neighborhood of the reflecting surface and at distances of 1, 10, and 100 m.

6-7. Alternating, sinusoidal flow of heat through a thick copper bar strikes an infinite heat sink (temperature cannot be changed). The amount of heat flowing in the incident wave is 100 watts/m² at the heat sink. Sketch the variation of temperature and heat flow at distances of a sixteenth-wavelength, an eighth-wavelength, 3 sixteenth-wavelengths, and a quarter-wavelength, using phasor diagrams as an aid. Express these distances in meters if the wave has a frequency of 10 cps.

6-8. A certain telephone cable has the following characteristics at 10 kc:

$\alpha = 2.43$ db/mile
$\beta = 0.59$ radian/mile
$Z_0 = 155 - j73$ ohms

This line is terminated in an impedance of $155 + j73$ ohms. If the line is 10 miles long and is required to have an output of 10 mv, what must be its input voltage? You may use either formulas or phasor diagrams.

6-9. The telephone cable of Prob. 6-8 has the following characteristics at 50 kc:

$\alpha = 3.53$ db/mile
$\beta = 2.60$ radians/mile
$Z_0 = 134 - j20$ ohms

The line is terminated in an impedance of $200 + j350$ ohms. For 10 miles of length and 10 mv out, what must be the input voltage?

6-10. A quarter-wavelength insulator has a loss of 0.1 db/wavelength. If the characteristic impedance is 50 ohms, what is the terminal impedance? Calculate the reflection from this impedance in a 50-ohm line insulated with this imperfect insulator. Is this a reasonable device to use?

6-11. The quarter-wavelength insulator of Prob. 6-10 is to be used over a band of frequencies. It has a bandwidth, which may be associated with a Q, just as with lumped resistance, capacitance, and inductance. Assuming that the device operates at a center frequency of 3,000 Mc, what is its Q? Over what frequency range would you consider using it? Justify your answer by calculating reflection vs. frequency.

6-12. Derive the expression for input impedance on a lossy line in terms of hyperbolic functions.

6-13. A radio-frequency transmission line has a loss of 3 db in its 5 wavelengths, when properly terminated. If 100 watts is to be dissipated in the load, how much more power must be transmitted down the line for the same output if its standing-wave ratio is increased to 4.0?

6-14. A radio-frequency transmission line has a loss of 0.5 db/wavelength at a certain frequency. If the line is 3 wavelengths long, what is the loss for a received power of 50 kw when matched and when mismatched so that the standing-wave ratio is 2.0? 10.0?

6-15. Determine the general expression for the ratio of the power required to be transmitted on a lossy mismatched line an eighth-wavelength long to that for a matched line of the same length, assuming that Z_0 is resistive. Find its value for reflection coefficients of 0.5 and 0.9 for Z_R resistive and $\alpha\lambda = 0.5$.

6-16. Do the computations of Prob. 6-15 for a line a quarter-wavelength long.

6-17. The electrical properties of sea water are important, since many radio-communication paths lie over sea water and radar is important to ships and aircraft at sea. The conductivity of sea water is nominally 4 mhos/m, although it may be as low as 1 mho/m in the Arctic and as high as 8 mhos/m in the Red Sea. Its dielectric constant is $81\epsilon_0$ up to 3,000 Mc, where it starts to decrease: (*a*) At what frequency does sea water change from a poor conductor to a poor dielectric? (*b*) Calculate α and β for a frequency of 1 Mc. (*c*) What is the wavelength in the sea at 1 Mc?

6-18. For fresh-water lakes, the conductivity varies, depending on impurities. Sometimes it is 0.005 mho/m. The dielectric constant is the same as for sea water. Perform the calculations of Prob. 6-17 for fresh water.

6-19. At 10 kc the magnetic field intensity in the surface of sea water due to a wave traveling along its surface is 1 μamp/m. What is its intensity at a depth of 30 m? Calculate the power flow per square meter at the surface and at 30 m depth.

6-20. Use a table in a handbook to calculate the frequency of transition from conductor to dielectric for five other materials.

6-21. Calculate the skin depth in phosphor bronze, whose *resistivity* (reciprocal of conductivity) is given as 5×10^{-6} ohm-cm. Repeat for carbon, whose resistivity is 3,000 ohm-cm. Use frequency of 10 Mc.

6-22. Calculate the resistance at 10 Mc for copper wire whose diameter is 0.1 in.

6-23. Gold and silver plating are often used to improve the resistance of surfaces at high frequencies. What is the effect of a 3-mil thickness of gold on the resistance of the wire in Prob. 6-22? Gold has a conductivity of 4.1×10^7 mhos/m. Why do you think gold is used, in view of your calculation?

6-24. Determine the phase and group velocities at 1 Mc for waves in fresh water of conductivity 0.005 mho/m and relative dielectric constant 81.

6-25. Calculate the group velocity for the transmission line of Example 2-2, at a frequency of 1,000 cps.

6-26. Discuss the concept of signal velocity as applied to a heat-conduction wave generated by a step function.

7. Lumped-constant Models—Artificial Lines

It is frequently easier to make a model of a distributed parameter structure using lumped elements (resistors, inductors, capacitors, and conductors in the electrical case) than it is to work with the actual structure itself. A lumped circuit in a relatively small box may simulate an actual telephone or power line hundreds of miles long. In analyzing heat conduction, it is frequently easier to use an electrical model (or a nonelectrical model for that matter) than to make actual measurements of heat flux and temperature. For example, it is very difficult to make measurements of the heat flux in a transistor but relatively easy to model transistor heat conduction with an electrical analog.

In addition to their uses in simulating the performance of actual traveling waves, artificial transmission lines find direct application in pulse generators, such as the one described in Example 4-2. Furthermore, they are frequently used to insert a time delay for some signal, as in an oscilloscope, where the signal presented on the screen must be delayed long enough for the sweep to get started.

Sometimes the reverse process works also. It is frequently useful to consider from a transmission-line point of view lumped-constant systems which approximate the performance of traveling-wave systems.

7-1. Introduction to Iterative Structures

A lumped-constant model for a traveling wave on a transmission line or in space is an iterative structure; that is, it is made up of a number of identical sections tied together, and the impedance at many specified points on the model is the same, provided that the model is properly terminated. Consider Fig. 2-6, which shows several possible approximations for a transmission line. Any one of these could be used as the basis for an artificial transmission line made up of large numbers of identical sections connected together.

An electric traveling wave is usually simulated by a ladder network. Such a network is illustrated in Fig. 7-1. The reason that it is called a ladder network is obvious from the configuration shown there. The network in Fig. 7-1*a* is terminated in the *mid-series* manner; that is, each end section has half as much series impedance as that between adjacent shunt elements. Figure 7-1*b* shows the same sort of network terminated in a *mid-shunt* manner; here the shunt admittances have been halved at the ends (so the shunt impedances are doubled). Although the mid-shunt- and mid-series-terminated lines are different at the ends, they are indistinguishable in the middle of the circuit model.

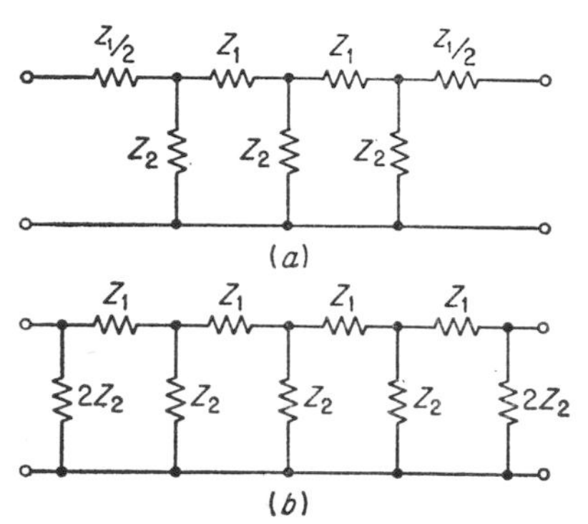

FIG. 7-1. Circuit models of transmission lines: (*a*) mid-series-terminated transmission-line model; (*b*) mid-shunt-terminated transmission-line model.

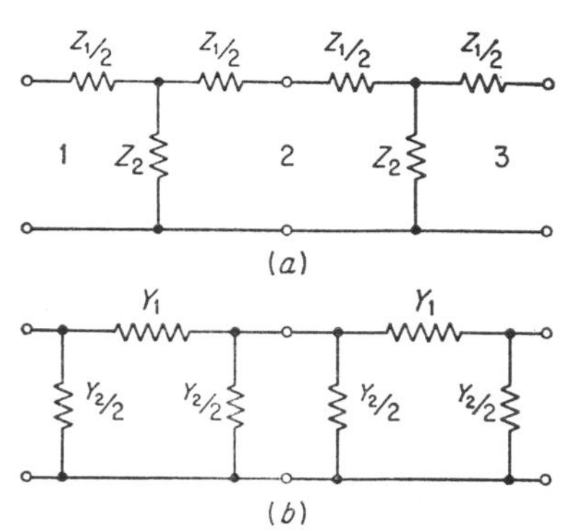

FIG. 7-2. T sections and π sections on artificial lines: (*a*) T sections; (*b*) π sections.

When the transmission line is terminated in the mid-series manner, it is normally considered to be made up of T sections. This is indicated in Fig. 7-2*a*, where two of the T sections of the circuit model of a transmission line are shown complete. Note that the series impedance per section is split evenly between the left- and right-hand sides of the shunt impedance. Figure 7-2*b* shows two sections with mid-shunt termination; such a model is considered to be made up of π sections. The T and the π both refer to the basic shape of the network impedance configuration, as indicated in the figure. For the π section, it is usually more convenient to talk in terms of admittances than impedances, so Fig. 7-2*b* shows the admittances. Note that, for the π section, half of the shunt admittance appears at the left end and half at the right end, whereas the series admittance is all in one piece. The admittances for the π section are, of course, given by

$$Y_1 = \frac{1}{Z_1} \qquad Y_2 = \frac{1}{Z_2} \tag{7-1}$$

Sometimes other types of circuit model are used. For example, an L section was used in Chap. 2 for deriving the properties of the transmission line. Since the characteristic impedance is the same for both ends of either type of symmetrical section (T or π), the circuit models for complete lines use such sections to simplify calculation.

Figure 7-3 shows a T section set up as a model for part of a transmission line. Here the series impedance is given by the series impedance per unit length on the transmission line times the length the model represents, and a similar method is used for the shunt admittance. Thus,

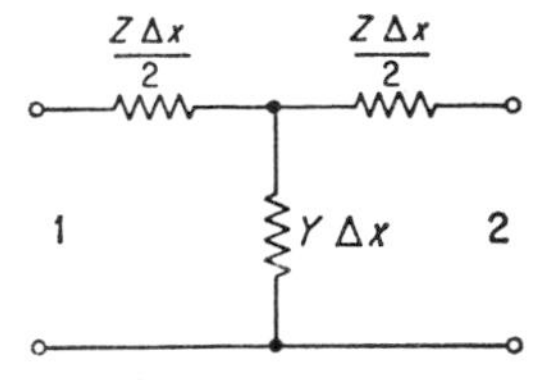

FIG. 7-3. T-section model for part of a transmission line.

$$Z = R + j\omega L \quad \text{ohms/m}$$
$$Y = G + j\omega C \quad \text{mhos/m} \tag{7-2}$$

This is an approximate model. Section 7-3 describes a method for obtaining an *exact* model for a section of line at a *single* frequency.

A model made with the approximate sections is a reasonably good representation for a transmission line, provided that the voltage drop between points 1 and 2 is small compared with the voltage across the line at either end of the section and provided that the loss in current through the shunt admittance is small compared with the total current flowing in the line. That is,

$$|V_2 - V_1| \ll |V_1| \qquad |I_2 - I_1| \ll |I_1| \tag{7-3}$$

It is possible to set up lumped equivalents for all the types of traveling wave discussed earlier. Only two examples will be considered at this point, the acoustic wave in a liquid and the thermal-conduction wave.

One of the telegrapher's equations for the acoustic wave is

$$\frac{\partial p}{\partial x} = -\rho_v \frac{\partial u_x}{\partial t} \tag{3-15}$$

Since p is measured in newtons per unit area, for a lumped-constant system we should multiply by some area ΔA to obtain a force. When this is done, Eq. (3-15) becomes

$$\frac{\partial f_x}{\partial x} = -\rho_v \, \Delta A \, \frac{\partial u_x}{\partial t}$$

Here it has been assumed that ΔA is in the YZ plane, so the resulting force is in the x direction. This is still in the form of an equation for a distributed-constant system. To make it appropriate for a lumped-constant system, we convert the partial derivative on the left to a difference divided by a length Δx and then multiply through by Δx on both sides of the equation. This results in the difference equation appropriate to the discussion of the lumped model:

$$f_x(x + \Delta x) - f_x(x) = -m \frac{\partial u_x}{\partial t} \tag{7-4}$$

Here the values of force are given at two different coordinates in the actual medium through which the wave is traveling, and m is the mass of a part of the lumped system, this mass being given by

$$m = \rho_v \Delta v = \rho_v \Delta A \Delta x$$

where Δv is the volume element simulated.

The other telegrapher's equation for a sound wave in a liquid is

$$\frac{\partial u_x}{\partial x} = -K \frac{\partial p}{\partial t} \tag{3-24}$$

In the mechanical terminology this equation is usually written in terms of a displacement rather than a velocity. If we let X represent the displacement in the x direction of any particular point from its rest value, Eq. (3-24) may be integrated with respect to time, yielding

$$\frac{\partial X}{\partial x} = -Kp = -K \frac{f_x}{\Delta A}$$

Again, we may convert the space derivative into a ratio of differences and multiply both sides of the equation by the difference in the x coordinate, Δx. The result is

$$X(x + \Delta x) - X(x) = -K \frac{f_x \Delta x}{\Delta A} \tag{7-5}$$

or, in more familiar form,

$$f_x = \frac{\Delta A}{K \Delta x} [X(x) - X(x + \Delta x)]$$

Equations (7-4) and (7-5) are dynamic equations corresponding to the electrical equations in Chap. 2 used in deriving the telegrapher's equations for the transmission line. Equation (7-4) is a D'Alembert equation associated with the inertia of a mass, and Eq. (7-5) is a "spring-stretching equation." When we make approximations comparable to those used in the transmission-line derivations of Chaps. 2 and 3, these equations represent the mass-spring system shown in Fig. 7-4*a*. Here we assume that, as far as Eq. (7-4) is concerned, the total change in force across the mass in Fig. 7-4*a* is small compared with the total force applied. This corresponds to the assumption for the transmission-line derivation that the current through the shunt elements may be neglected, as compared with the total current in the line, in setting up the voltage-drop equation.

When this assumption is made, Eq. (7-4) describes the acceleration of the mass, and Eq. (7-5) describes the pull due to the spring being stretched.

When a large number of these sections are connected together, the spring-mass system shown in Fig. 7-4*b* results, and this corresponds directly to the electrical system shown in Fig. 7-1.

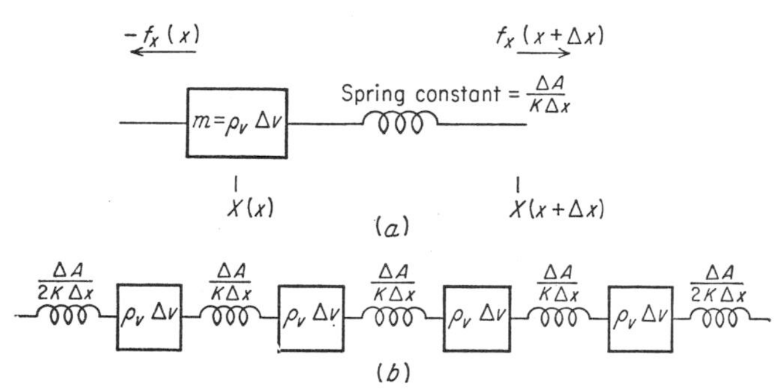

FIG. 7-4. Lumped-constant equivalent of acoustic waves in liquid: (*a*) single-section model; (*b*) four-section model.

The lumped equivalent of the thermal wave may be seen from examination and modification of the telegrapher's equations for the thermal wave. The first telegrapher's equation for the thermal wave is

$$\frac{\partial \tau}{\partial x} = -\frac{1}{k}q_x \tag{3-36}$$

Following the usual procedure, we turn the spatial derivative into a difference and multiply both sides by the differential distance. That is,

$$\tau(x + \Delta x) - \tau(x) = -\frac{\Delta x}{k\,\Delta A}q_{xA} = -R_t q_{xA} \tag{7-6}$$

Here the rate of heat flow per unit area, q_x, has been replaced by a total heat flow across an area ΔA, q_{xA}. The thermal resistance R_t has been used in accordance with its normal definition.

The second telegrapher's equation for thermal conduction is

$$\frac{\partial q_x}{\partial x} = -S\rho_v\frac{\partial \tau}{\partial t} \tag{3-38}$$

Changing the space derivative to a difference ratio and substituting q_{xA} for q_x with the appropriate multiplication by ΔA, this becomes

$$q_{xA}(x + \Delta x) - q_{xA}(x) = -S\rho_v\,\Delta v\frac{\partial \tau}{\partial t} = -C_t\frac{\partial \tau}{\partial t} \tag{7-7}$$

where C_t is thermal capacity. Equation (7-6) refers to a region of negligible thermal capacity but high thermal resistance, whereas Eq. (7-7) refers to a region of high thermal capacity and small thermal resistance. In Fig. 7-5 results are shown first for a single section and then for several sections connected together. In setting up the model in Fig. 7-5, it is necessary to neglect the change in heat-flow rate across the thermal resistor and the change in temperature across the thermal capacitor. Thus, in Fig. 7-5*a* the thin section represents a region of high thermal resistance, whereas the bulky section has low thermal resistance but large heat-storage capacity. When a number of sections are connected together, they take the form shown in Fig. 7-5*b*.

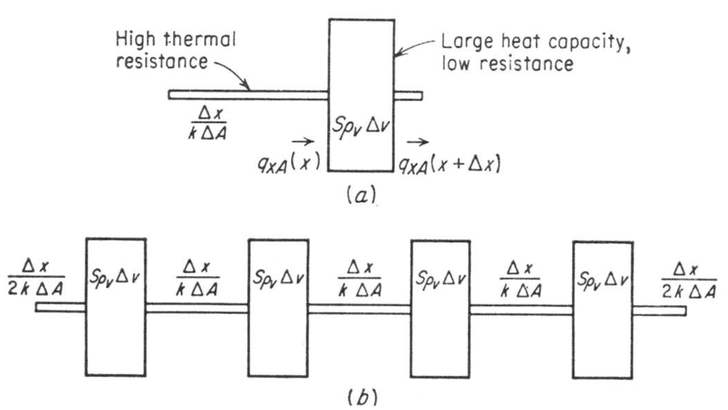

FIG. 7-5. Lumped-constant equivalent of thermal wave: (*a*) single-section lumped thermal model; (*b*) multiple-section lumped thermal model.

Lumped-constant equivalents could be made for the other types of acoustic waves, for diffusion, and for electromagnetic waves in space. However, they are not detailed here, since the examples given illustrate the basic process.

7-2. Iterative Impedance and Propagation Constant

When a circuit model is used to represent a transmission line, a characteristic impedance is developed which repeats at comparable points (such as the ends of T or π sections); this is known as the *iterative impedance*. Similarly, a propagation constant can be developed which is directly analogous to the propagation constant on the transmission line. Consider the situation for the T section shown in Fig. 7-6.

If a line is properly terminated, then

$$Z_R = Z_0$$

Not only is this true, however, but input impedance is the same as the output impedance for a properly terminated line, so that

$$Z_s = Z_0 = Z_R$$

The characteristic impedance for the T section is different from that for the π section; hence we write for the T section

(7-8) $$Z_s = Z_{0T} = Z_R$$

Analysis of Fig. 7-6 shows that, when these conditions are met, the input impedance is given by

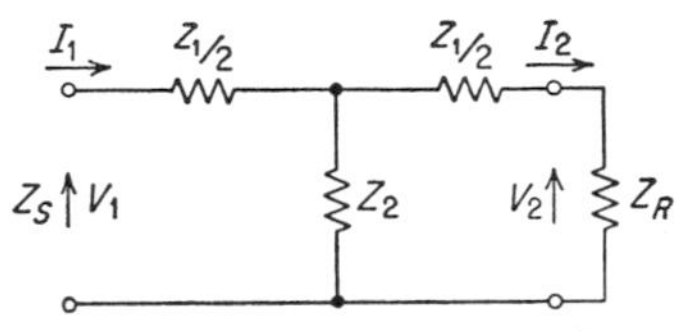

FIG. 7-6. T section for impedance analysis.

$$Z_{0T} = \frac{Z_1}{2} + \frac{Z_2(Z_1/2 + Z_{0T})}{Z_1/2 + Z_2 + Z_{0T}}$$

Solving this equation for Z_{0T}, we find

(7-9) $$Z_{0T} = \sqrt{Z_1 Z_2 + \frac{Z_1^2}{4}} = \sqrt{Z_1 Z_2}\sqrt{1 + \frac{Z_1}{4Z_2}}$$

On a transmission line the characteristic impedance is

$$Z_0 = \sqrt{\frac{R + j\omega L}{G + j\omega C}} = \sqrt{\frac{Z\,\Delta x}{Y\,\Delta x}} = \sqrt{Z_1 Z_2}$$

where

$$Z = R + j\omega L \qquad \text{ohms/m}$$
$$Y = G + j\omega C \qquad \text{mhos/m}$$

and for a section representing a length Δx,

$$Z_1 = Z\,\Delta x$$
$$Z_2 = \frac{1}{Y\,\Delta x}$$

Hence, for the T section, the characteristic impedance is, in terms of the characteristic impedance for the transmission line it represents,

(7-10) $$Z_{0T} = Z_0\sqrt{1 + \frac{Z_1}{4Z_2}}$$

A similar analysis for a π section shows that the characteristic impedance for the π section is given by

(7-11) $$Z_{0\pi} = \frac{Z_0}{\sqrt{1 + Z_1/4Z_2}}$$

The implications of this are considered in Sec. 7-3. Before treating them, the propagation constant expressions are developed.

Let the propagation factor $\gamma\,\Delta x = \gamma_T$ for the T section. Using Fig. 7-6, we can write

(7-12) $$V_2 = V_1 e^{-\gamma_T}$$

Therefore, we may write

$$\gamma_T = \log_e \frac{V_1}{V_2}$$

In the circuit of Fig. 7-6, V_2 can be found in terms of V_1. Thus,

$$V_2 = I_2 Z_{0T} = \frac{\begin{vmatrix} Z_1/2 + Z_2 & V_1 \\ -Z_2 & 0 \end{vmatrix}}{\begin{vmatrix} Z_1/2 + Z_2 & -Z_2 \\ -Z_2 & Z_1/2 + Z_2 + Z_{0T} \end{vmatrix}} Z_{0T}$$

$$= \frac{V_1 Z_2 Z_{0T}}{(Z_1/2 + Z_2)(Z_1/2 + Z_2 + Z_{0T}) - Z_2^2}$$

$$\gamma_T = \log_e \frac{Z_1 Z_2 + Z_1^2/4 + Z_1 Z_{0T}/2 + Z_2 Z_{0T}}{Z_{0T} Z_2}$$

so that

$$\gamma_T = \log_e \left(1 + \frac{Z_1}{2Z_2} + \frac{Z_{0T}}{Z_2}\right) \tag{7-13}$$

Comparable expressions for characteristic impedance as a ratio of velocity to force, of temperature to heat flow, or of other appropriate quantities could be developed for the nonelectric waves. Likewise, comparable propagation constants could be developed. However, separate derivations are not used, since the analogous quantities can be substituted in the electrical derivations.

7-3. Approximate and "Exact" Models

The iterative structures described in Sec. 7-2 may be used as approximate models for an actual transmission line by simply substituting in the lumped-constant iterative structure the values of the R, L, G, and C corresponding to a particular length of line. At a single frequency it is possible to use different impedances that make the phase shift and characteristic impedance for a model correspond exactly to those of the original traveling wave. Either the approximate or the "exact" models may often be simplified and the number of components reduced by network transformations, provided that it is not necessary to observe voltages and currents at various points along the line. Network transformations are frequently employed in the design of delay lines, such as those used in the radar modulators discussed in Chap. 4. Since these changes to the networks are properly a subject in network theory rather than in traveling waves and since they are quite involved, they will not be discussed further here.[1]

[1] See, for example, E. A. Guillemin, "Synthesis of Passive Network," John Wiley & Sons, Inc., New York, 1957.

Most of this section deals with approximate models. For an approximate model, we use the following relations for a length of line Δx:

$$R_1 = R\,\Delta x \qquad L_1 = L\,\Delta x \qquad G_2 = G\,\Delta x \qquad C_2 = C\,\Delta x$$

Comparable relations may be developed for nonelectric waves.

The characteristic impedance for a T section was given as

$$Z_{0T} = Z_0 \sqrt{1 + \frac{Z_1}{4Z_2}} \tag{7-9}$$

The second term under the square-root sign can be written in terms of shunt admittance instead of shunt impedance, in which case it becomes

$$\frac{Z_1}{4Z_2} = \frac{Z_1 Y_2}{4}$$

where $Y_2 = 1/Z_2$. Utilizing the values for R, L, G, and C set forth above, this becomes

$$\frac{Z_1}{4Z_2} = \frac{(R + j\omega L)(G + j\omega C)(\Delta x)^2}{4} = \frac{(\gamma\,\Delta x)^2}{4}$$

Hence, we may write the T-section characteristic impedance as

$$Z_{0T} = Z_0 \sqrt{1 + \frac{(\gamma\,\Delta x)^2}{4}} \tag{7-14}$$

For good simulation, the T-section characteristic impedance should be close to the characteristic impedance for the transmission line, which means that

$$|\gamma\,\Delta x| \ll 1$$

We may expand the square root in Eq. (7-14) by the binomial expansion, the first two terms of which yield a reasonable approximation for this condition:

$$Z_{0T} \approx Z_0 \left[1 + \frac{(\gamma\,\Delta x)^2}{8}\right] \tag{7-15}$$

Thus the amount of error is directly expressible in terms of the length of line simulated by a single section.

The propagation constant was given by Eq. (7-13), which is in the form

$$\log_e (1 + x)$$

If each section represents a short enough distance, then the phase shift and attenuation in that section will be small, and we can utilize the series expansion

$$\log_e (1 + x) = x - \frac{x^2}{2} + \frac{x^3}{3} - \cdots$$

Thus, provided that the terms in Eq. (7-13) corresponding to x in this series are small, we may write

$$\gamma_T = \frac{Z_{0T}}{Z_2} + \frac{Z_1}{2Z_2} - \frac{1}{2}\left(\frac{Z_1^2}{4Z_2^2} + \frac{Z_1Z_{0T}}{Z_2^2} + \frac{Z_{0T}^2}{Z_2^2}\right) + \cdots$$

Each of the terms of this expression may be developed in terms of $\gamma\,\Delta x$ as follows:

$$\frac{Z_{0T}}{Z_2} = \frac{\sqrt{Z_1Z_2}}{Z_2}\sqrt{1 + \frac{Z_1}{4Z_2}} = \sqrt{\frac{Z_1}{Z_2}}\sqrt{1 + \frac{Z_1}{4Z_2}} = \gamma\,\Delta x\sqrt{1 + \frac{(\gamma\,\Delta x)^2}{4}}$$
$$\approx \gamma\,\Delta x\left[1 + \frac{(\gamma\,\Delta x)^2}{8} - \frac{(\gamma\,\Delta x)^4}{32} + \cdots\right]$$
$$\frac{Z_1}{2Z_2} = \frac{Z_1Y_2}{2} = \frac{(\gamma\,\Delta x)^2}{2}$$
$$\frac{Z_1^2}{8Z_2^2} = \frac{(\gamma\,\Delta x)^4}{8}$$
$$\frac{Z_1Z_{0T}}{2Z_2^2} = \frac{(\gamma\,\Delta x)^3}{2}\sqrt{1 + \frac{(\gamma\,\Delta x)^2}{4}} = \frac{(\gamma\,\Delta x)^3}{2}\left[1 + \frac{(\gamma\,\Delta x)^2}{8} - \cdots\right]$$
$$\frac{Z_{0T}^2}{2Z_2^2} = \frac{(\gamma\,\Delta x)^2}{2}\left[1 + \frac{(\gamma\,\Delta x)^2}{4}\right]$$

Combining these, we find that the terms in $(\gamma\,\Delta x)^2$ cancel out. Thus,

$$\gamma_T = \gamma\,\Delta x - \frac{(\gamma\,\Delta x)^3}{24} - \frac{(\gamma\,\Delta x)^4}{4} + \cdots$$

which reduces to,

$$\gamma_T = \gamma\,\Delta x\left[1 - \frac{(\gamma\,\Delta x)^2}{24} + \cdots\right] \tag{7-16}$$

Comparing this expression and that for the characteristic impedance (7-15), it would appear that the limiting factor would ordinarily be the characteristic impedance rather than the phase shift. However, the characteristic-impedance effect is the same regardless of the number of sections, whereas errors in phase shift and attenuation build up from section to section. The total propagation factor for N sections of line having length

$$l = N\,\Delta x$$

is

$$\gamma_T N = \gamma\,\Delta x\,N\left[1 - \frac{(\gamma\,\Delta x)^2}{24} + \cdots\right] \tag{7-16a}$$
$$= \gamma l - N\frac{(\gamma\,\Delta x)^3}{24} + \cdots$$

When the error becomes large enough, the higher-order terms in $\gamma\,\Delta x$ must not be discarded in Eq. (7-16). If, however, the model of the transmission line is sufficiently good that the total error, as indicated in (7-16a), is small, and not simply the error per section, only the terms shown here need be used.

Some of the problems associated with simulating transmission lines are best shown by examples. The following two examples illustrate the problems and their solutions.

Example 7-1. Consider a lossless line whose length is 20 wavelengths. First let us determine what happens if we try simulation by section such that

$$\Delta x = 0.1\lambda$$

Since we are dealing with a lossless line, this means that

$$\gamma\,\Delta x = j\beta\,\Delta x = j0.2\pi$$

Hence,

$$Z_{0T} = Z_0\sqrt{1 + \frac{(\gamma\,\Delta x)^2}{4}} \approx Z_0\left(1 - \frac{0.04\pi^2}{8}\right)$$
$$\approx Z_0(1 - 0.049)$$

Therefore the error in the characteristic impedance is 4.9 per cent when 0.1-wavelength sections are used. This error could, of course, be reduced considerably by using shorter sections.

Now let us consider the error in propagation constant and phase shift. Using Eq. (7-16), we find

$$\gamma_T = \gamma\,\Delta x\left(1 + \frac{0.04\pi^2}{24}\right) = \gamma\,\Delta x(1 + 0.0163)$$

This means that the error in phase shift per section is only 1.63 per cent.

In the total 20 wavelengths there are 200 sections. Hence, the total phase shift is $200\beta\,\Delta x$. Thus,

$$\text{Total phase shift} = 40\pi(1 + 0.0163) = 40\pi + 2.05$$

Hence, the error, which is only 1.63 per cent per section, totals up to about 120°, or $2\pi/3$ radians. It seems unreasonable to allow such a large error, and therefore a smaller length for each section must be considered.

Suppose that the total phase-shift error is to be kept within 36° (0.2π radians). The total phase error will be given by

$$\frac{(\beta\,\Delta x)^2}{24}\,40\pi = 0.2\pi$$

Using the relation

$$\beta = \frac{2\pi}{\lambda}$$

we may solve this to find

$$\left(\frac{\Delta x}{\lambda}\right)^2 = \frac{0.2 \times 24}{40(2\pi)^2} = \frac{0.12}{(2\pi)^2}$$

so

$$\frac{\Delta x}{\lambda} = 0.05513$$

Hence, the total number of sections is

$$N = \frac{20\lambda}{0.05513\lambda} = 362.7$$

But the number of sections must be an integer, so N must be 363, resulting in

$$\frac{\Delta x}{\lambda} = \frac{20}{363} = 0.05509$$

Thus, to keep the phase error within reasonable limits, it is necessary to use almost twice as many sections as originally postulated. The characteristic impedance in this case becomes

$$Z_{0T} = Z_0[1 - \tfrac{1}{4}(2\pi \times 0.05509)^2] = Z_0[1 - 0.03] = 0.97Z_0$$

So the error in the characteristic impedance is only 3 per cent.

Example 7-2. It is desired to simulate a 30-km length of the telephone toll cable used in Example 2-2. Impedance simulation must be within 2 per cent at 1,000 cps, and the over-all phase shift at 1,000 cps must be within 0.1 radian (5.7°). We wish to find the number of sections and their characteristics. This toll cable is a lossy one, so the calculations are not so simple as in Example 7-1. Here we have

$$\begin{aligned}\text{Total phase shift and attenuation} &= \gamma_T N \\ &= \gamma l \left[1 - \frac{(\gamma\,\Delta x)^2}{24}\right]\end{aligned}$$

Substituting from Example 2-2, we find

$$\begin{aligned}\gamma_T N &= (0.076 + j0.083)30\left[1 - \frac{(\gamma\,\Delta x)^2}{24}\right] \\ &= (2.28 + j2.49)\left[1 - \frac{(0.076 + j0.083)^2(\Delta x)^2}{24}\right] \\ &= (2.28 + j2.49)[1 + (6.2 \times 10^{-5} - j5.25 \times 10^{-3})(\Delta x)^2] \\ &= 2.28 + 1.31 \times 10^{-2}(\Delta x)^2 + j[2.49 - 1.18 \times 10^{-2}(\Delta x)^2]\end{aligned}$$

To satisfy the requirement on error in phase shift, we observe that the phase-shift error is the second term of the imaginary part of the above equation. Hence,

$$1.18 \times 10^{-2}(\Delta x)^2 = 0.1$$

or

$$(\Delta x)^2 = \frac{10}{1.18}$$
$$\Delta x = 2.92 \text{ km}$$

Therefore, 10 sections of this length simulate 29.2 km of the line. The characteristic impedance is well within the limits set for it. The parameters for the individual sections are

$$L_1 = 0.69 \times 2.92 = 2.02 \text{ mh}$$
$$R_1 = 52.2 \times 2.92 = 153 \text{ ohms}$$
$$G_2 = 0.623 \times 2.92 = 1.82\ \mu\text{mhos}$$
$$C_2 = 0.038 \times 2.92 = 0.111\ \mu\text{f}$$

The result is shown in Fig. 7-7. To simulate 30 rather than just 29.2 km in this manner, it would be necessary to use 11 somewhat shorter sections; that is, Δx should be $\frac{30}{11}$ or 2.72 km. The individual components of the artificial sections are reduced accordingly.

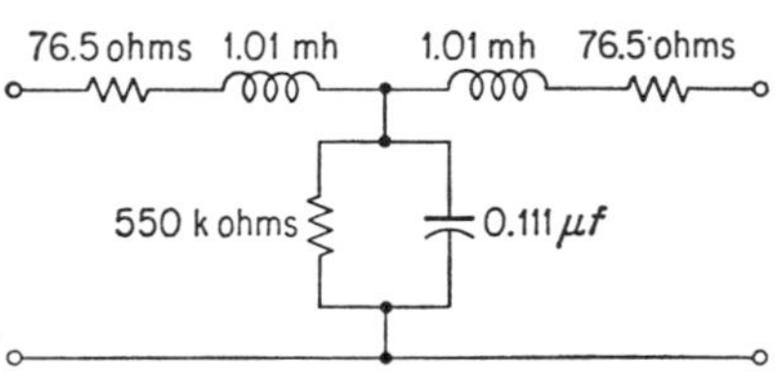

FIG. 7-7. One section of a 30-km simulated line.

A comparable example may be considered for an acoustic wave utilizing exactly the same techniques that are used for the transmission line. Such an example follows.

Example 7-3. Consider an acoustic wave traveling through a circular tube of water 10 m long with a 10-cm radius. With this diameter, the friction at the walls should have a small effect, and the wave may be considered as essentially plane—a finite piece of an infinite plane wave.

For water (see Example 3-5),

$$\rho_v = 10^3 \text{ kg/m}^3$$
$$K = 4.78 \times 10^{-10} \text{ sec}^2/\text{kg}$$

For this example, the cross-sectional area ΔA is

$$\Delta A = (0.1)^2\pi = 0.01\pi$$

Hence, the mass shown in Fig. 7-4 is

$$m = \rho_v \, \Delta v = 10^3 \times 0.01\pi \, \Delta x = 10\pi \, \Delta x \qquad \text{kg}$$

The spring constant has been shown to be

$$\frac{\Delta A}{K \, \Delta x} = \frac{10^{-2}\pi}{4.78 \times 10^{-10} \, \Delta x} = \frac{10^8\pi}{4.78 \, \Delta x}$$

The characteristic impedance is given by

$$Z_0 = \frac{U}{F} = \frac{1}{\Delta A}\sqrt{\frac{K}{\rho_v}} = 2.20 \times 10^{-5} \text{ m/newton-sec}$$

The propagation constant for the acoustic wave proper is given by

$$\gamma = j\omega\sqrt{\rho_v K}$$

In this case, γ becomes

$$\gamma = \frac{j\omega}{1{,}450}$$

Using the electrical analogy, the propagation constant for the T-section equivalent is given by

$$\gamma_T = j\frac{\omega \, \Delta x}{1{,}450}\left[1 - \frac{1}{24}\left(\frac{\omega \, \Delta x}{1{,}450}\right)^2\right]$$

If we wish the phase error to be 0.1 radian or less, the error part of this expression may be equated to the 0.1. Thus,

$$\frac{\omega l}{1{,}450}\frac{1}{24}\left(\frac{\omega \, \Delta x}{1{,}450}\right)^2 = 0.1$$

Solving for $(\Delta x)^2$, we have

$$(\Delta x)^2 = 24\left(\frac{1{,}450}{\omega}\right)^3 10^{-2}$$

If the acoustic wave under consideration is at a frequency of 925 cps so that $\omega/4 = 1{,}450$, this results in

$$\Delta x = 0.0611 \text{ m or } 6.11 \text{ cm}$$

Thus, in the 10 m total length, there are 163 sections.

For each section there are a mass and a spring. The mass is

$$m = 10\pi \times 0.0611 = 1.92 \text{ kg}$$

The spring constant is given by

$$\frac{10^8\pi}{4.78 \times 0.0611} = 1.075 \times 10^9 \text{ newtons/m}$$

This is a rather stiff spring. The effect can be achieved with a steel rod, however, and the rod need not be particularly thick. The spring constant for a thin rod is given by

$$\text{Spring constant} = \frac{Y_0(\text{area})}{\text{length}}$$

Thus, if we specify either the cross-sectional area or the length, we may determine the other for the piece of steel acting as the spring. Suppose we specify that the length of the spring is 5 cm. Then we have for the area, using the value for Young's modulus for steel found in Table 3-5,

$$\text{Area} = \frac{1.075 \times 10^9 \times 5 \times 10^{-2}}{2.1 \times 10^{11}} = 2.55 \times 10^{-4}\ \text{m}^2$$

If the spring has a circular cross section, its radius is therefore given by

$$\sqrt{\frac{2.55 \times 10^{-4}}{\pi}} = 0.90\ \text{cm}$$

Suppose that the masses are cylinders of copper with a 5-cm radius. The volume associated with 1.92 kg (using the density of copper from Table 3-5) is

$$1.92 = \rho_v\ \Delta v = 8.9 \times 10^3\ \Delta v$$

Hence,

$$\Delta v = \frac{1.92 \times 10^{-3}}{8.9} = 2.15 \times 10^{-4}\ \text{m}^3$$

For a 5-cm-radius cylinder we can find the length by dividing this volume by the area of a 5-cm-radius circle. The result is

$$\Delta x = 2.74\ \text{cm}$$

Hence, the lumped-constant equivalent for the 10-m-long pipe at 925 cps with 0.1 radian total phase-shift error is made up of 163 sections of the type shown in Fig. 7-8. Since the lumped equivalent is as long as the original, it may not be worthwhile to use as a "model."

Examples 7-1 and 7-3 indicate the great many sections that are required when the traveling-wave medium being simulated is many wavelengths long. The line of Example 7-2 had only a few wavelengths, and therefore the simulation could be made with a more reasonable number of sections, even though the phase-shift limitation was fairly stringent there also.

The examples above will give better performance for the lower frequencies than for the design frequency. Hence they represent broadband simulation of the traveling wave. Single-frequency simulation may be made exact by making the T- or π-section characteristic impedance exactly equal to the characteristic impedance of the transmission

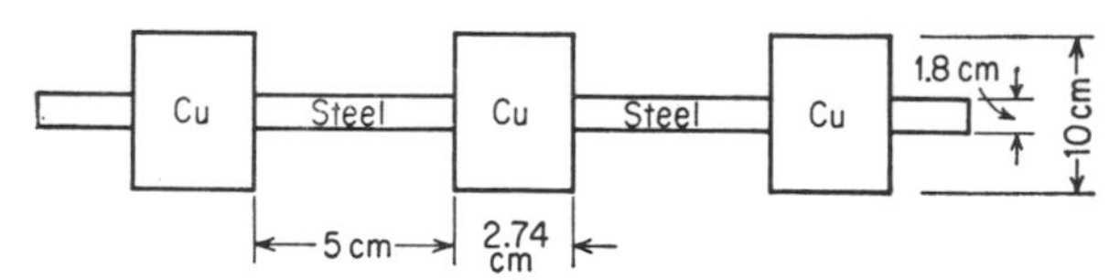

FIG. 7-8. Lumped-constant equivalent for acoustic wave in large pipe.

line and making its phase shift and attenuation exactly equal to those of the transmission line. Thus, from (7-9), we set

$$Z_0 = Z_{0T} = \sqrt{Z_1 Z_2 + \frac{Z_1^2}{4}}$$

and, from (7-13),

$$e^{\gamma\,\Delta x} = e^{\gamma_T} = 1 + \frac{Z_1}{2Z_2} + \sqrt{\frac{Z_1}{Z_2} + \frac{Z_1^2}{4Z_2^2}}$$

These equations may be solved simultaneously for Z_1 and Z_2 at a particular frequency. When this is done, the simulation is exact at the particular frequency where the impedances are evaluated.

Details of the solution of these equations are left as a problem. The results are

$$Z_1 = 2Z_0 \frac{e^{\gamma\,\Delta x} - 1}{e^{\gamma\,\Delta x} + 1} \tag{7-17}$$

and

$$Z_2 = \frac{Z_0^2 - Z_1^2/4}{Z_1} \tag{7-18}$$

The exact simulation at a single frequency represented by these equations is frequently used in modeling power lines, for which the frequency is normally constant.

Example 7-4. Consider the telephone line described in Example 7-2. Suppose that 11 sections are used, as required in that example, but that now the sections give exact simulation at 1,000 cps. Then,

$$\Delta x = \tfrac{30}{11} = 2.727 \text{ km}$$
$$\gamma\,\Delta x = (0.076 + j0.083)2.727 = 0.207 + j2.26$$
$$Z_0 = 345 - j319 \text{ ohms}$$

For the model, from (7-17),

$$Z_1 = 2(345 - j319)\,\frac{e^{0.207}e^{j0.226} - 1}{e^{0.207}e^{j0.226} + 1} = 940\underline{/-42.6^\circ}\,\frac{0.20 + j0.275}{2.20 + j0.275}$$
$$= 144\underline{/4.3^\circ} = 144 + j10.8 \text{ ohms}$$

From (7-18),

$$Z_2 = \frac{Z_0^2}{Z_1} - \frac{Z_1}{4}$$

To the accuracy possible here, this shows that

$$Z_2 = -j1{,}522 \text{ ohms}$$

This should be compared with the values computed from the *approximate* method. Thus,

$$R\,\Delta x = 52.2 \times 2.727 = 142 \text{ ohms}$$
$$\omega L\,\Delta x = 6.9 \times 10^{-4} \times 2\pi \times 10^3 \times 2.727 = 11.8 \text{ ohms}$$
$$\frac{1}{\omega C\,\Delta x} = \frac{1}{2\pi \times 10^3 \times 3.8 \times 10^{-8} \times 2.727} = \frac{10^5}{65.0} = 1{,}540 \text{ ohms}$$

For this example, therefore, the exact and approximate values are about the same.

If the 1,000-cps line is modeled by only two sections, the results are quite different. Then

$$\gamma\,\Delta x = (0.076 + j0.083)15 = 1.14 + j1.245$$
$$Z_1 = 940\underline{/-42.6^\circ}\,\frac{e^{1.14}e^{j1.245} - 1}{e^{1.14}e^{j1.245} + 1}$$
$$= 940\underline{/-42.6^\circ}\,\frac{+j2.95}{2.00 + j2.95}$$
$$= 780\underline{/-8.5^\circ} = 770 - j122 \text{ ohms}$$

Thus, in this case the series reactance is *capacitive.*

7-4. Electrical Models for Nonelectric Waves

Frequently, electrical models for nonelectric waves may be devised most simply by (1) making the transmission-line analog for the nonelectric wave and (2) making the iterative circuit model for the transmission line.

The easiest way to do this for an acoustic wave is simply to let the numerical value of velocity equal the numerical value of voltage and the numerical value of pressure equal the numerical value of current. Then the impedance and propagation constants come out the same. This works reasonably well for an acoustic wave in air; however, the values for impedance for acoustic waves in liquids and solids are quite impractical when applied to the electrical model. The values for thermal waves are also quite impractical.

When the simple approach does not work, it is necessary to scale the values used for pressure, velocity, voltage, and current; that is, instead of making pressure in newtons per square meter equal the number of amperes, it may be desirable to make this equal the number of milliamperes or kiloamperes. The procedure can best be explained by reference to examples.

Example 7-5. *Electrical Simulation of Acoustic Wave in Air.* Here we shall consider the situation discussed in Example 3-3. The characteristic impedance, for the A analogy, is given by

$$Z_0 = 0.00235 \text{ ohm}$$

The density ρ_v corresponds to the capacitance per unit length C. Therefore

$$C = 1.29 \text{ farads/unit length}$$

The quantity $1/\gamma_g p_a$ corresponds to the inductance per unit length L. Hence

$$L = \frac{1}{1.40 \times 10^5} = \frac{10^{-6}}{0.14} = 7.15\ \mu\text{h/unit length}$$

Obviously, such values are unrealistic for the electrical model. A capacitance of 1.29 farads is *tremendous.* If the unit length is so large that only 1.29 μf is required, then only 7.15 $\mu\mu$h is permitted, and this is absurdly small. The solution is in *scaling.*

Suppose we let v correspond to u_{xn} in hundredths of millimeters per second and let i correspond to p in newtons per square meter. Then, in our new units,

$$\frac{v}{i} = Z_0 = \frac{u_{xn}}{p} = 10^5 \times 0.00235$$
$$= 235 \text{ (hundredths of mm/sec)/(newtons/m}^2\text{)}$$

To see what happens to L and C, consider

$$\frac{\partial u_x}{\partial x} = -\frac{1}{\gamma_g p_a}\frac{\partial p}{\partial t} \tag{3-14}$$

$$\frac{\partial p}{\partial x} = -\rho_v \frac{\partial u_x}{\partial t} \tag{3-15}$$

Since the number of the new units of u_x is 10^5 times the number of meters per second, the equations now read

Analogous equations

$$\frac{\partial u_{xn}}{\partial x} = -\frac{10^5}{\gamma_g p_a}\frac{\partial p}{\partial t} \qquad \frac{\partial v}{\partial x} = -L\frac{\partial i}{\partial t}$$

$$10^5\frac{\partial p}{\partial x} = -\rho_v\frac{\partial u_{xn}}{\partial t} \qquad \frac{\partial i}{\partial x} = -C\frac{\partial v}{\partial t}$$

Hence, analogous to L is

$$\frac{10^5}{\gamma_g p_a} = 7.15 \times 10^5\ \mu\text{h/unit length}$$
$$= 0.715\ \text{henry/unit length}$$

and analogous to C is

$$\frac{\rho_v}{10^5} = 1.29 \times 10^{-5}\ \text{farad/unit length}$$
$$= 12.9\ \mu\text{f/unit length}$$

The velocity of propagation on this line is

$$v_p = \frac{1}{\sqrt{LC}} = \frac{1}{\sqrt{0.715 \times 12.9 \times 10^{-6}}} = 328\ \text{unit lengths/sec}$$

The unit of length may be chosen as anything appropriate. Actually, since simulation with lumped constants is the goal, it does not matter.

Suppose we wish to simulate 5 m of $1{,}000/2\pi$ cps sound in air, with sections corresponding to about 0.1 radian phase shift per section. The total number of sections is

$$N = \frac{\beta 5}{0.1} = \frac{\omega 5}{0.1 v_p}$$

For air, $v_p = 330$ m/sec, so

$$N = \frac{10^3 \times 5}{10^{-1} \times 3.3 \times 10^2} = 152\ \text{sections}$$

If we use the same frequency in the model,

$$\gamma_T = j\beta_T = \frac{j\omega\,\Delta x}{v_p} = j0.1$$

Hence,

$$\Delta x = \frac{0.1}{10^3}v_p = \frac{0.1 \times 328}{10^3} = 3.28 \times 10^{-2}\ \text{unit length}$$

So

$$L_1 = L\,\Delta x = 0.715 \times 3.28 \times 10^{-2} = 2.34 \times 10^{-2}\ \text{henry} = 23.4\ \text{mh}$$
$$C_2 = C\,\Delta x = 12.9 \times 3.28 \times 10^{-2} = 0.424\ \mu\text{f}$$

The simulated wave is made up of 152 sections with L_1 and C_2 as shown and with voltage corresponding to particle velocity in meters per second times 10^5. If criteria other than the 0.1 radian per section had been used, the results would have been somewhat different.

Example 7-6. *Acoustic Wave in Water.* This is the same as Example 3-5 for the acoustic wave. Here the characteristic impedance is

$$Z_0 = 6.9 \times 10^{-7} \text{ m}^3/\text{newton-sec, or ohm}$$

Obviously this is a very low characteristic impedance for an electric transmission line. The problem may be solved, however, by making the voltage correspond to the velocity in hundredths of a micron per second rather than in meters per second. In this case, we have a scale factor of 10^8:

$$Z_0 = \frac{u_x \text{ (m/sec} \times 10^8)}{p \text{ (newtons/m}^2)} = 69$$

This is not an unreasonable value for an electric transmission line. The same result might have been achieved by expressing u in centimeters per second and p in newtons per square kilometer.

The values of L and C are $10^8 K$ for L and $\rho_v/10^8$ for C. Thus

$$\begin{aligned} L &= 4.78 \times 10^{-2} \text{ henry/unit length} \\ &= 47.8 \text{ mh/unit length} \\ C &= \frac{10^3}{10^8} = 10 \text{ } \mu\text{f/unit length} \end{aligned}$$

The method of Example 7-6 may be used to find the length per section and values for L_1 and C_2.

With acoustic waves, a scale factor was required only for the magnitude of the voltage-velocity or current-pressure analogs. The scale factors for parameters resulted from these, and the time factor remained the same. With heat problems, the times involved are long, and a time scale factor is frequently called for. Example 7-7 shows the approach in this situation.

Example 7-7. *Modeling a Thermal Wave.* Suppose that, in the city described in Example 3-10, it is desired to model heat flow into the ground to learn how to take into account irregular temperature variations. One year in the heat problem is to be represented in the model by 1 sec. Other values are to be chosen as reasonable. The model is to be good enough so that 0.1 radian for a period of 1 day is represented by one section.

The pertinent equations are

		Analogous equation
(3-36)	$\dfrac{\partial \tau}{\partial x} = -\dfrac{1}{k} q_x$	$\dfrac{\partial v}{\partial x} = -Ri$
(3-38)	$\dfrac{\partial q_x}{\partial x} = -S\rho_v \dfrac{\partial \tau}{\partial t}$	$\dfrac{\partial i}{\partial x} = -C \dfrac{\partial v}{\partial t}$

If the time scale is changed by a factor equal to the number of seconds in a year (3.15×10^7), Eq. (3-38) becomes

$$\frac{\partial q_x}{\partial x} = -\frac{S\rho_v}{3.15 \times 10^7} \frac{\partial \tau}{\partial t'}$$

where t' is the time in years. Hence, if degrees centigrade are analogous to volts and watts per square meter to amperes and if we assume $k = 2.0$ and $D_t = 4 \times 10^{-7}$, we find

$$S\rho_v = \frac{k}{D_t} = \frac{2}{4} \times 10^7 = 5 \times 10^6$$

For a period of 1 day in "real time," the model time is $\frac{1}{365}$ sec. Thus, the frequency is 365 cps. To find the distance corresponding to 0.1 radian, note that

$$\gamma = \beta(1 + j) = \sqrt{\frac{\omega S\rho_v}{2k}}\,(1 + j) \tag{3-44}$$

In this case, the model factor is also pertinent, so

$$\beta = \sqrt{\frac{\pi \times 365 \times 5 \times 10^6}{3.15 \times 10^7 \times 2.0}} = \sqrt{91.2} = 9.56$$

Now

$$\beta\,\Delta x = 0.1$$

So

$$\Delta x = \frac{0.1}{9.56} = 0.01046 \text{ unit length}$$

If q_x in watts per square meter is analogous to i in amperes and if τ in degrees centigrade is analogous to v in volts, the capacitance per section is

$$C_2 = \frac{S\rho_v\,\Delta x}{3.15 \times 10^7} = \frac{5 \times 10^6}{3.15 \times 10^7} 1.05 \times 10^{-2} = 1.66 \times 10^{-3} \text{ farad}$$

a very large capacitance indeed.

If the current in amperes corresponds to q_x in kilowatts per square meter, the value of $\partial q_x/\partial x$ for a given $\partial \tau/\partial t$ is reduced by 10^3.

Thus we have

$$\frac{\partial q_{xkw}}{\partial x} = -\frac{S\rho_v}{3.15 \times 10^{10}} \frac{\partial \tau}{\partial t'}$$

and

$$C_2 = 1.66 \ \mu\text{f}$$

a reasonable value.

This scale also makes the numerical value of q_x smaller in (3-36). Thus

$$\frac{\partial \tau}{\partial x} = -\frac{10^3}{k} q_{xkw}$$

so

$$R_1 = R \, \Delta x = \frac{10^3}{2.0} = 1.05 \times 10^{-2} = 5.25 \text{ ohms}$$

Of course, the scale factors chosen are arbitrary, and different ones might also be appropriate.

Note that the procedure followed in all the examples involves the same basic steps:

1. Where time matters, choose an appropriate factor.
2. Determine the distance per section in unit lengths.
3. Check to see whether analogies give reasonable electrical parameters without further scaling.
4. If not, apply scale factors to bring results within reason.

Another method, not illustrated by example, involves *first* making the nonelectric Z_0 reasonable for an electric line by scaling, and *then* making a lumped model from the direct electrical analog of this "prescaled" nonelectrical problem.

7-5. Nonelectrical Lumped-constant Models

In Sec. 7-4 lumped-constant electrical models of nonelectric waves were discussed. In Sec. 7-2 lumped-constant nonelectrical models of nonelectric waves were discussed. There the same types of nonelectrical quantities were used in the model as in the original—that is, force and velocity in modeling an acoustic wave, and temperature and heat flow in modeling a thermal wave. It is not necessary, however, to restrict our modeling of nonelectric waves to the same type of parameters or to electric circuits. The tables of Chap. 3 which summarize the various waves (Tables 3-10 to 3-12) can be the source of many types of analogies.

For example, any of the acoustic waves may be modeled by any lumped-constant system having two kinds of energy storage and no dissipation. We have already considered electric and magnetic energy storage (C and L), as well as storage in inertia and the spring constant of masses and springs in linear motion. Inertia of rotation and torsional springing could also be used. With a suitable device for separating kinetic and potential energy and eliminating loss, even flow of liquids could be used to simulate an acoustic wave.

A lumped-constant thermal model could readily be applied to a diffusion problem or to an electromagnetic wave in a conductor. Likewise, a lumped-constant diffusion model may be applied to thermal problems or lossy transmission lines in cables.

Since there is no really general approach that can be easily stated, the use of one kind of nonelectrical model is shown by an example. This is a liquid-diffusion model of a thermal problem. The thermal problem could as well be an electromagnetic wave in a conductor.

Example 7-8. *Lumped Diffusion Analog of a Thermal Wave.* First consider what a lumped diffusion model of a diffusion process itself must look like. Consider the diffusion equations

Analogous equations

$$\frac{\partial c}{\partial x} = -\frac{1}{D} m_{vx} \qquad \frac{\partial \tau}{\partial x} = -\frac{1}{k} q_x = -\frac{1}{S\rho_v D_t} q_x \tag{3-53}$$

$$\frac{\partial m_{vx}}{\partial x} = -\frac{\partial c}{\partial t} \qquad \frac{\partial q_x}{\partial x} = -S\rho_v \frac{\partial \tau}{\partial t} = -\frac{k}{D_t}\frac{\partial \tau}{\partial t} \tag{3-54}$$

The first of these (Fick's law) is a resistive, or loss, type of equation. In "lumped-constant fluid flow," such an equation can be used to

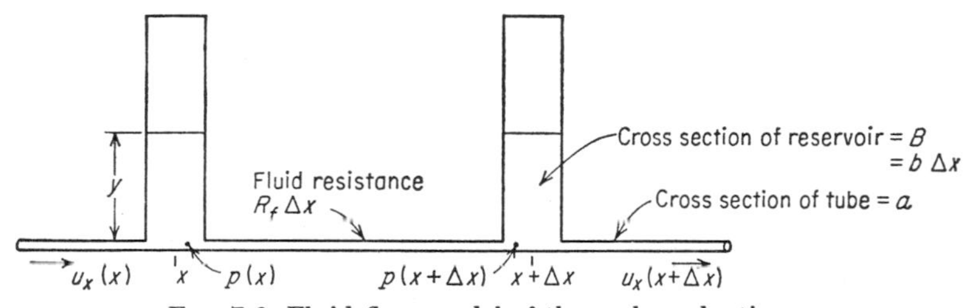

FIG. 7-9. Fluid flow model of thermal conduction.

describe the pressure drop in a thin tube due to the viscosity of the fluid. The second equation is simply a continuity equation relating net outflow rate to rate of change of storage of the molecules, whose concentration is represented by c. With a single liquid, such an equation describes the storage and flow rate of a reservoir.

Figure 7-9 illustrates a series combination of these two "lumped elements." Here the pressure in the thin tubes and at the bottom of the reservoirs corresponds to concentration. Either an actual mass-flow rate or a velocity may be used to correspond to m_{vx}. In the model discussed in this example, velocity corresponds to m_{vx}.

Let us consider the thermal quantities to be modeled. In any thermal problem we must model τ, q, k, and $S\rho_v$, where these are all as defined in Chap. 3. We shall model τ with the pressure in the liquid p, and q_x with the velocity of the liquid entering and leaving the reservoirs.

In our model, the first equation has to do with the viscous flow through the tubes, and the second with storage in the reservoir.

The pressure drop in a thin tube is linearly related to the mean flow velocity, provided that the conditions for laminar flow prevail. These are present for low enough velocities in thin enough tubes, the criterion being a Reynolds number less than 2,320.† It is assumed here that the model is so set up that this condition is met and flow is laminar.

The "Ohm's law" for a pipe is called the Hagen-Poiseuille law. For the situation shown in Fig. 7-9, it may be stated as

$$p(x + \Delta x) - p(x) = -R_f u_x(x)\,\Delta x \tag{7-19}$$

where p is the pressure at the bottom of the reservoir and in the tube, R_f is the fluid resistance per unit length, and u_x is the mean velocity in the tube. The fluid resistance in a length Δx is given by

$$R_f\,\Delta x = \frac{8\pi\mu\,\Delta x}{a} \tag{7-20}$$

where a is the cross-sectional area of the tube, μ is the viscosity of the liquid, and Δx is the length of the tube. The viscosity is in mks units, kilograms per meter-(second)^2.

Note that it is assumed here that there is no pressure drop in going from one side of the reservoir to the other—a typical assumption for traveling-wave models.

Equation (7-19) can be seen to correspond directly with (3-53) and (7-6). In differential-equation form it is

Analogous equation

$$\frac{\partial p}{\partial x} = -R_f u_x \qquad\qquad \frac{\partial \tau}{\partial x} = -\frac{1}{k}\,q_x \tag{7-19a}$$

The continuity equation for the reservoir (neglecting inertial effects in the reservoir itself) is

$$au_x(x + \Delta x) - au_x(x) = -\frac{d(By)}{dt} = -B\,\frac{dy}{dt}$$

† R. A. Dodge and M. J. Thompson, "Fluid Mechanics," chap. 8, McGraw-Hill Book Company, Inc., New York, 1937. See also J. M. Coulson and J. F. Richardson, "Chemical Engineering," vol. I, chap. 3, McGraw-Hill Book Company, Inc., New York, 1954.

where au_x is the volume rate of flow in the tube, B is the cross section of the reservoir, and y is the depth in the reservoir. It has been assumed that there is no pressure drop across the reservoir.

Although y is a perfectly good indicator, we are using p as the intensity variable, so we must relate p and y. The pressure at the bottom of the reservoir is just the weight of the fluid per unit area, so

$$p = \rho_v g y$$

where ρ_v is density and g is gravitational acceleration, and the continuity equation may be written

$$u_x(x + \Delta x) - u_x(x) = -\frac{B}{a}\frac{dy}{dt} = -\frac{B}{\rho_v g a}\frac{dp}{dt} \tag{7-21}$$

If we consider the amount of reservoir area per unit length to be

$$b = \frac{B}{\Delta x}$$

(7-19) may be written as a telegrapher's equation, like (3-54):

Analogous equation

$$\frac{\partial u_x}{\partial x} = -\frac{b}{\rho_v g a}\frac{\partial p}{\partial t} \qquad \frac{\partial q_x}{\partial x} = -S\rho_v\frac{\partial \tau}{\partial t} \tag{7-21a}$$

Hence the analogies are

τ corresponds to p

q_x corresponds to u_x

$\frac{1}{k}$ corresponds to R_f

$\frac{b}{\rho_v g a}$ corresponds to $S\rho_v$

For a numerical example, let us again use the parameters of Example 3-10, in which the depth for burying water pipes was determined. A time scale is called for, but the fluid analog cannot move so fast as the electrical analog. Thus, let us scale 4 days in "real time" per minute on the model. The time scale is

$$24 \times 4 \times 60 = 5.76 \times 10^3$$

Thus, the second heat equation becomes

$$\frac{\partial q_x}{\partial x} = -\frac{S\rho_v}{5.76 \times 10^3}\frac{\partial \tau}{\partial t'}$$

We may find the cross section of the capillary tube from the first heat equation. Thus, if u_x in meters per second is analogous to q_x

in watts per square meter and if p in newtons per square meter is analogous to τ in degrees centigrade,

$$R_f = \frac{1}{k} = \frac{1}{2.0} \frac{\text{newton-sec}}{\text{m}^4}$$

To find the reservoir size required, use (7-21a), as modified by the time factor. Then

$$\frac{S\rho_v}{5.76 \times 10^3} = \frac{b}{\rho_v g a}$$

$S\rho_v$ on the left refers to the heat problem:

$$S\rho_v = 5 \times 10^6$$

ρ_v on the right is that of the fluid in the model. Assuming water, ρ_v is 10^3 kg/m^3. Thus, we have

$$\frac{b}{a} = \frac{5 \times 10^6}{5.76 \times 10^3} \times 10^3 \times 9.8 = 8.5 \times 10^6$$

The method of Example 7-7 may be used to find the number of unit lengths involved, with the same result:

$$\Delta x = 0.01046 \text{ length}$$

Thus, for the section,

$$R_{f1} = R_f\,\Delta x = \frac{0.01046}{2} = 5.23 \times 10^{-3} \text{ newton-sec/m}^3$$

By (7-18),

$$R_{f1} = \frac{8\pi\mu l}{a}$$

where l is the length of the tube.

Since both l/a and b/a are specified, a may be chosen arbitrarily. However, the fluid-resistance expression assumes lamellar flow, and l should be reasonably long. Suppose a is 10^{-4} m^2. Then,

$$l = \frac{R_{f1}a}{8\pi\mu} = \frac{5.23 \times 10^{-3} \times 10^{-4}}{8\pi \times 10^{-3}} = 2.09 \times 10^{-5} \text{ m}$$

since μ is 10^{-3} in mks units. Obviously this is much too short to be practical, and scaling must increase the value of R_f, preferably by about 10^4.

This result is possible by expressing u_x in meters per second as analogous to q_x in watts per square centimeter. When this is done,

$$\frac{\partial\tau}{\partial x} = -\frac{10^4}{k}\,q_{x,\text{cm}}$$

where $q_{x,\mathrm{cm}}$ is heat flux in watts per square centimeter. Now

$$R_f = \frac{10^4}{k} = 5 \times 10^3$$

If we use $a = 10^{-4}$ m^2,

$$l = 2.09 \times 10^{-1} \text{ m} = 20.9 \text{ cm}$$

This is reasonable. The effect on the reservoir is seen by the heat equation

$$10^4 \frac{\partial q_{x,\mathrm{cm}}}{\partial x} = -\frac{S\rho_v}{5.76 \times 10^3} \frac{\partial \tau}{\partial t'}$$

Hence,

$$\frac{b}{a}\left(\frac{1}{\rho_v g}\right) = \frac{S\rho_v}{5.76 \times 10^7} = \frac{5 \times 10^6}{5.76 \times 10^7}$$

$$\frac{b}{a} = 8.5 \times 10^2$$

So

$$b = 8.5 \times 10^2 \times 10^{-4} = 8.5 \times 10^{-2} \text{ m}^2$$

The total reservoir area per section is thus

$$B = 8.5 \times 10^{-2} \times 1.046 \times 10^{-2} = 8.9 \times 10^{-4} \text{ m}^2 = 8.9 \text{ cm}^2$$

As a final check, consider the reservoir height. The total temperature range at the surface is specified (for the fundamental) as 38.9°C.

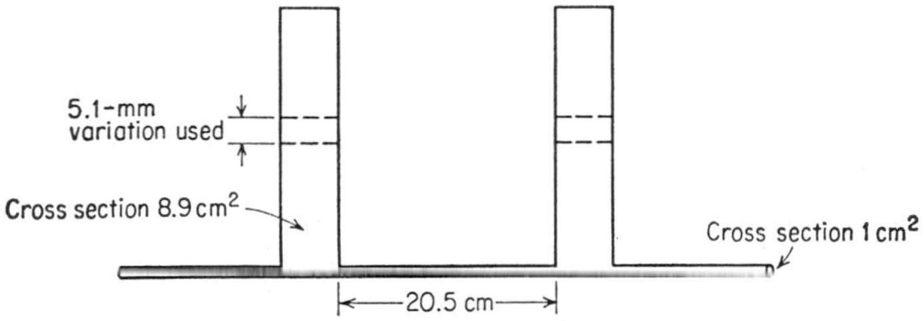

FIG. 7-10. Fluid model of heat flow into ground. Each reservoir and one capillary represent 1.046 cm in ground. One minute represents four days of heat flow.

With a 10° safety factor, about 50° variation should be provided for. So far we have assumed that p in newtons per square meter is analogous to τ in degrees centigrade; hence, a variation of 50 newtons/m^2 is called for. The pressure in water is just its height times the weight per unit volume:

$$p = y\rho_v g$$

where y is the height. Thus, the Δy called for is

$$\Delta y = \frac{\Delta p}{\rho_v g} = \frac{50}{10^3 \times 9.8} = 5.1 \times 10^{-3} \text{ m} = 5.1 \text{ mm}$$

This is easy to accommodate but difficult to measure accurately, so another scale factor may be in order—though it is not discussed here.

The final model is shown in Fig. 7-10. It will perform reasonably well, but the reservoirs appear a bit small.

7-6. Summary

Various types of iterative networks may be set up to simulate transmission lines and other traveling waves. Examples include electrical "ladder networks," spring-mass systems, thermal-resistance–thermal-capacity systems, and fluid systems. The basic modeling procedure involves converting the telegrapher's equations for the wave to be modeled to difference equations (actually the telegrapher's equation may have been derived from these very difference equations). A comparable set of equations is written for the lumped-constant system, and analogous quantities are compared either on a one-to-one basis or by scaling. The lumped-constant system need not be the same as the distributed system, provided that comparable equations can be found.

It is required that the same number of "energy-storage" and "dissipation" elements be present in analogous systems. Thus, for a lossless line or an acoustic wave, two types of energy storage must be present in the lumped model, and dissipation elements are not permitted. For a thermal or diffusion wave, or for an electromagnetic wave in a conductor, only one storage element is permitted, and a dissipation element is required.

In dealing with the lumped iterative networks, the impedance and propagation constants are similar to those for the actual distributed situation if the modeling does not attempt to simulate too long a section in one lumped section. For the electrical iterative network, in the T-section configuration,

$$Z_{0T} = Z_0 \sqrt{1 + \frac{Z_1}{4Z_2}} \tag{7-10}$$

$$Z_{0T} \approx Z_0 \left[1 + \frac{(\gamma \, \Delta x)^2}{8} \right] \tag{7-15}$$

$$\gamma_T = \log_e \left(1 + \frac{Z_1}{2Z_2} + \frac{Z_{0T}}{Z_2} \right) \tag{7-13}$$

$$\gamma_T \approx \gamma \, \Delta x \left[1 - \frac{(\gamma \, \Delta x)^2}{24} \right] \tag{7-16}$$

Here the difference between the lumped and distributed systems is shown by the second factors in (7-15) and (7-16).

Comparable equations are available for π sections.

Exact models can be made for traveling waves at a given frequency, by making Z_{0T} for the network equal Z_0 for the wave and γ_T for the network equal $\gamma\ \Delta x$ for the wave.

PROBLEMS

7-1. Show how to set up a lumped-constant mechanical model for a shear wave in a solid. Assume rectilinear motion.

7-2. Repeat Prob. 7-1 using torsional motion.

7-3. Show how to set up a lumped-constant mechanical model for a vibrating string.

7-4. Show how to set up a lumped-constant mechanical model for a vibrating membrane using qualitative arguments only.

7-5. Show how to set up a lumped-constant fluid model for charge-carrier diffusion, neglecting recombination and generation.

7-6. Derive the expression for the π-section characteristic impedance (electrical).

7-7. Derive the expression for the π-section propagation constant (electrical).

7-8. Discuss reflections on model transmission lines.

7-9. Show a mechanical model for an acoustic wave different from that developed in Sec. 7-1.

7-10. It is desired to make a lumped-constant electrical simulation of a telephone line with the following characteristics:

R = 100 ohms/mile
L = 1 mh/mile
C = 0.1 μf/mile
G negligible

This line is to be used in a 100-kc carrier system. The error in over-all phase shift should not exceed 30° in simulating a 50-mile length of the line. Determine the parameters and the number of sections for an appropriate artificial line. Calculate the resultant error in Z_0 at 100 kc.

7-11. It is desired to make a lumped-constant simulation of the following telephone line:

R = 47.5 ohms/mile
L = 0.910 mh/mile
G = 150 μmhos/mile
C = 0.06 μf/mile

This line is used in a carrier system having a maximum frequency of 50 kc. Repeater stations are 10 miles apart. The simulation of a section between repeaters should have, at most, an error in phase shift of 30°. Determine the parameters and the number of sections for an appropriate model. Calculate the resultant error in Z_0 at 50 kc and at 10 kc.

7-12. A power line is operated at 60 cps, where it has the following parameters:

R = 0.10 ohm/mile
L = 1.0 mh/mile
C = 0.01 μf/mile
G negligible

If the line is 20 miles long, determine its exact T-section equivalent.

7-13. Determine the exact π-section equivalent for a length of line less than a quarter-wavelength long.

7-14. Determine the exact T-section equivalent for the line of Prob. 7-12 if it is 50 miles long. Compare with the approximate T-section values of parameters.

7-15. Set up an electric-circuit model for the acoustic wave in an organ pipe 2 m long (a half-wavelength) and 5 cm in diameter. Assume that the maximum phase-shift error may be 10°.

7-16. Set up a lumped-constant mechanical model for the organ pipe of Prob. 7-15.

7-17. Set up an electric-circuit model for heat flow in a masonry wall 20 cm thick. Assume that your model represents the flow through 1 m^2 of the wall. The phase-shift error at 1 cycle/day should not exceed 5°.

7-18. Set up a lumped-constant thermal model for the wall of Prob. 7-17, using copper for the thermal capacitors and asbestos for the thermal resistors.

7-19. Set up an electrical model for diffusion of electrons into germanium through an area of 1 mm^2. Assume that the phase shift at 10^4 cps should not exceed 15° at a depth of 1 mm in the germanium. This is at 30°C.

7-20. Set up a water capillary-reservoir model for the heat flow of Prob. 7-17.

7-21. Set up an electrical model for diffusion of indium into germanium, through a 1-mm^2 surface, at 700°C. The phase shift for a wave with a period of half an hour should not exceed $\pi/4$ radians at a depth of 10^{-6} m.

7-22. Set up a spring-mass lumped analogy for the transmission line of Prob. 7-10, using dashpots for resistors.

7-23. Set up a spring-mass lumped analogy for the transmission line of Prob. 7-11, using dashpots for resistors.

7-24. Set up a lumped circuit model for transmission of 10-kc sine waves into sea water (conductivity, 4 mhos/m) to a depth of 20 m, with a maximum phase-shift error of 20°. Assume that the model represents transmission through 1 m^2.

7-25. Repeat Prob. 7-24, but with a lumped thermal model.

BIBLIOGRAPHY

Olson, Harry F.: "Dynamical Analogies," D. Van Nostrand Company, Inc., Princeton, N.J., 1943. (Provides an excellent description of lumped-constant analogies for electrical, acoustical, and mechanical systems.)

Schneider, P. J.: "Conduction Heat Transfer," chap. 13, Addison-Wesley Publishing Company, Reading, Mass., 1955. (A discussion of electrical modeling of thermal transmission.)

8. Transmission Lines for Communication and Power

Long-distance communication and transmission of power utilize electric traveling waves exclusively. Even the types of wave that were treated as lossless in Chap. 3 actually have sufficient loss to be impractical for truly long-distance transmission of information or of large amounts of power. The transmission line is one of the most efficient devices known for this purpose, and it is treated in this chapter with that use in mind. Information-transmission problems and the United States telephone system are discussed along with a few of the problems associated with power transmission. The chapter does not treat nonelectrical problems.

8-1. Information Transmission

Information is transmitted from one place to another in various forms. The telegraph was the first electric information-transmitting device. Today, telegraph lines are in wide use, but they are employed almost exclusively for printing telegraphs (teletypewriters). The telephone was the next electric information-transmitting device developed. Both telegraph and telephone were originally restricted to wire transmission, but the development of radio expanded the utility of both. The telephone and telegraph systems in the United States are extremely extensive, and their improvement has led to a great many advances, not only in traveling-wave work, but in all fields of electrical engineering.

Information is also transmitted electrically by facsimile photographs and by television. Telemetering finds wide application—from presenting measurements by earth satellites to the control of oil pipelines. With the oil pipelines, the various flow rates and pressures are transmitted electrically to a central location, and control signals are sent back which operate valves and pumps. Various types of business and scientific data are transmitted in digital form for use by computing facilities in central locations.

These are the principal types of information-transmitting systems. Others could be named, but they are less important.

Information theory has received wide discussion since Shannon published his original paper on the subject.[1] Information theory deals with measurement of the efficiency with which various information-transmission systems utilize the available channels and with coding to obtain optimum utilization. Information is defined in terms of the statistics of signals. It can be shown that a given channel can transmit information at a faster rate if the noise in the channel is small than if the noise is large. Thus, the signal-to-noise ratio is important in information theory, as in other phases of information transmission. It can also be shown that voice and television as normally transmitted are extremely wasteful and inefficient methods of information transmission.

Since information theory deals largely with the selection of optimum codes for transmitting information and the evaluation of existing systems, it will not be considered further here. We shall assume that the coding system is already known and that the problem is one of providing a transmission facility which will reproduce with a minimum of distortion the input signal at a distant location.

Regardless of actual information content, certain specified bandwidths are required by different systems of communication. High-speed automatic telegraph requires almost 100 cps bandwidth. Telephone requires, as an absolute minimum, something like 300 to 2,500 cps, and present-day standards call for 200 to 3,500 cps. Telephone lines used for broadcast program material have an upper bandwidth limitation of about 7,500 cps. Facsimile depends on the speed at which the information is to be transmitted, but ordinary television, with its relatively fixed speed, requires bandwidths in excess of 4 Mc. Some telemetering applications require even higher bandwidths, because of the extremely fast phenomena to be monitored.

The bandwidth used by a communication channel is dependent on the modulation of the carrier. It is customary to transmit information by varying some characteristic of a sinusoid (amplitude, phase, frequency). The sinusoid whose properties are varied is the *carrier*, and the process by which they are varied is known as *modulation*. Various modulation methods require bandwidths from the same amount as the information (single-sideband amplitude modulation) to many times that bandwidth (wideband frequency modulation, pulse-code modulation, etc.).

[1] C. E. Shannon, A Mathematical Theory of Communication, *Bell System Tech. J.*, vol. 27, pp. 399–423, 623–656, 1948.

In general, signal-to-noise ratios are set in terms of allowable error rates for automatic communication systems or intolerable interference for systems such as telephone and television, depending on perception by human senses. The required signal-to-noise ratio is very much a function of the grade of service desired. For a high-grade telephone service the signal-to-noise ratio should be about 40 db. For other purposes it may be more or less.

8-2. The United States Telephone System

A telephone system utilizes the elements shown in Fig. 8-1 in various amounts and combinations. The telephone user operates the *local,* or *subscriber, station,* which is connected to his local *exchange* through *local lines,* also known as *loops.* At the exchange, switching facilities are available which permit connection with other subscribers attached to the

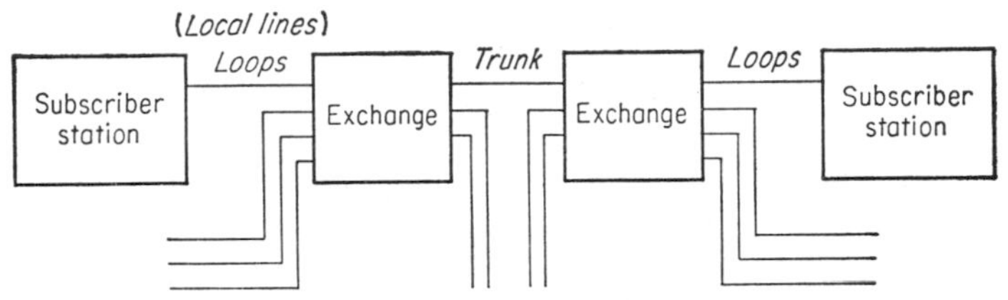

FIG. 8-1. Elements of a telephone system.

same exchange or with subscribers at distant points. In the case of a distant subscriber, a call is placed between exchanges by *trunk lines* (sometimes known as *toll lines* if the exchanges are in different cities), and the call is then connected from the remote exchange through local loops to the called subscriber station.

Each subscriber station consists of a number of elements: (1) the microphone (transmitter in telephone terminology), (2) the earphone (receiver), (3) a bell or other signaling device, and (4) impedance matching and balancing equipment. The latter equipment is usually in the base of a cradle-type phone and in the wall box of other types of phone.

Loops and short trunks within a city are usually very much shorter than the shortest wavelength involved, so the analysis of their performance is quite simple and does not even involve traveling waves. Usually a single-section lumped model suffices for such analysis. The toll trunks which go between cities, however, are real transmission lines, and a study of their operation involves a full knowledge of traveling waves. Variations with frequency of the attenuation and the velocity of propagation have a severe effect on toll trunks and must be compensated for in some fashion.

In general, the toll-trunk transmission lines have sufficient attenuation that amplification is necessary along the way. The first application of the original triode vacuum tube was to long-distance telephony. A special set of terminology has been developed over the years to describe the equipment required for long-distance telephony, including the amplifiers.

The amplifiers used in long-distance telephony are called *repeaters* when they are between *terminal stations.* Many configurations of repeaters have been developed; some of them, by a clever balancing scheme, utilize only one amplifier for signals going in both directions on a pair of wires, and others use separate amplifiers for the conversations going in opposite directions, which may be on either one or two pairs of wires. The designs of the networks that go with these telephone amplifiers can be quite involved. The basic amplifiers themselves, however, are relatively simple, except that they must be much more stable than most other amplifiers or must at least be capable of automatic gain control. To understand the need for this, it should be noted that some carrier systems have coast-to-coast attenuations, at the higher-frequency end, of thousands of decibels. Since a change on the order of 3 db is readily perceived by the human ear, a change of a very small fraction of a per cent in the over-all attenuation in such a system causes a very noticeable change in the resulting output level. Hence, the gain must be controlled very accurately.

It is interesting to examine the system which has been set up for switching long-distance telephone calls in the United States. An over-all system has been established for both operator and subscriber dialing of long-distance (*toll*) calls.

The system is based on a large number of various types of *switching center.* A local exchange (*end office*) is called a *tributary* of a *toll center.* Several toll centers and *toll points* are connected to a single *primary center,* which itself is one of several connected to one *sectional center.* The sectional centers are connected to *regional centers.* Figure 8-2 shows the standard switching pattern for calls between two regions.

Two basic types of circuits are shown in the figure. The dash lines represent *high-usage groups* of circuits, and the solid lines represent *final groups.* High-usage groups are made available wherever there is sufficient direct traffic between two cities to require them, regardless of the routing for the final group. The automatic equipment selects a circuit from the most direct group if one is available. If none is available, it takes an alternate route. In practice, only a very small percentage of calls go by a circuitous route, but in theory, at least, even calls between, say, the Los Angeles regional center and the Sacramento regional center might have to be routed through points in the Middle West.

Figure 8-3 illustrates the alternate routings possible for a call from Hibbing, Minn., to Davenport, Iowa. The most direct route would be by high-usage circuits from Hibbing to Minneapolis and from Minneapolis to Davenport. If the Hibbing to Minneapolis circuits were all

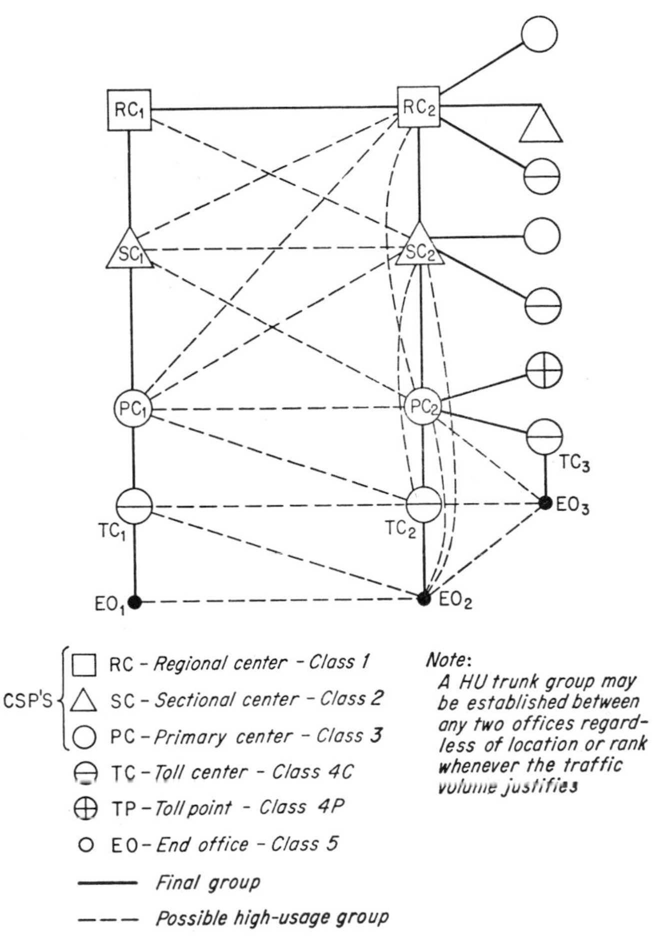

FIG. 8-2. Standard switching pattern for distance dialing between two regions. (*Courtesy American Telephone and Telegraph Company; by permission.*)

busy, the final groups from Hibbing to Duluth and Duluth to Minneapolis might be used. Likewise, if the high-usage groups from Minneapolis to Davenport were in use, the routing might involve either high-usage groups from Minneapolis to Des Moines and a final group from Des Moines to Davenport or a final group from Minneapolis to Chicago and a high-usage group from Chicago to Davenport (not shown). If all shorter alternates were unavailable, the call could go from Hibbing to Duluth to Minneapolis to Chicago to Des Moines to Davenport.

The nationwide plan for control switching points is shown in Fig. 8-4 as it was in 1951. Later versions of this are somewhat different but have not been publicly released.

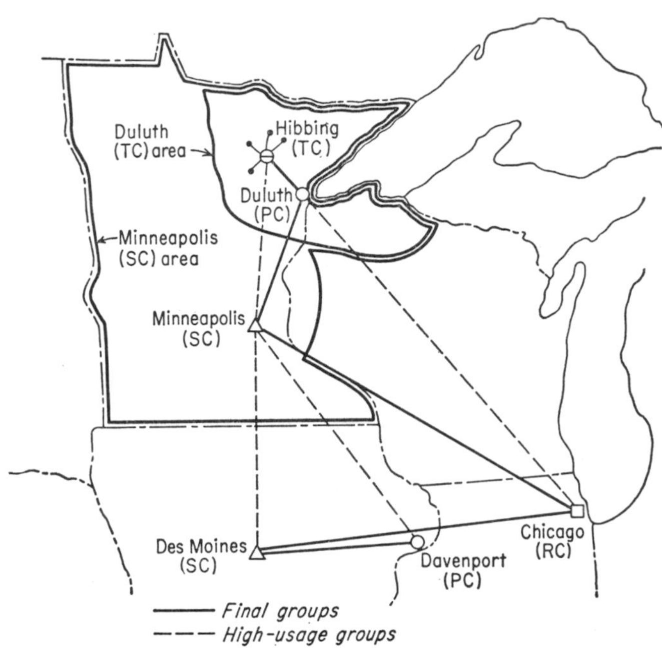

FIG. 8-3. Alternative routing possibilities between Davenport, Iowa, and Hibbing, Minn. (*Courtesy American Telephone and Telegraph Company; by permission.*)

Nearly all switching into alternate routes is done automatically at the various switching centers. It is controlled by a computer, which uses a collection of metal cards with holes punched in them, through which a light shines, to show the various routes for transmission between a given pair of points and the order in which the alternates should be tried.

8-3. Problems of Speech and Picture Transmission

Many of the problems of speech and picture transmission have to do with the wide range of frequencies involved. For example, satisfactory speech transmission requires a bandwidth of from 200 to 3,300 cps, or a range of almost 1 to 17. Music requires a considerably wider range for high-fidelity reproduction, say, from 30 to 15,000 cps, a 1-to-500 range. Television is worst of all, since it requires from 30 cps to 4.5 Mc, or a range of 1 to 150,000 in frequency. To keep a transmission system

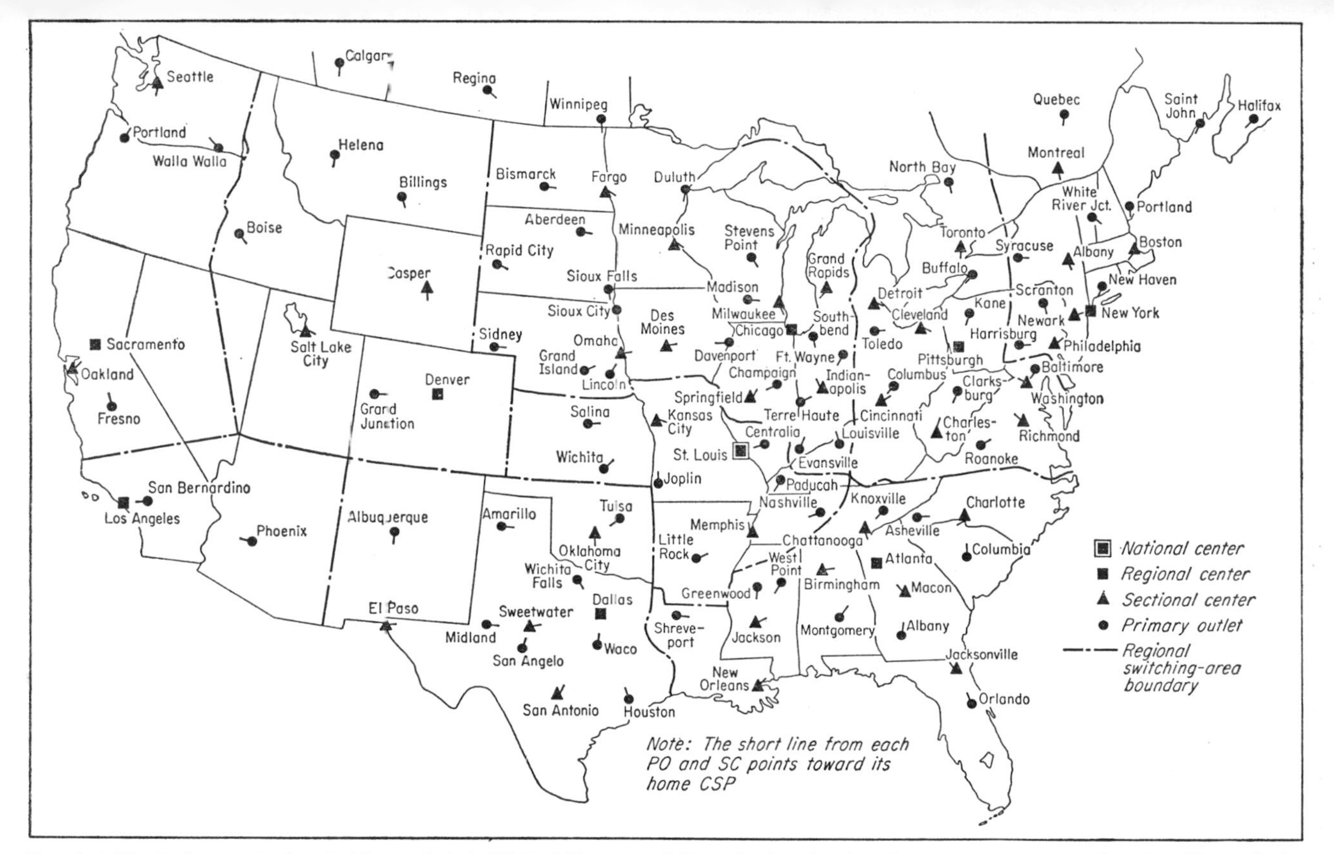

FIG. 8-4. Tentative control switching points in United States and Canada showing homing arrangements and regional boundaries (December, 1951). Later plans dc not include a national center. (*From J. J. Pilliod, Fundamental Plans for the Toll Telephone Plant, Elec. Eng., vol. 71, p. 785, September, 1952; by permission.*)

distortion-free over such a wide range is very difficult. Fortunately, only the frequencies down to 3,500 cps need to be precisely reproduced, as the lower frequencies are important mainly for synchronizing pulses which do not require as faithful reproduction as the picture itself.

Two methods can be used for keeping distortion to a minimum: the line may have its parameters adjusted (by *loading*), or *equalizers* may be inserted at intervals. These add attenuation in those parts of the range where the line attenuation is the least, so that the over-all attenuation is essentially constant. Equalizers are also used to correct for errors in velocity of propagation.

In general, there are two types of distortion for which compensation is necessary: *amplitude-frequency* distortion and *phase-frequency* distortion. The former is due to an amplitude-frequency characteristic (or an *attenuation-frequency* characteristic) other than that desired. Ordinarily the desired characteristic across a speech or television band is constant attenuation. Amplitude-phase distortion occurs when the phase shift in the transmission line is not directly proportional to frequency; this usually manifests itself as delay distortion. An extreme case would cause the high-frequency components to arrive at a significantly different time from the low-frequency components. This was the type of distortion first encountered in transatlantic telegraph cables. In television it can cause extremely severe complications.

The attenuation observed in practice on telephone-type circuits varies from less than 0.03 db/mile to greater than 10 db/mile. Minimum attenuation occurs on open-wire lines at relatively low voice frequencies. Maximum attenuation occurs in coaxial cables at high carrier telephone frequencies, between 4 and 10 Mc.

To get some idea of the magnitude of the compensation problem on a long coaxial cable, suppose that such a cable is used over a range of 50 kc to 3 Mc. In this frequency range, the attenuation in decibels per mile varies from 1.4 to 10. Suppose further that this is a 4,000-mile coast-to-coast circuit. The over-all attenuation, therefore, varies from 5,600 to 40,000 db. Recalling that the number of decibels divided by 10 is the exponent to which 10 must be raised to obtain the power ratio, it can be seen that the over-all attenuation in the cable corresponds to a power ratio of from 10^{560} to $10^{4,000}$. This is indeed a fantastic variation, as well as a fantastic attenuation. Obviously, if the signal-to-noise ratio is to be maintained at a reasonable level, repeaters must be close together on such a system, and the equalization must be extremely severe. Thus, the equalizers themselves must produce an attenuation at the lowest frequency of 34,400 db over the entire coast-to-coast path.

Since an attenuation of about 40 db is the maximum that can be tolerated between repeaters, the repeaters on the system just described would have to be spaced 4 miles apart; thus there would be 1,000 in 4,000 miles. This is, indeed, the spacing used on the Type L3 coaxial carrier system operated by the American Telephone and Telegraph Company.

Repeater spacing does not need to be nearly so close with open-wire voice-frequency circuits. In fact, the repeaters can be spaced more than 200 miles apart on such circuits.

Considering the high cost of repeater stations, it would seem that the open-wire circuits might predominate. In fact, however, they are becoming obsolete. The reasons are that they are much more subject to weather conditions than are cable circuits and more likely to receive and cause interference, so that the number of channels which may be used is very limited. One problem encountered with the open-wire system is the hum coupled in from power lines, not only at 60 cps but also at the various high-frequency harmonics set up by the generators. Another problem is that the frequency range is limited, because at higher frequencies the lines tend to radiate and to receive like antennas, thus both causing and receiving interference. As a result, carrier frequencies on open-wire lines are restricted to a maximum of about 150 kc.

This limitation precludes the use of the open-wire line for television and restricts the number of voice channels which may be carried to about 12 two-way channels per pair of conductors. The coaxial cable used on the L3 systems carrics 1 television channel and 600 telephone channels or 1,860 one-way telephone channels. Thus, even though expensive repeaters are required at frequent intervals, the cost per channel is considerably less, provided that the larger number of channels is required on a particular path.

In addition to the extremes of open-wire and coaxial cable, a large proportion of the toll telephone circuits is in ordinary cables. Carrier systems in such cables are exemplified by the American Telephone and Telegraph Company's older K carrier, which uses one pair in each direction for 12 channels, with 17-mile repeater spacing and circuits up to 3,000 miles in total length, and by the newer N carrier, with 12 two-way channels on one pair and repeater spacing of 8 miles for total distances up to 200 to 400 miles.

A great deal of the toll-circuit mileage today is carried by radio relay instead of by cable or open-wire systems. The telephone engineer treats the radio path between two radio stations in much the same way as he does the coaxial-cable path between two repeater stations. The radio relays must be well within the line of sight, which ordinarily places them at a separation of about 30 miles. The radio-relay systems have a channel capacity comparable with that of the coaxial cable, with a considerable potential for expansion.

As examples consider American Telephone and Telegraph's TD-2,

TH, and TJ systems. The TD-2 was the first widely used long-haul radio system. It operates in the band from 3,700 to 4,200 Mc and carries 3,000 two-way channels on the 5 working channels (1 protection). The TH system operates in the band from 5,925 to 6,425 Mc, carrying about 10,800 channels on 6 working carriers (2 protection). The TJ is a short-haul radio system with 720 two-way channels on 3 working carriers (3 protection) in the 10,700-to-11,700-Mc band.

The problems of radio relay are traveling-wave problems, but they are somewhat different from those that have been treated for transmission lines and plane waves. Some idea of a few of the problems is given in Chap. 10, where plane-wave reflections are considered at other than normal incidence. When the stations are within line of sight of each other, the principal problems associated with the microwave relay are *multipath* problems; that is, the signal may follow both the direct path and a path via a ground reflection point. Since these paths are of different lengths, the time delays involved are also different, so that the signal received is made up of two components with different delays. These signals are superimposed by the receiver, with adverse effects similar to those which would result if two different lengths of cable were connected between stations. Occasionally, radio relay also involves attenuation problems due to the presence of moisture, either in the form of rain and snow or in the form of clouds between the stations.

In summary, then, bandwidths of from less than 100 cps for telegraph signals to greater than 4 Mc for television are required by communication systems. The lowest attenuation is encountered with open-wire lines, but they have largely been replaced by cables and microwave systems. Because of their limited capacity, ordinary cables with wires close together are installed today only where the traffic load does not justify a coaxial-cable or microwave system. Where the traffic load is high enough, coaxial cables capable of carrying up to 1,860 one-way channels per pair of conductors may be used. Similar channel capacity is available with microwave systems. Both microwave-relay and coaxial-cable systems are also frequently used with smaller traffic loads, where this is economically justified.

8-4. Carrier Systems

The problem of getting more communication per pair of wires has been studied for a long time. In the very early days of the telegraph, numerous mechanical multiplexing schemes were developed which permitted more than one telegraphic message to be sent simultaneously along the same pair of wires. Some of Edison's early inventions had to do with this problem. Soon after the introduction of long-distance telephony, the same problem arose. Various schemes were developed whereby, for example, two pairs of wires could carry three conversations by use of "phantom" circuits. These bridging arrangements are still widely used for short-distance multiplexing. When repeater amplifiers were first introduced, a great deal of effort was put into keeping the conversations on one pair of wires and arranging to use one or two amplifiers to serve the two directions of transmission on a single pair of wires. Later it became feasible to use one pair of wires to carry a number of conversations, but it was difficult to make this work in both directions, so separate pairs of wires were used for the two directions of transmission. In systems being installed today, this is almost always the case.

The change was brought about by the use of carrier systems, which are the basis for the modern toll telephone plant. With carrier systems each individual telephone conversation is used to modulate a different carrier frequency, and all conversations are transmitted upon the same pair of wires. In the early systems, three carrier frequencies above the voice-frequency range were used, and the voice-frequency channel provided a fourth channel. Modern carrier systems for use with cables and open wires have about 12 conversations per pair of wires, and, as indicated above, modern coaxial-cable and microwave systems have hundreds and even thousands of conversations per path.

Not only does the carrier system reduce the number of wires, but at repeater stations only one repeater amplifier (admittedly more complex than a voice-frequency amplifier) is required for each *set* of conversations rather than one for each conversation.

Although the discussion here is quite brief, carrier telephony is a subject that can be studied at great length. Probably the best single source of information on carrier telephony is the *Bell System Technical Journal.* Papers on carrier telephony, of course, have also appeared in numerous other journals, including, in this country, those of the American Institute of Electrical Engineers.

A summary of carrier systems by various manufacturers in this country and elsewhere is presented in the Federal Telephone Handbook.[1]

The Type C carrier of the American Telephone and Telegraph Company, manufactured by the Western Electric Company, is now obsolete for new installations, although it is still in wide use. It is described here because it is simple enough to be explained understandably in a relatively brief discussion, whereas modern systems are much more complex. The principles involved are applicable, with complications, to all the more advanced systems.

The Type C system is used on open-wire lines and has a total capacity

[1] "Reference Data for Radio Engineers," 4th ed., pp. 833–837, International Telephone and Telegraph Corporation, New York, 1956.

of four channels, three modulated onto carriers and one at voice frequency. The entire system operates in the frequency band between 0 and 30 kc. In this band, in addition to the voice frequency, there are three channels assigned for transmission in one direction (referred to as east-west) and three for transmission in the other direction (referred to as west-east). The basic principle of such a system is illustrated in Fig. 8-5, which shows the three channels of the carrier system at two terminals. Three voice-frequency channels come in from the left at the top and are used to modulate oscillators at three different frequencies above the voice-frequency band. By the use of an appropriate modulator and appropriate filters for each channel, only a single voice sideband results, so

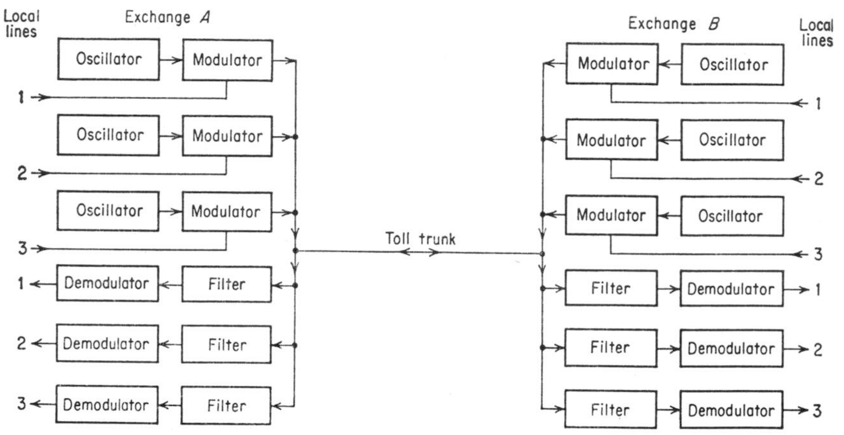

FIG. 8-5. Basic principle of carrier telephony.

that, in effect, the modulation system merely translates the 250-to-2,750-cps speech band up to a band of, say, 30,250 to 32,750 cps. Since each of the oscillators is at a different frequency, the bands into which the three speech channels are translated do not overlap. They are all combined and sent on the toll trunk to the exchange at the right-hand end. At that point the incoming signal is passed through three filters, each one of which will pass only the range of frequencies associated with a particular voice channel. After passing through the filters, the composite signal containing three channels going one way and three channels going the other way has been separated into a series of single channels going in one direction only. At the right-hand terminal the three channels going in the right-hand direction are detected (demodulated) separately and sent on to the subscriber lines at that exchange. The signals from those subscriber lines going toward the left receive the same kind of treatment and eventually end up at the subscriber line at the left-hand exchange.

The Type C carrier system does not use the 200-to-3,300-cps bandwidth considered a minimum by modern standards. Instead, it uses a frequency band from 250 to 2,750 cps. Typical frequency assignments for such a system are shown in Fig. 8-6. The allocations shown are for the Type CS carrier system, which is one of a number of versions of the Type C carrier. The diagram indicates the carrier frequencies themselves and the allotted passbands for the voice channels. The east-west channels have the carrier below the sideband carrying the intelligence, whereas the west-east channels have the carrier above the intelligence passband. Other versions of the Type C carrier reverse this system for carrier location.

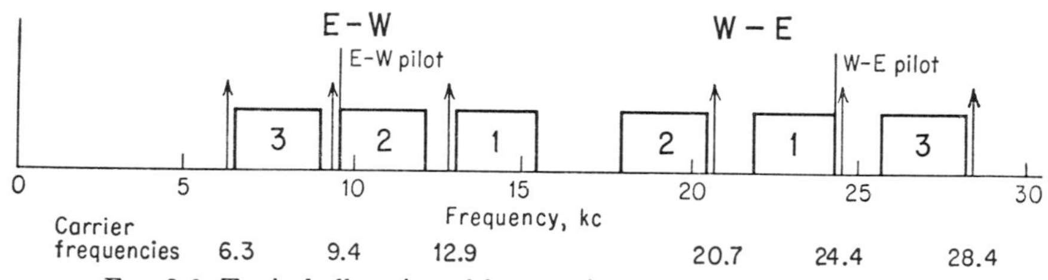

FIG. 8-6. Typical allocation of frequencies for Type C carrier system.

In addition to the carriers and the voice passbands, *pilot frequencies* are noted for the two directions. The pilot frequencies are associated with *pilot carriers*, which are constant-amplitude signals sent along the line to aid in gain control. In the repeater itself, separate narrow-band filters isolate the pilot-carrier frequency from the other frequencies in the system and adjust the gain of the repeater amplifiers so that the level at the output of each repeater is always constant, even though the input level may vary. This is necessary because of changes in gain due to weather conditions, temperature of the wires, and aging of the amplifiers. On the more advanced systems, amplification is frequently carried out in groups of channels, each of which has its own pilot carrier. Thus, not only is the over-all gain compensated, but the gain-frequency characteristic of the repeater is adjusted as it changes.

Whereas Fig. 8-5 illustrates the principle of the carrier system, Fig. 8-7 shows a somewhat more detailed block diagram of a terminal; here the complete system for a single channel in both directions is indicated. At the left-hand side of the terminal equipment is a *hybrid coil*, which is, in effect, a balancing transformer that permits the amplified incoming signal to be connected to the local line without having it go on through the amplifier in the other direction. If its balancing were not carried out, there would be an echo path between the two ends of the system, and signals once started would continue to circulate; hence oscillation would be possible.

Taking the signal going from the channel 1 local line toward the outgoing toll line, observe that the first step is to modulate the carrier frequency with the voice frequency, thereby translating the original voice-frequency band up to one of the bands indicated in Fig. 8-6. This signal is immediately passed through a filter which eliminates undesired components created in the modulation process. At this point the signal is mixed with the output of similar modulation and filter setups for the other two channels. After passing through a transformer, the pilot carrier is added to the three voice channels, and the combination is amplified in the transmitting amplifier and passed through the transmitting level pad. This serves to adjust the over-all gain and also makes

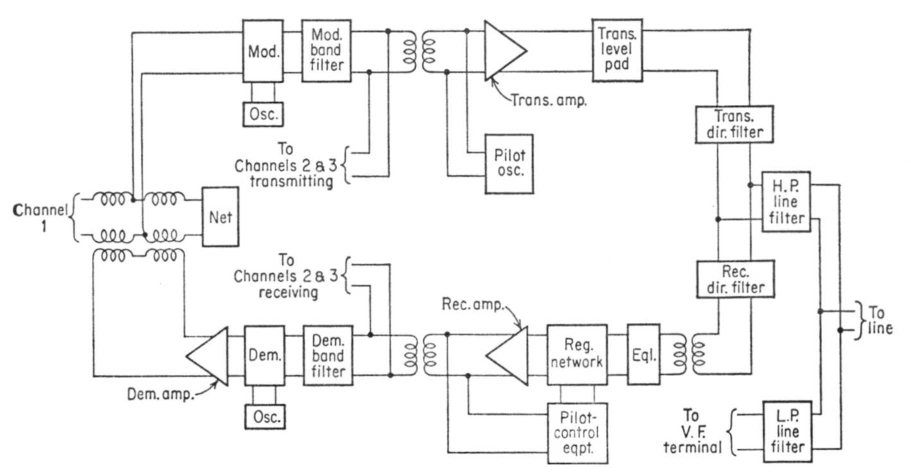

Fig. 8-7. Block diagram for Type C carrier system. (*From "Principles of Electricity Applied to Telephone and Telegraph Work," p.* 280, *American Telephone and Telegraph Company, New York, January,* 1953; *by permission.*)

it possible to transmit with the amplitude-frequency characteristic desired. The output passes through the transmitter filter, which transmits a band appropriate for all the channels in the one direction and rejects the band appropriate for channels in the other direction. The signal then goes through a high-pass line filter, which eliminates any voice-frequency components that may be present, and on to the toll line.

A signal coming in from the toll line goes through the same high-pass filter to eliminate voice-frequency components, which may also be present in this system, and then through the receiver directional filter. The transmitter directional filter stops the signal from the line and prevents it from going on toward the transmitter amplifier. After passing through the receiver directional filter and the transformer, the signal goes through an equalizer, which serves to compensate for variations in attenuation and velocity of propagation in the line. It then goes through a regulating network, which adjusts the gain to the proper level, and on through the receiver amplifier. The pilot-control equipment compares the output of the amplifier at the pilot frequency with what it should be and adjusts the regulating network accordingly. The incoming voice signal, after passing through a transformer, goes to the demodulating band filters. The channel 1 signal passes through the channel 1 filter and is rejected by the channel 2 and channel 3 filters; likewise, the channel 2 signal passes through its filter and is rejected by the others; etc. After being separated, the channel 1 signal is demodulated (translated back down to voice frequency) and passed through a final amplification stage, which provides a signal at its output at an appropriate level for the local line. It then passes through the hybrid coil to the line. Since the Type C carrier not only utilizes the three carrier channels but also the voice-frequency channel, a filter is provided at the toll line to separate the voice frequencies.

Equalization is necessary on most telephone lines. Figure 8-8 shows an example of the equalization required for the Type C carrier system. Note how the line loss varies with frequency—from about 13 to about 28 db, in the particular case chosen. In addition to the line loss, there is a filter loss due to the filters that separate the upper and lower groups of channels. The loss of the equalizer which must compensate for the uneven loss in the line and filters is indicated at the top of the diagram. The net result for the lower-group channels is loss L_1 and for the upper-group channels is loss L_2. By using amplifiers with different gains, it is possible to bring the over-all levels to the same point at either end of the line.

Equalization is very simple in the Type C system. In more complicated carrier systems, however, it becomes extremely difficult. Equalization and filtering problems in the telephone system have led to the development of much of modern network theory.

The Type C carrier was replaced or supplemented on open wires by the Type J carrier and is now being replaced by Type O. Many of the early open-wire lines were themselves replaced by cables with the Type K carrier. The Type J system provides 12 two-way voice channels with a bandwidth ranging from 200 to 3,300 cps in a frequency range of 36 to 140 kc. The carrier spacing is 4 kc. The channels are in separate groups for each direction, with the west-east channels from 36 to 84 kc and the east-west channels from 92 to 143 kc. Repeaters on this system are usually spaced about 50 miles.

The Type K system sends 12 two-way channels over *two* 19-gauge *pairs* in a cable. Because of the characteristics of the multiconductor cable, it is not feasible to operate such a system to as high a carrier frequency as the open-wire system. Hence, the east-west conversations

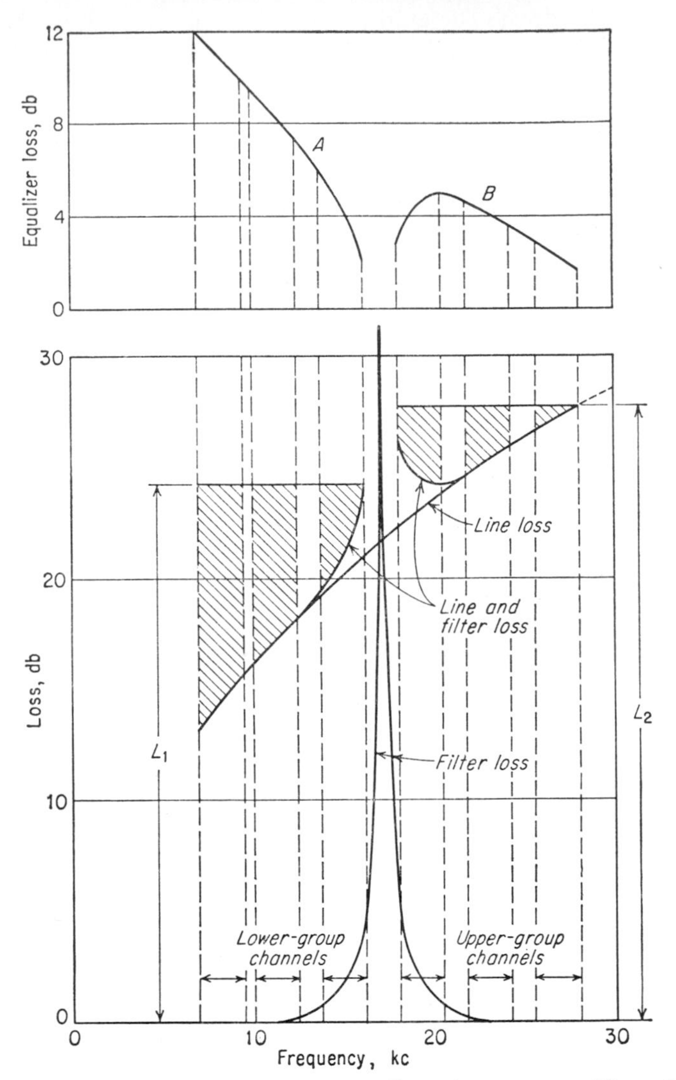

FIG. 8-8. Attenuation equalization for Type C carrier system. (*From "Principles of Electricity Applied to Telephone and Telegraph Work," p. 190, American Telephone and Telegraph Company, New York, January,* 1953; *by permission.*)

go on one pair and the west-east on another. The frequency band used is 12 to 60 kc, and repeaters are spaced, on the average, 17 miles.

The Type C carrier is being largely replaced for new construction by the Type O carrier. This is a carrier system with the lower band starting at 2 kc and the upper band reaching 156 kc. It can provide up to 16 high-grade transmission channels in each direction on one pair of open wires, but without a voice-frequency channel. Type N systems, rather than Type K, are now used in large quantities for short-haul systems on wires in cables.

A coaxial-cable system for long-haul purposes is the L3 system.[1] This system uses a coaxial cable that has much better frequency-response characteristics than the multiwire cables. Usually six coaxial lines are in one large cable with a number of regular pairs present also. Each coax carries 1,860 one-way telephone channels, or 600 telephone channels and one 4.2-Mc television channel. The range of operation is from 308 kc to 8.32 Mc. In a complex system such as the L3, it is necessary to group the individual channels prior to the final assembly of all channels onto one coax. Thus, the channels are grouped by 12s, and by 60s. An over-all discussion of the L3 system would be much too lengthy to include here, but the reference cited presents an interesting discussion and refers to a group of other papers describing in detail the various aspects of the system.

It will be noted that the Type C system is the only one described which operates all the way down to voice frequency. It is usually not considered economical to operate a multichannel carrier system over the wide range of frequencies that would be required if a voice-frequency band were included along with the carrier frequencies. Thus, with the Type J system, a frequency range of 715 to 1 would be required if it operated down to 200 cps. With Type L3, a range of 4,248 would be required. It is difficult enough to equalize such systems over the range that they do use, and the cost of equalization over this increased range would be much greater than the added benefit from the few channels that could be placed in the lower-frequency region. It is possible, however, to operate a Type C system over the same pair of wires as a Type J system, since the lowest frequency of the Type J system is higher than the highest frequency of the Type C system.

Frequency allocations for American carrier telephone systems are shown in Fig. 8-9. Some systems other than those described are shown in this figure. These are carrier systems used for short-haul purposes, a number of which have been developed in recent years. The principal difference between these systems and the long-haul systems is that compromises have been made in the design that permit the use of less expensive equipment which is perfectly adequate for a system of a few hundred miles but is not adequate for a transcontinental system.

The microwave-relay systems are not shown on the allocation diagram of Fig. 8-9. However, the basic principles involved in wire carrier

[1] C. H. Elmendorf, R. D. Ehrbar, R. H. Klie, and A. J. Grossman, The L3 Coaxial System, *Trans. AIEE, Part I (Communications and Electronics)*, p. 395, September, 1953.

telephony are also utilized in the microwave systems. In general, most microwave systems of a transcontinental type utilize carrier frequency allocations somewhat similar to those of the Type L system. Some local microwave systems use smaller numbers of channels.

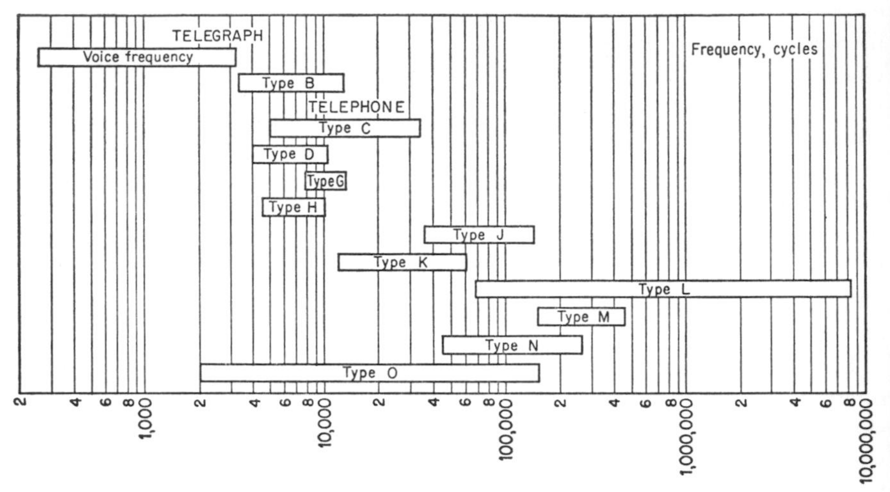

FIG. 8-9. Frequency allocations for American carrier telephone systems. (*From "Principles of Electricity Applied to Telephone and Telegraph Work," p. 227, American Telephone and Telegraph Company, New York, January,* 1953; *by permission.*)

8-5. Distortion, Equalizing, and Loading

Both in Chap. 2 and in this chapter examples have been given of telephone lines for which distortion is present. Even the low-loss lines treated at the end of Chap. 2 are not distortionless, since their attenuation is proportional to their resistance per unit length, and this resistance, because of skin effect, is a function of frequency. Figure 8-8 illustrates the problem of attenuation equalization in the Type C open-wire carrier system. Equalization is usually considerably more of a problem in cable systems.

Since in the distortionless system both attenuation and velocity of propagation are independent of frequency, we must have

$$\alpha = \text{constant}$$
$$\beta = \omega(\text{constant})$$

The propagation constant for any transmission line is, of course, given by

$$\gamma = \sqrt{(R + j\omega L)(G + j\omega C)}$$

This may be rewritten as

$$\gamma = \sqrt{RG\left(1 + \frac{j\omega L}{R}\right)\left(1 + \frac{j\omega C}{G}\right)}$$

To make the attenuation independent of frequency except for the variation of R and G with frequency, the two complex factors must be made equal, so that their product represents a perfect square. They are equal when

$$\frac{L}{R} = \frac{C}{G} \tag{8-1}$$

In this case, the propagation constant is given by

$$\gamma = \sqrt{RG\left(1 + \frac{j\omega L}{R}\right)^2} \tag{8-2}$$

Hence,

$$\alpha = \sqrt{RG} \tag{8-3}$$

$$\beta = \omega L \sqrt{\frac{G}{R}} = \omega L \sqrt{\frac{C}{L}} = \omega \sqrt{LC} \tag{8-4}$$

Thus, when Eq. (8-1) holds, we have the "distortionless line," and, in fact, the phase-shift constant β is the same as it is for a lossless line.

In telephone cables the insulation is normally good enough so that G is low compared with C. On the other hand, the inductance in a cable is quite low, and the resistance cannot be made as low as desirable without unduly increasing the cost of the cable. Hence the cause of distortion in telephone cables is usually that the inductance-to-resistance ratio is much smaller than the capacitance-to-conductance ratio.

In the early days of the telephone, the problem of distortion seemed insurmountable. However, Prof. Michael Pupin implemented the use of *loading coils*[1] at intervals to increase the effective inductance per unit length of the transmission line. Pupin's loading coils came just before the development of the vacuum tube, which, as an amplifier, made possible true long-distance telephony. It would not have been possible at that time, however, without loading coils.

Loading coils are quite effective in reducing the distortion on voice-frequency circuits, as shown in Example 8-1. However, they also have the effect of turning the transmission line into a low-pass filter whose cutoff frequency is quite close to the upper end of the voice-frequency range. Since this is necessary, if the loading is to do a significant amount of good, loading is not possible with carrier telephone systems,

[1] Suggested earlier by Heaviside.

because the frequencies used are far above the cutoff frequency for the filter. With carrier systems the compensation for distortion must be carried out by the equalizers in the repeater stations and terminals.

The problems of loading are best illustrated by an example.

Example 8-1. Consider a voice-frequency circuit in a toll cable on a pair of 19-gauge wires. The parameters for this pair are as follows:

$R = 86.0$ ohms/mile
$L = 1$ mh/mile
$G = 1.4$ μmhos/mile
$C = 0.062$ μf/mile

These are the values at 1,000 cps and 20°C. Actually these values are functions of both frequency and temperature, but they are assumed constant here to simplify the examples. Calculating attenuation and velocity of propagation at the lower and upper ends of the voice-frequency bands and in the middle, we find the following results. At 200 cps, the attenuation is 0.0579 neper/mile, and the velocity of propagation is 2.16×10^4 m/sec. At the mid-frequency of 1,000 cps, the attenuation has increased to 0.126 neper/mile, and the velocity of propagation has more than doubled, to 4.6×10^4 miles/sec. At the upper end of the voice-frequency band, at 3,300 cps, the attenuation has increased again to 0.212 neper/mile, and the velocity of propagation has almost doubled again, to 7.72×10^4. If we consider 100 miles of this line, the attenuation varies from 5.79 nepers at the low end of the band to 21.2 nepers at the high end of the band. This is really a large range of attenuation. Likewise, the time of travel for 200 cps is $\frac{1}{216}$ sec, and at 3,300 cps it is $\frac{1}{772}$ sec. Obviously, either loading or extensive equalization is required.

Examining the appropriate ratios, we find

$$\frac{G}{C} = 22.6 \qquad \text{but} \qquad \frac{R}{L} = 86{,}000$$

It is therefore obvious that some sort of increase in L is called for. If we apply a loading coil every mile (or at some other interval, with an amount of inductance per mile averaged to the value which would be appropriate for 1-mile spacing) and call this inductance L_1, then we must have

$$\frac{R}{L + L_1} = 22.6$$

or

$$L + L_1 = \frac{86}{22.6} = 3.8 \text{ henrys/mile}$$

The inductance of the cable itself is negligible compared with this total, and the inductance of the loading coil per mile will be 3.8 henrys.

Suppose we insert a coil of this magnitude every mile. The loaded transmission line then amounts to a T section whose series inductance is 3.8 henrys and whose shunt element and series resistance are those of the transmission line itself. Using (8-3) and (8-4), we find

$$\alpha = \sqrt{RG} = \sqrt{86 \times 1.4 \times 10^{-6}} = 11.0 \times 10^{-3} \text{ neper/mile}$$

which is significantly less than even the lowest value without loading. The phase shift is given by

$$\beta = \omega \sqrt{LC} = 2\pi f \sqrt{3.8 \times 6.2 \times 10^{-8}} = 3.05 \times 10^{-3} f$$

and

$$v_p = \frac{1}{4.86 \times 10^{-4}}$$
$$= 2.05 \times 10^3 \text{ miles/sec}$$

This is not practical, however, because of the filter properties of the T section. Recall that the characteristic impedance for a T section is given by

$$(7\text{-}14) \qquad Z_{0T} = Z_0 \sqrt{1 + \frac{(\gamma\, \Delta x)^2}{4}}$$

At 200 cps,

$$\gamma\, \Delta x = 0.011 + j0.61$$

so that

$$(\gamma\, \Delta x)^2 \approx -0.37$$

Thus, even at the lowest frequency, the impedance is changed significantly. The impedance actually goes to zero where

$$(\gamma\, \Delta x)^2 = -4$$

Neglecting the effect of α, this occurs where

$$3.05^2 \times 10^{-6} f^2 = 4$$
$$f = 655 \text{ cps}$$

This is the cutoff frequency of the equivalent low-pass filter. For higher frequencies the characteristic impedance is almost pure imaginary, and no power can be transmitted. Since this is hardly above the lower end of the voice-frequency band, it is obvious that such loading cannot work satisfactorily. The loading coils could be placed closer together, but it is apparent that they would have to be placed a great deal closer together if the cutoff frequency were to be extended to the upper end of the voice-frequency band. The result would be uneconomic.

Fortunately, almost the same result may be obtained with a great deal lower value of loading inductance, which does not result in such a low cutoff frequency. One of the actual loading-coil arrangements used on this line has an inductance of

$$L_1 = 0.150 \text{ henry/mile}$$

The resistance of these coils is 12 ohms/mile. Thus, the ratio of resistance to inductance is given by

$$\frac{R + R_1}{L + L_1} = \frac{98}{0.151} = 650$$

This is a long way from 22.6, but it is certainly a great deal better than that for the unloaded line. In fact, the results with this loading are almost as good as with the optimum loading.

If we perform the same calculations that were performed before, namely, those for the attenuation and the velocity of propagation at 200, 1,000, and 3,300 cps, we find that the attenuation varies only from 0.0313, at 200 cps, to 0.0315, at 3,300 cps, and that the velocity of propagation varies only from 1.01×10^4 miles/sec to 1.04×10^4 miles/sec in the range from 200 to 3,300 cps. Thus, these coils almost perfectly compensate for the distortion in the line, and at the same time reduce the attenuation to a value only half that without loading at 200 cps. The results are summarized in Table 8-1.

TABLE 8-1. LOADED-VS.-UNLOADED-LINE COMPARISONS

Frequency, cps	Attenuation		Velocity of propagation, miles/sec	
	Unloaded	Loaded†	Unloaded	Loaded†
200	0.0579	0.0313	2.16×10^4	1.01×10^4
1,000	0.126	0.0315	4.6×10^4	1.035×10^4
3,300	0.212	0.0315	7.72×10^4	1.04×10^4

† Loaded line uses 150 mh/mile.

If coils of this size are spaced at a distance of 0.5 mile, we find that $\beta\,\Delta x = 3{,}300(2\pi/10{,}400)\frac{1}{2} = 1$. Hence,

$$\frac{(\gamma\,\Delta x)^2}{4} = -0.25$$

and the variation in impedance over the passband will be relatively small. Thus, this type of loading may be readily achieved with coils of 0.075 henry spaced at intervals of $\frac{1}{2}$ mile.

For transatlantic telegraph cables, loading coils at close intervals are not very practical. In the 1920s, when permalloy became available, a solution was found for the loading problem for these cables. Modern submarine telegraph cables are "continuously loaded"; that is, their inductance per unit length is increased over the entire length rather than by adding lumped inductances. This increased inductance is obtained by wrapping the center conductor of the coaxial cable with permalloy tape having a very high permeability. By reducing delay distortion, this method increases the capacity of a telegraph cable from a few tens of words per minute to several hundred, and actually to more than a thousand.

At the present time telephone cables are also in use on a transoceanic basis. These cables have repeaters at frequent intervals, however, and equalization is carried on in the repeater. The transatlantic telephone cables are carrier systems, and considerable effort is currently being expended on increasing their capacity without increasing their number (because of the high cost per cable). Early in 1960 they had 36 high-grade circuits, plus some order wire circuits, with an increase to 48 planned for the immediate future.

8-6. Power Lines and Power Systems

Long-distance power transmission is carried out on three-phase high-voltage lines. The designers of such lines are much more concerned with efficiency than with basic traveling-wave considerations. When the three-phase lines are balanced and the line parameters are known, power-line problems may be treated (on a phase-to-neutral basis) just like all other traveling-wave problems. When the loads are not balanced, it is customary to resolve transmission into phasor components rotating in different directions, and with different magnitudes and angular velocities, by the method of symmetrical components.[1]

As indicated in Chap. 2, transmission lines for power are invariably quite short in terms of a wavelength. Therefore, any single power transmission line is readily simulated with a single section of artificial line. When steady state is considered, the exact model described in Chap. 7 is more appropriate than the approximate model.

[1] J. G. Tarboux, "Introduction to Electric Power Systems," chap. 10, International Textbook Company, Scranton, Pa., 1944.

Power-system problems are usually relatively complicated because of interconnections between numerous generators, loads, and lines. Practical solutions to these problems are nearly always obtained by computer methods. The a-c network analyzer consists of a number of lumped elements which may be connected in such a way as to simulate various lengths of transmission lines and various generators and loads. Problems are then worked out on the network analyzer by a trial-and-error method. Power-system problems are also worked out by digital computers, and, in fact, some power systems utilize digital computers for the automatic and most economic dispatching of power from a number of generating stations.

In order to utilize the network analyzer and in order to calculate power-system problems most easily, it is desirable to use an equivalent T section or an equivalent π section. Equations (7-9) and (7-13) give the expressions for the characteristic impedance of a T section and its propagation constant. These were solved simultaneously to obtain the values for Z_1 and Z_2:

$$Z_1 = 2Z_0 \frac{e^{\gamma \Delta x} - 1}{e^{\gamma \Delta x} + 1} \tag{7-17}$$

$$Z_2 = \frac{Z_0^2 - Z_1^2/4}{Z_1} \tag{7-18}$$

The exact π-section equivalent is different from the T section, and either can be used.

For a short power transmission line it is customary to neglect the shunt element and simply use the exact series resistance and inductance of the line as a part of the circuit. When the line becomes long enough so that this is not a good approximation, the expressions of (7-17) and (7-18) must be used.

It is not always possible to use these simple expressions for power-system computations. One of the big problems in power systems is the transmission of pulses induced by lightning and of transients associated with such faults as short or open circuits. Since the Fourier analysis of these transients contains many higher-frequency components than those at 60 cps, a single-frequency equivalent cannot be used. In such a case, either a multiple-section artificial line is used, or the transients are calculated on the basis of their performance on the actual line.

Many of the most significant problems of power systems are associated with protection of the lines against overvoltage and overcurrent. Truly tremendous circuit breakers have been developed to interrupt short-circuit faults of the order of millions of kilovolt-amperes. Lightning arresters are also important, of course. It is important to keep adequate control of the voltage on a long power line, for open-circuiting the load can result in an increased voltage at that point, because of the Ferranti effect, described in Chap. 5. These excessive voltages can break down the insulation on transformers.

Two examples are given here of the interconnections found in power systems. The first is the small power system operated by the Public Service Company of New Mexico in central and northern New Mexico. The second is a portion of the very large system operated by the Bonneville Power Administration in Washington, Oregon, Idaho, and Montana.

The power system of the Public Service Company of New Mexico is typical of systems that serve one major city and a number of smaller ones nearby. The basic transmission voltage of the system is 115 kv line to line on three phase. Transmission from the main substations to distribution substations located near the various loads is at 46 kv. These subsidiary stations drop the voltage to the distribution level of 12,600 volts, and pole transformers drop this to 440, 220, and 120 volts.

The 115-kv transmission system is shown in Fig. 8-10. There are two major generating stations and two minor ones in the system. The major stations are Reeves, north of Albuquerque, and Person, south of the city. The minor stations are Prager, near downtown Albuquerque, and a station in Santa Fe. The center of the system is the West Mesa switching substation, located on Albuquerque's west side. At each of the generating stations are 115- to 46-kv transformers. A major substation is at Sandia Base, on the east side of Albuquerque. From the substations the 46-kv lines radiate throughout the city. The city is completely looped by 115-kv lines. There are arrangements for breaking the loop and delivering power by any one of the possible paths should one line fail. This is a typical situation for supplying a city of a few hundred thousand.

Also radiating from the West Mesa substation are lines 75 miles to the west to the uranium milling and mining center of Grants and 60 miles to the northeast to Santa Fe. A line owned by the AEC runs from Santa Fe to Los Alamos (about 40 miles), and the Public Service Company operates a line from Santa Fe across to Las Vegas, a distance of about 50 miles. The Santa Fe and Los Alamos peak load is about 30,000 kva, and the Las Vegas load is about 10,000 kva. It should be noted that power-system loads are always given in kilovolt-amperes rather than in kilowatts, since the line capacity and transformer capacity are determined by the current which flows, even though a good part of it may be reactive. In general, power systems are operated with the current lagging the voltage at a power factor on the order of 0.9. The power factor is simply the cosine of the angle by which the current lags the voltage. If the power factor of the load is less than 0.9, it is usually corrected with capacitors (which draw current leading the voltage by 90°).

The Public Service Company of New Mexico, like many other power companies, has facilities for interconnecting its system with other systems. Part of the interconnection is shown in Fig. 8-10. The steam generating plant of the Plains Electric Transmission and Generation Co-operative, Incorporated, is located about 25 miles north of Albuquerque, and there is a tie between this and the West Mesa substation of the Public Service Company. The Plains Co-operative supplies power to a number of REA cooperative distribution systems in the state. One such system lies to the north of Los Alamos, Santa Fe, and Las Vegas, and a dashed line on

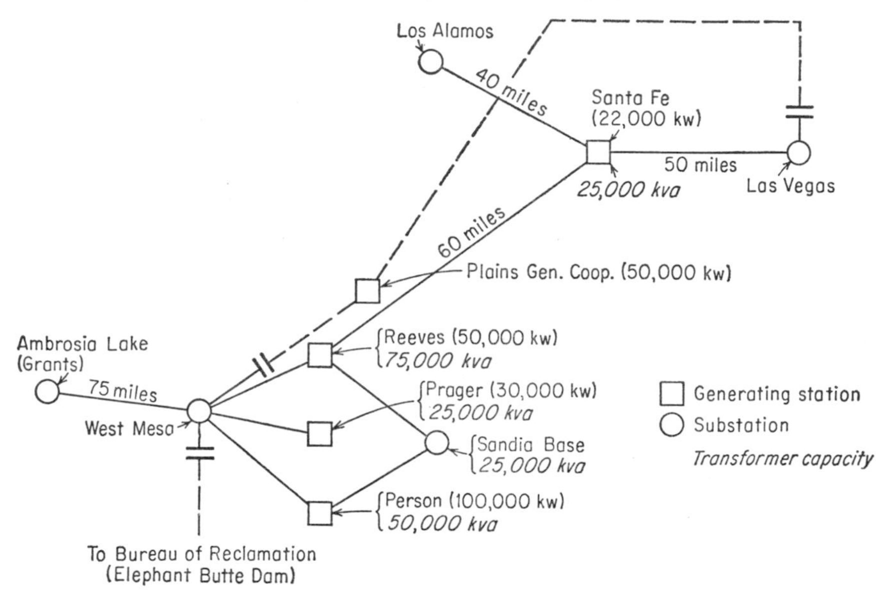

FIG. 8-10. Public Service Company of New Mexico power system.

Fig. 8-10 indicates the transmission from the Plains station through this cooperative system. This loop covers a number of small communities and ties in with the Public Service Company of New Mexico again at Las Vegas. Thus, although the load for Las Vegas alone is about 10,000 kva, the line serving Las Vegas must be able to handle 20,000 kva, because an extra 10,000 kva of load is occasionally fed northward into the cooperative. The cooperation works both ways, and at times the Plains generating plant supplies power to the Albuquerque system.

Another interconnection is with the Bureau of Reclamation system centered on a hydroelectric plant at Elephant Butte Dam, 150 miles to the south. This is actually a tie with a larger system, since the Bureau of Reclamation system ties in with other utilities to the south and east. As with the Plains system, power may be sent in either direction, depending on the demand.

Figure 8-11 shows a partial one-line diagram for a much larger system, that operated by the Bonneville Power Administration. Many of the substations for the Bonneville system have a larger transformer capacity than the entire generating capacity of the Public Service Company of New Mexico. The Bonneville Power Administration system is a public system containing elements operated by the Bureau of Reclamation and

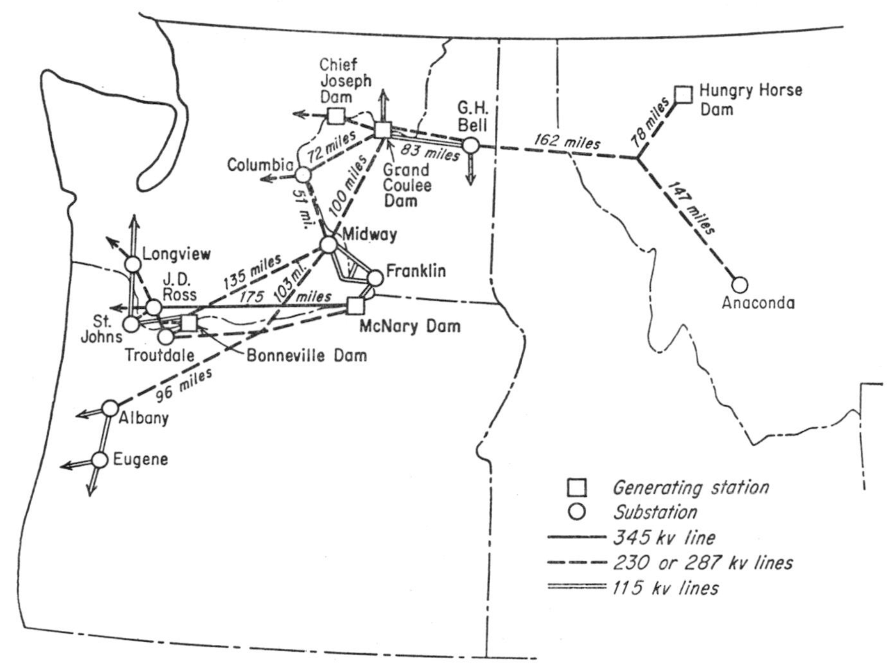

FIG. 8-11. Partial system diagram of Bonneville Power Administration, 1957. Generating capacity in kilowatts: Grand Coulee—1,944,000; McNary—962,000; Bonneville—600,000; Chief Joseph—400,000; Hungry Horse—300,000. Substation transformer capacity in kilovolt-amperes: J. D. Ross—585,000; G. H. Bell—400,000; Troutdale—300,000; St. Johns—262,500; Longview—175,000; Albany—162,000; Midway—165,000; Columbia—150,000; Anaconda—150,000; Eugene—64,500; Franklin—24,000.

others operated by the Corps of Engineers of the United States Army. It feeds power into numerous private and public distribution systems, some of which have their own generating plants also. Bonneville is a hydroelectric system based upon the Columbia River, as shown in Fig. 8-11.

In the diagram of Fig. 8-11, a great many of the large substations and all but one of the smaller substations of the Bonneville system (Franklin) have been omitted. If all the major substations had been shown, the diagram would have been too complex to follow. All the major generating stations on the line in 1957 are shown, although there are some smaller dams which are not shown, and the new station at The Dalles, though quite large, is not included because it was not in operation in 1957. The substations shown are a cluster in the vicinity of Portland, Ore. (J. D. Ross, St. Johns, and Troutdale); a few stations in west central Oregon (Albany and Eugene); a number of the stations between the McNary Dam generating plant and the Grand Coulee and Chief Joseph Dams generating plants; the major G. H. Bell station near Spokane, Wash.; and the Anaconda station at the extreme eastern side of the system in Montana. All the dams shown are on the Columbia River, except the Hungry Horse Dam in Montana.

Three basic transmission voltages are used in this system, as well as some lower voltages for short, low-capacity lines (not shown). The highest voltage in use is 345 kv; it is on a line to the J. D. Ross substation near Portland, Ore., from the McNary Dam generating plant. Most of the major transmission lines in the system are operated at 230 kv and are being changed over to 287 kv. A number, as indicated, are operated at 115 kv.

When large blocks of power are to be transmitted a long distance, it is necessary to go to higher and higher voltages for efficiency of transmission. A number of European systems are using voltages in excess of 500 kv, and tests are under way with such systems in this country.

It can be seen from the diagram of Fig. 8-11 that many of the lines in a system like Bonneville are significant portions of a quarter-wavelength, so that the Ferranti effect can be quite noticeable and losses can be quite large if the line is not operated efficiently.

Some of the substations in this system are principally for the purpose of switching and changing from 230 to 115 kv. Others reduce the voltage to a level suitable to send out to distribution substations. A station of the latter type is represented in Fig. 8-12. This is the Franklin substation in southeastern Washington. The diagram of the station is rather complicated even when the single-line form is shown. It should be realized that each of the single lines actually represents a three-phase transmission line. The heavy horizontal lines represent major bus bars, those at the top being at 115 kv and the one at the bottom being at 12.5 kv. The 115-kv bus bar is directly connected to the McNary Dam generating station. Through the two lines which follow different paths to the Midway substation, it is connected to the rest of the system. Examination of the many paths by which power may flow from McNary Dam (the closest) and the other dams in the system to the Franklin substation indicates the complexity of analysis of a power system.

The switching arrangement for the 115 kv is made completely flexible: any one of the lines may be taken out of the circuit; by utilizing the two bus bars, two pieces of the system may be kept separate; and any one of the circuit breakers may be removed from the circuit if it should fail. These flexible arrangements for switching are necessary to ensure reliability, which, of course, is extremely important in a power system. Note

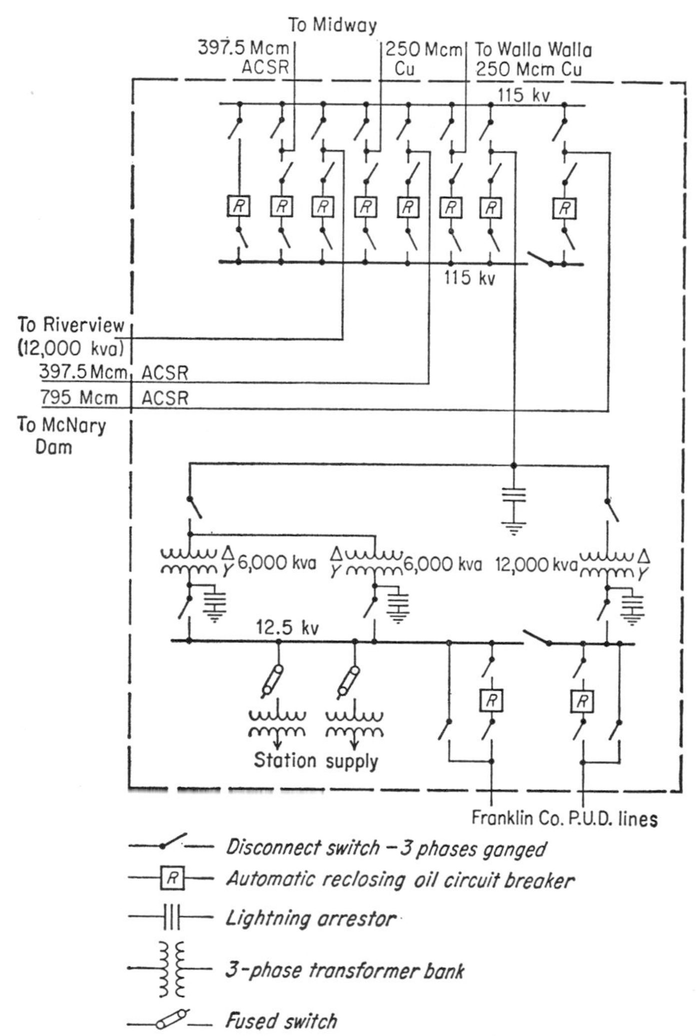

FIG. 8-12. Simplified single-line diagram of Franklin Substation of Bonneville Power Administration.

also the flexibility allowing separation of the various transformers shown in the bottom of the diagram; thus, only those transformers which need be used are on the line at a particular time, and failure of any of the transformers would not prohibit use of the others.

The circuit breakers shown are all of the oil-bath automatic-reclosing type. If a fault occurs on the system and a transient heavy current arrives at the substation, the circuit breaker on that particular circuit will open. After a short interval, it will reclose if the fault has cleared. If the fault has not cleared, it will reclose only momentarily, open again, and continue trying for a specified number of trials.

Note that lightning arresters are used on both sides of the transformer bank to protect it thoroughly. The numbers shown on the 115-kv lines refer to the cross section of the conductors, in thousands of circular mils, and to the type of wire used—copper (Cu) and aluminum-cable steel-reinforced (ACSR).

An over-all system plan does not exist for power as it does for telephone because of the large number of separate power systems in the country which need not be connected together. There are a large number of small telephone systems in this country also, but they must all fit the same basic plan, so that long-distance service may be provided from one system to the next. With power systems this is not necessary. In general, however, power systems in a given vicinity may be interconnected, as indicated in the discussion of the Public Service Company of New Mexico. The Bonneville system interconnects with many other systems—both public and private.

As an illustration of the type of power-line calculations which can be carried out for a single, relatively simple line, consider the Albuquerque–Santa Fe line of the Public Service Company of New Mexico.

Example 8-2. A 115-kv three-phase line extends from the Reeves station in north Albuquerque to Santa Fe. The distance is slightly less than 60 miles, but we shall assume here that it is exactly 60 miles. The line is carried on wooden towers and has three conductors, each of which is an ACSR cable of 477,000 circular mils. The line is spaced 12 ft between conductors, with all conductors in the same horizontal plane.

It can be shown that the parameters of this line are:

$R = 0.216$ ohm/mile

$X_L = 0.755$ ohm/mile

$B_C = 5.6 \times 10^{-6}$ mho/mile

The propagation constant per mile of line is therefore given by

$$\begin{aligned}\gamma &= \sqrt{(0.216 + j0.755)(j5.6 \times 10^{-6})} \\ &= 10^{-3} \sqrt{0.782\underline{/74.0^\circ} \times 5.6\underline{/90^\circ}} \\ &= 2.10 \times 10^{-3}\underline{/82.0^\circ} \\ &= 0.291 \times 10^{-3} + j2.08 \times 10^{-3}\end{aligned}$$

For the 60 miles then,

$$\alpha x = 1.75 \times 10^{-2} \text{ neper}$$
$$\beta x = 0.125 \text{ radian} = 7.12^\circ$$

It is interesting to note that the attenuation is only 0.152 db, so that this would be a very low loss communication line indeed. Nevertheless, as shown later, the amount of power lost in this line is 770 kw for each of the three phases.

Let us assume that this line is operating at its maximum capacity of 50,000 kva at 0.9 power factor and that the voltage at Santa Fe is 115 kv. We shall compute the voltage at Albuquerque and the power lost in the line as well as the kilovolt-amperes delivered to the line at Albuquerque.

The voltage from one-phase wire to neutral is given by

$$\frac{115}{\sqrt{3}} \text{ kv}$$

Hence, at Santa Fe the current is given by the number of volt-amperes divided by this voltage, or

$$I = \left(\frac{5 \times 10^7}{3}\right)\left(\frac{\sqrt{3}}{1.15 \times 10^5}\right) = 251\underline{/-26^\circ}$$

Although we could calculate the performance of this line on our normal basis, using reflection coefficients, it is more customary to compute power-line performance on the basis of equivalent circuits. The exact T-section equivalent may be obtained from Eqs. (7-17) and (7-18). Note, first, that

$$e^{\gamma \Delta x} - 1 = 0.0097 + j0.127$$

and that the characteristic impedance is

$$Z_0 = 10^3 \sqrt{\frac{0.755 - j0.216}{5.6}} = 375\underline{/-8.0^\circ} \text{ ohms}$$

Hence, from Eq. (7-17), we have

$$\begin{aligned} Z_1 &= 2 \times 375\underline{/-80^\circ} \times \frac{0.0097 + j0.127}{2.01 + j0.127} \\ &= 375\underline{/-8.0} \times \frac{0.127\underline{/85.6^\circ}}{1.01\underline{/3.6^\circ}} \\ &= 47.1\underline{/74.0^\circ} \\ &= 13.0 + j45.5 \text{ ohms} \end{aligned}$$

Z_2 may be obtained from (7-18) and Z_1. In this case, we may neglect the second term in Eq. (7-18), so that

$$Z_2 \approx \frac{Z_0^2}{Z_1} = \frac{3.75^2 \times 10^4 \underline{/-16.0^\circ}}{47.1\underline{/74.0^\circ}}$$
$$= -j2.99 \times 10^3 \text{ ohms}$$

The resulting T-section equivalent is given in Fig. 8-13. To calculate the performance of the line, we shall work back, in Fig. 8-13, from

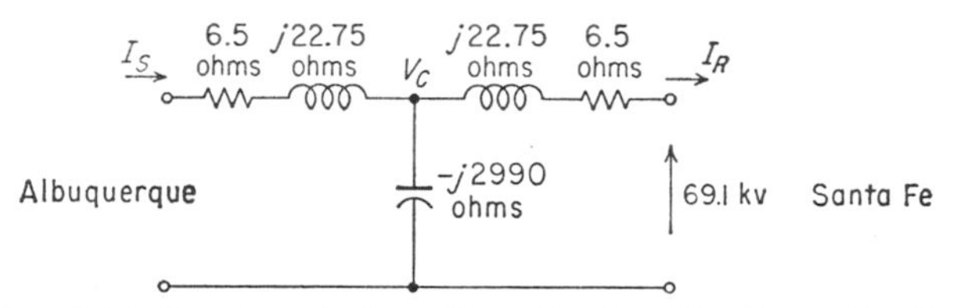

FIG. 8-13. Exact equivalent T section for 60-mile 115-kv line.

the Santa Fe end. The voltage at the center is given by

$$\begin{aligned} V_c &= 66.3 + 23.55\underline{/74.0^\circ} \times 0.251\underline{/-26^\circ} \\ &= 66.3 + 5.91\underline{/48.0^\circ} = 66.3 + 3.95 + j4.33 \\ &= 70.25 + j4.33 \text{ kv} \end{aligned}$$

The current into the capacitance is given by the voltage across the capacitance divided by its reactance, or

$$I_c = \frac{70.25 + j4.33}{-j2{,}990} = -1.46 \times 10^{-3} + j0.0235 \text{ ka}$$

Hence, the sending-end current is the sum of I_c and I_R, or

$$\begin{aligned} I_s &= 0.225 - j0.110 - 0.001 + j0.0235 \\ &= 0.224 - j0.0865 \\ &= 0.240\underline{/-21.2^\circ} \text{ ka} \end{aligned}$$

The sending-end voltage is given by the sum of the voltage across the capacitance and the drop in the initial series leg, so that it is

$$\begin{aligned} V_s &= 70.25 + j4.33 + 23.55\underline{/74.0^\circ} \times 0.240\underline{/-21.2^\circ} \\ &= 73.66 + j8.83 = 74.1\underline{/6.8^\circ} \text{ kv} \end{aligned}$$

It can therefore be seen that the total drop in the line is about 8 kv; that is, the Albuquerque end must supply 8 kv more than the desired voltage at Santa Fe under this particular loading condition. The capacitance of the line actually corrects somewhat for the lagging power factor at the load, so that the angle by which the sending-end current lags the sending-end voltage is only 28.0°, as compared with 32.4° lag neglecting capacitance.

The power loss may be determined from the I^2R loss in the two resistances. Thus the total power loss is given by

$$240^2 \times 6.5 + 251^2 \times 6.5 \text{ watts/phase} = 787 \text{ kw/phase}$$

The efficiency of the line is therefore given by

$$\text{Efficiency} = \frac{5 \times 10^4 \times 0.9 - 3 \times 7.87 \times 10^2}{5.0 \times 10^4 \times 0.9}$$

$$= \frac{42.65}{45} = 94.9\%$$

Thus, this is a quite efficient transmission line. Such efficiency is common for power transmission lines at 60 cps, yet even power transmission at radio frequencies is frequently on much less efficient lines. Telephone and other *signal* lines usually have extremely low efficiency.

8-7. Summary

Power and information are normally transmitted over long distances by electric transmission lines (and radio, in the case of information). The amount of bandwidth required for communication purposes varies from about 100 cps, for printing telegraph, to 4.5 Mc, for television. Once the method of coding and modulation has been established (which may increase the required bandwidth), the communication engineer has the job of providing a reasonably distortion- and noise-free channel of this bandwidth.

The telephone system in the United States is the world's largest computer, in the sense that it utilizes automatic switching over a major portion of its circuits. The switching plan involves *high-usage groups* of channels that directly connect points between which there is a reasonable amount of direct traffic and *final groups* of circuits for alternate routing. The routing is selected by a computer to be the most economical available at a particular time.

Nearly all long-distance telephone circuits utilize *carrier*, in which the information is modulated onto a carrier frequency above the voice-frequency band. Early carrier systems also utilized voice frequency. The modern L3 coaxial carrier system is capable of carrying 1,860 one-way telephone messages on a single coaxial line. Because of the extreme attenuations encountered on this line and the need to equalize attenuation at the various frequencies, it is necessary to have repeater stations every 4 miles and a very accurate automatic-gain-control system.

On telephone lines, attenuation and velocity of propagation are not independent of frequency. On carrier systems their effects must be compensated for by *equalizers* at the repeater stations and terminals. On voice-frequency lines *loading coils* are inserted at intervals to make the line more nearly distortionless (to make attenuation independent of frequency and phase shift linearly proportional to frequency). Such a system cannot be used for carrier because the loading coils act to convert the line into a low-pass filter whose upper cutoff frequency is well below the maximum frequency required by the carrier system.

Power transmission lines are always short enough so that their steady-state behavior may be represented by a single exact equivalent T or π section. The complicated interconnections ordinarily found in power systems make it almost a necessity to utilize a network analyzer as an analog computer to solve the detailed stability problems of the system. Many of the problems of power systems have to do with protection against overvoltage or overcurrent and against the effects of transients.

Power is transmitted over long distances at extremely high voltages, with 115 and 230 kv being common and with higher voltages in use in some places. Power transmission is always by three-phase lines.

Most power systems interconnect several generating stations and several substations, and various types of connections are possible in case of failure of some component of the system.

A typical power line has an efficiency greater than 94 per cent. The attenuation of a power line is therefore only a small fraction of a decibel. This is one of the big distinctions between power and communication lines, since attenuations of the order of 30 to 40 db are not uncommon between repeater stations on a telephone system, and system attenuations in a radio system may run as high as 200 db.

PROBLEMS

8-1. Use the library to determine the characteristics of one of the carrier telephone or telegraph systems not discussed in Chap. 8. Write a brief report describing the system. Include discussion of repeater spacing, frequency allocation, repeater and terminal equipment, equalization, etc.

8-2. Use the library to determine the characteristics of a telephone microwave-relay system. Write a brief report describing the system. Include discussion of repeater spacing, frequency allocation (both subcarrier and main carrier), repeater and terminal equipment, cross talk, etc. Compare this system with the wire systems described here.

8-3. A 19-gauge telephone-cable pair has the following parameters:

R = 84.0 ohms/mile
L = 0.001 henry/mile
C = 0.061 μf/mile
G = 1.0 μmho/mile

Determine the amount of loading required to make this line distortionless. If the loading coils were spaced at 1-mile intervals, what would be the cutoff frequency of the loaded line? Find the attenuation and the velocity of propagation.

8-4. In practice, the line of Prob. 8-3 is loaded with 31-mh coils at intervals of 1.135 miles. Compare the attenuation and velocity of propagation of the line so loaded with the unloaded line at 200, 1,000, and 3,000 cps. The loading-coil resistance is 3.2 ohms/mile.

8-5. Another loading arrangement for the line of Prob. 8-3 uses 44-mh coils at intervals of 1.135 miles. Compare the attenuation and velocity of propagation with those for the unloaded line at 200, 1,000, and 3,000 cps. The loading-coil resistance is 3.4 ohms/mile.

8-6. A 16-gauge telephone-cable pair has the following parameters:

$R = 48.5$ ohms/mile
$L = 0.001$ henry/mile
$C = 0.061\ \mu$f/mile
$G = 1.5\ \mu$mhos/mile

Perform on this line the calculations indicated for Prob. 8-3.

8-7. Perform the calculations of Prob. 8-4 on the line of Prob. 8-6.

8-8. Perform the calculations of Prob. 8-5 on the line of Prob. 8-6.

8-9. Using the library, write a brief report on the basis for telegraph bandwidths.

8-10. Using the library, write a brief report on the basis for telephone bandwidths.

8-11. Using the library, write a brief report on the reason for television bandwidths.

8-12. Determine the location of control switching points for your area, and prepare a sketch showing alternate routings for a call between two cities in different regions.

8-13. Determine the parameters of the power system in your area, and prepare a sketch like that of Fig. 8-10 or Fig. 8-11.

8-14. Examine a nearby small substation and sketch a one-line diagram for it.

8-15. A power line with the same per-mile parameters as that of Example 8-2 is 100 miles long. It is operated with a load of 40,000 kva with 0.85 lagging power factor. Determine the regulation (no-load–to–full-load voltage change) if the receiving-end voltage is 120 kv line to line.

8-16. Determine the efficiency of the line of Prob. 8-15.

8-17. A heavy power line is operated at 287 kv, line to line. It is made of 1,000,-000-circular mil aluminum-cable steel-reinforced, with an effective spacing of 32 ft. Use a handbook to determine its resistance, series reactance, and shunt susceptance per mile.

8-18. A certain power line has the following parameters:

$R = 0.100$ ohm/km
$X_L = 0.500$ ohm/km
$B_C = 10^{-5}$ mho/km

Determine the exact equivalent T section for this line if it is 200 km long.

8-19. For the line of Prob. 8-18, determine efficiency with a load of 200,000 kva at 0.9 lagging power factor. Determine regulation of the line. Line-to-neutral voltage is 230 kv at the load.

9. Impedance Charts and Matching, Resonators, and Waveguides

Many specialized traveling-wave techniques have been developed for radio-frequency transmission lines and waveguides and for acoustic resonators. With these lines and resonators, relatively short distances in meters are long in wavelengths; hence, the use of traveling-wave techniques is particularly important. The techniques used are applicable to any wave for which losses are small.

At high radio frequencies the losses due to standing waves become extremely important. It is necessary to apply matching networks to transmission lines so that long lines will be terminated in their characteristic impedance, or as close to it as possible. Frequently the matching networks themselves utilize short- or open-circuited lengths of transmission lines as reactances.

To assist in the calculations of the type carried out in Chaps. 5 and 6, special charts have been developed which greatly speed working with impedance, standing waves, and matching. These are discussed in this chapter.

Transmission lines and their acoustic equivalents are frequently used as resonators, serving the same purpose as lumped-constant resonant circuits do at lower frequencies. Frequently these resonators take the form of cavities, which are simply resonant lengths of waveguides. The techniques used in treating such resonators are also applicable in wave mechanics.[1]

Waveguides themselves are discussed here. These are hollow pipes which guide electromagnetic transmission through space in the same manner that wire transmission lines guide waves. The difference is that,

[1] For a discussion of this analogy, see William Shockley, "Electrons and Holes in Semiconductors," chap. 14, D. Van Nostrand Company, Inc., Princeton, N.J., 1950.

with a waveguide, it is difficult to distinguish the separate conductors. Waveguides are extensively used, particularly at frequencies above 1,000 Mc.

9-1. Impedance Charts

Several types of special charts have been developed to assist in impedance calculations. The most popular one is the Smith chart.[1] In the Smith chart impedances and admittances on the transmission line are represented by points inside a unit circle. By simple rotation on the chart, the effect of the position on the line may be determined.

To see how a Smith chart is developed, consider Eq. (5-29):

$$Z = Z_0 \frac{1 + \Gamma_R e^{-j2\beta d}}{1 - \Gamma_R e^{-j2\beta d}} \tag{5-29}$$

This, of course, is the lossless form of the equation. When loss must be considered, this becomes

$$Z = Z_0 \frac{1 + \Gamma_R e^{-2\alpha d} e^{-j2\beta d}}{1 - \Gamma_R e^{-2\alpha d} e^{-j2\beta d}} \tag{6-8}$$

Here we shall assume a lossless line, but the effect of loss may be easily taken care of by modifying the magnitude of the reflection coefficient used in the lossless case. Thus,

$$\Gamma_{R,\mathrm{eff}} = \Gamma_R e^{-2\alpha d}$$

Working with these equations involves considerable numerical calculations of complex quantities. It was to simplify these calculations that the Smith chart and other impedance charts were developed.

It is necessary to separate the reflection coefficient into its magnitude and associated angle in the complex plane. Thus,

$$\Gamma_R = K e^{jk} \tag{9-1}$$

Writing (5-29) in this form, we obtain

$$Z = Z_0 \frac{1 + K e^{j(k-2\beta d)}}{1 - K e^{j(k-2\beta d)}} \tag{9-2}$$

Defining

$$\theta = k - 2\beta d \tag{9-3}$$

the expression for the impedance becomes

$$Z = Z_0 \frac{1 + K e^{j\theta}}{1 - K e^{j\theta}} \tag{9-4}$$

[1] P. H. Smith, Transmission Line Calculator, *Electronics*, p. 29, January, 1939; and An Improved Transmission Line Calculator, *Electronics*, p. 130, January, 1944.

The Smith chart is constructed by plotting the ratio of the actual impedance to the characteristic impedance on a diagram whose polar coordinates are K for the radial direction and θ for the angle. K varies between 0 and 1; thus, the entire diagram is contained within a circle of unit radius. This situation is shown in Fig. 9-1. It should be noted that there is one and only one position on this chart for each

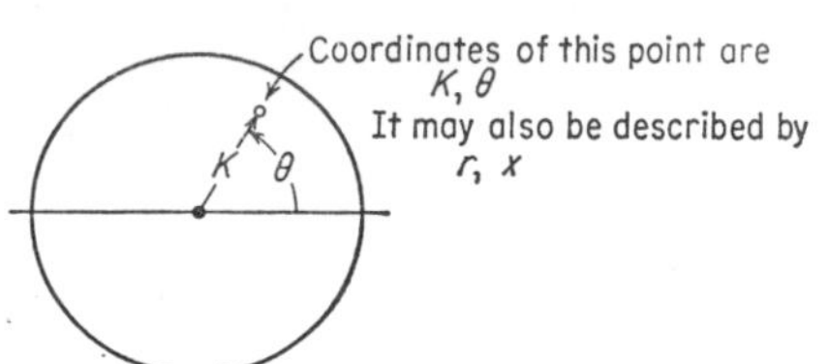

FIG. 9-1. Polar coordinate basis for Smith chart.

$$z = \frac{Z}{Z_0}$$

For convenience in interpretation, it is sometimes desirable to utilize rectangular rather than polar coordinates. Let us therefore define

$$w = u + jv = Ke^{j\theta} \tag{9-5}$$

In these terms, (9-4) becomes

$$z = \frac{1 + w}{1 - w}$$

Hence,

$$w = \frac{z - 1}{z + 1} = \frac{r + jx - 1}{r + jx + 1} \tag{9-6}$$

where $r = \dfrac{R}{Z_0}$, the resistive component of the impedance at any point on the line

$x = \dfrac{X}{Z_0}$, the reactive component of the impedance at any point on the line

These definitions apply when Z_0 is a resistance. Even if it is not, r is still the real part of z, and x the imaginary part, but the values R/Z_0 and X/Z_0 do not apply.

To see what the Smith chart looks like, we consider separately the cases of zero reactance (pure resistance), zero resistance (pure reactance), constant resistance with variable reactance, and constant reactance with variable resistance.

Consider a case where the impedance on the line is a pure resistance. Then $x = 0$, and

$$u + jv = \frac{r - 1}{r + 1} = u = K$$

Hence, the contour of zero reactance is a horizontal straight line, with a scale linear in the reflection coefficient. Of course, this was required by the original setup of the polar coordinates. It should be noted, however, that the scale is not linear in terms of r.

Now consider what happens at the point corresponding to infinite reactance. It can be seen that the numerator and denominator of the expression for w [Eq. (9-6)] approach the same value, so that w approaches 1. That is, infinite reactance corresponds to the point at $1 + j0$ at the extreme right-hand side of the circle of Fig. 9-1.

Now consider a pure reactance, that is, $r = 0$. Here

$$
\begin{aligned}
w = u + jv = \frac{jx - 1}{jx + 1} &= 1 \frac{\underline{/\arctan (x/-1)}}{\underline{/\arctan x}} \\
&= 1 \frac{\underline{/180^\circ - \arctan x}}{\underline{/\arctan x}} \\
&= 1 e^{j\pi} e^{-j2 \arctan x}
\end{aligned}
$$

Thus the impedance points corresponding to pure reactance lie on the unit circle which bounds the diagram, with their angle being determined by the size of the reactance. The angular scale is not a linear one. Note, however, that, when $x = 0$, $w = -1$; that is, the point of zero reactance and zero resistance is at the left-hand side of the circle. Note also that reactance equal to the characteristic impedance ($x = 1$) is given by $w = e^{j\pi}e^{-j\pi/2} = j$ at the top of the circle and that a pure negative reactance of the same size is located on the diagram at $-j$, or the bottom of the circle.

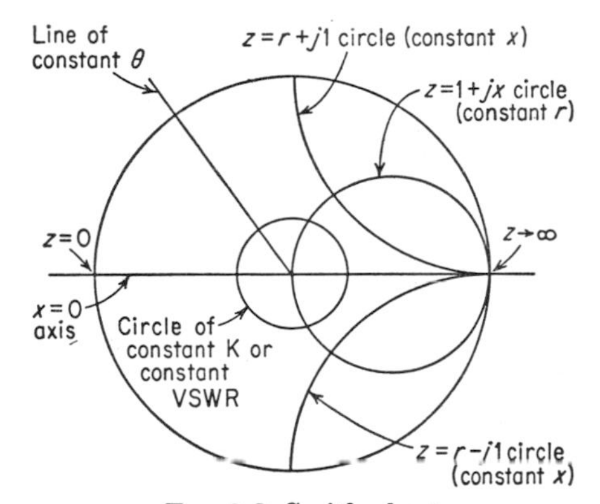

FIG. 9-2. Smith chart.

It can be shown that a contour of constant resistance and variable reactance is a circle, as in Fig. 9-2. The horizontal intercepts for this circle occur where $x = 0$ at $u = (r - 1)/(r + 1)$ and for infinite x at $u = 1$.

Hence the circle has a radius

$$
\frac{1}{2}\left(1 - \frac{r - 1}{r + 1}\right) = \frac{1}{r + 1}
$$

and the circle passes through the point $1 + j0$. It has its center on the horizontal axis at

$$
1 - \frac{1}{r + 1} = \frac{r}{r + 1}
$$

It is also possible to show that the contours of constant reactance are circles normal to the contours of constant resistance. All the circles of constant reactance pass through the point corresponding to infinite reactance, that is, the right-hand edge of the diagram.

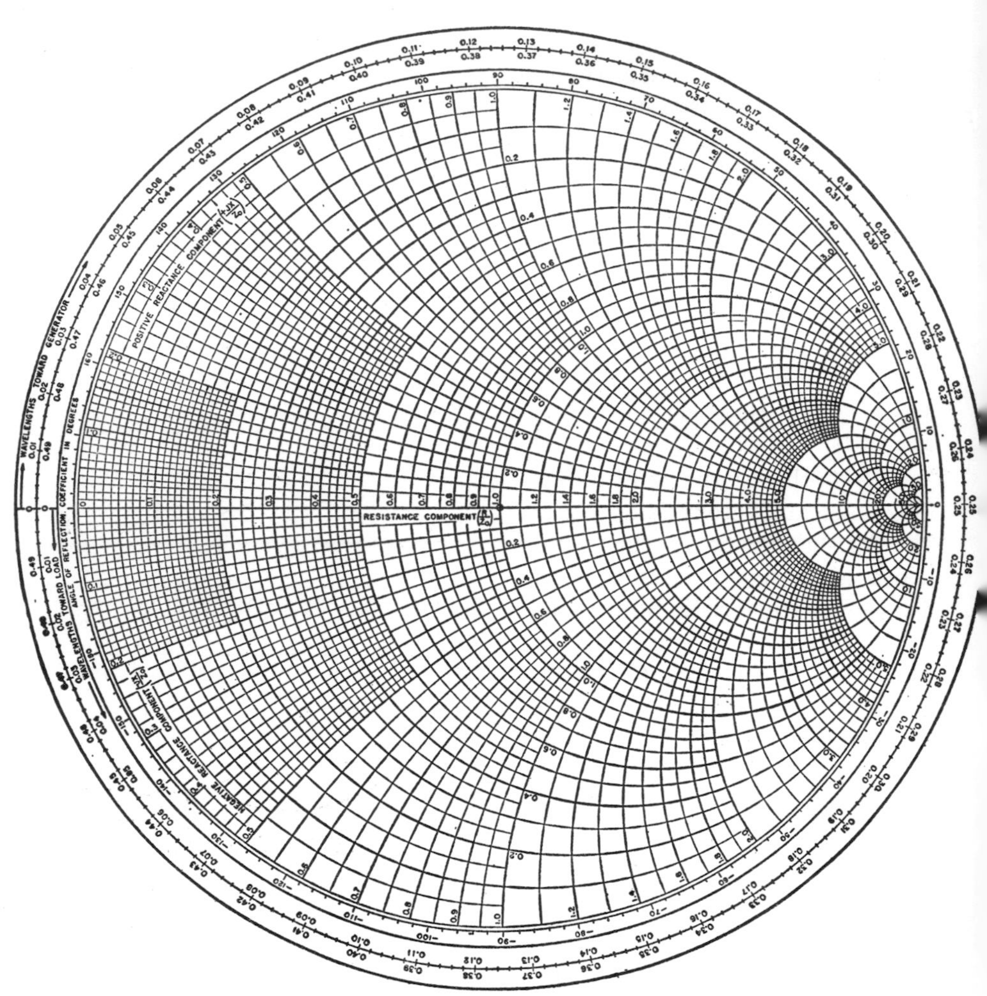

FIG. 9-3. Smith chart. (*From P. H. Smith, An Improved Transmission Line Calculator, Electronics, p.* 131, *January,* 1944; *by permission.*)

Figure 9-3 shows a complete Smith chart ready to use. Note that the minimum value of impedance and also the maximum value of impedance for a particular standing-wave ratio occur when the impedance is a pure resistance. The minimum value occurs where the circle of constant standing-wave ratio (and hence constant reflection coefficient) intercepts the horizontal axis to the left of the center, and the point of maximum impedance occurs where it intercepts to the right of the center. Naturally, the minimum voltage also occurs at the point where the impedance is minimum and the maximum voltage at the distance where the impedance is maximum.

The standard Smith chart shown can be purchased in quantity and therefore is useful for doing practical problems. It can also be purchased in the form of a slide rule.[1] Note that the scales for the angle θ are given in terms of distance rather than angle and that the angular scale around the edge of the chart corresponds to the angle of the reflection coefficient. The diagram was, of course, set up with the reflection coefficient as the radial coordinate, so that circles concentric with the center of the unit circle are circles of constant reflection coefficient. Since the standing-wave ratio is determined only by the magnitude of the reflection coefficient, these circles are also contours of constant standing-wave ratio. Note that the distances are given in wavelengths toward the generator and wavelengths toward the load, so that it is easy to determine which direction to advance as the position on the line is changed.

The use of the Smith chart is illustrated in the following examples.

Example 9-1. *Determining Input Impedance of a Line Terminated with Resistive Load.* Suppose that the characteristic impedance of a given line is

$$Z_0 = 50 \text{ ohms}$$

and that the load impedance is

$$Z_R = 90 + j0$$

Hence,

$$z_R = 1.8$$

Suppose, also, that the wavelength is 1 m and that the length of the line is 1.2 m. We wish to find the sending-end impedance.

We enter the chart at 1.8 on the horizontal axis. This is also the magnitude of standing-wave ratio (which can be readily shown to be equal to the value of resistance at the intersection of the constant-standing-wave-ratio circle with the resistance coordinate on the horizontal axis). Since one complete circuit of the diagram corresponds to a half-wavelength [because of the $2\beta L$ rather than βL appearing in Eqs. (5-29) and (9-2) to (9-4)], we must go completely around the diagram twice for 1 wavelength. Then we must go 0.2 wavelength farther. Since we start with a load impedance, we go

[1] Charts may be purchased from The Emeloid Co., Inc., Hillside 5, N.J., and the General Radio Co., West Concord, Mass. The slide rule is available only from The Emeloid Co.

in the direction toward the generator. The situation is shown in Fig. 9-4. Note that on the distance scale the initial reading is 0.25. It is on a line from the center through the receiving-end impedance to the right-hand side of the diagram. After going 0.2 wavelength farther, we are at 0.45. Since for a lossless line the standing-wave ratio remains constant as one travels down the line, we need merely move around the circle of constant standing-wave ratio from 0.25 to 0.45 and obtain the impedance from the intersection of this circle with the line from the center out to the 0.45 marker. This intersection is noted on Fig. 9-4. The corresponding values for resistance and reactance are read to be

$$z_s = 0.60 - j0.22$$

so

$$Z_s = 30 - j11$$

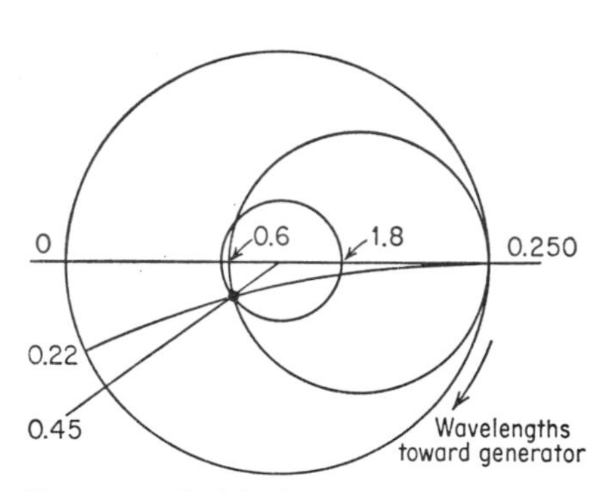

FIG. 9-4. Smith-chart example for Z_R resistive.

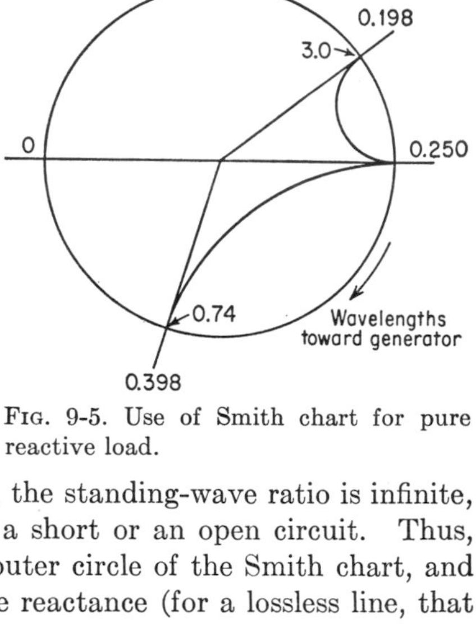

FIG. 9-5. Use of Smith chart for pure reactive load.

When the load is a pure reactance, the standing-wave ratio is infinite, just as it would be if the load were a short or an open circuit. Thus, computations are performed on the outer circle of the Smith chart, and the input impedance is always a pure reactance (for a lossless line, that is). This is shown in Example 9-2.

Example 9-2. *Determining Z_s for Pure Reactive Load.* Given

$$Z_R = j3.0Z_0$$
$$\text{Length} = 0.700\lambda$$

Find Z_s.

The impedance $j3.0$ is on the outer circle of the Smith chart, as shown in Fig. 9-5. The distance scale for this reactance reads 0.198 wavelength toward the generator. The angle of the reflection coefficient may also be read and is found to be 37°.

To find the input impedance, it is merely necessary to go a distance of 0.700 wavelength toward the generator from the 0.198. A distance 0.500 results in returning to the 0.198 point. An additional distance 0.200 must then be gone to 0.398. At 0.398, the impedance we read for the sending end is

$$Z_s = -j0.74Z_0$$

Example 9-3. *Determining Z_s for Complex Load Impedance.* Given

$$z_R = 1.8 - j0.5$$
$$\text{Length} = 2.11\lambda$$

Find z_s.

The point on the Smith chart corresponding to $1.8 - j0.5$ is located. A line from the center through this point intersects the

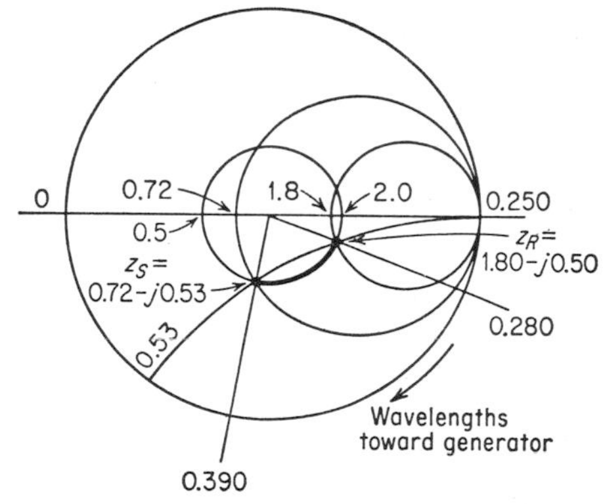

Fig. 9-6. Use of Smith chart for complex load impedance.

wavelength-toward-the-generator scale at 0.280. Rotating along a circle centered at $1 + j0$ (the center of the Smith chart) back to the horizontal axis on the right-hand side, we find the intersection of this circle at 2.0. Since this is a circle of constant standing-wave ratio or constant magnitude of a reflection coefficient, we know that the standing-wave ratio is 2.0, just as it would be if the load were a resistance twice the characteristic impedance.

To find the sending-end impedance, we go along this circle of constant standing-wave ratio a distance corresponding to the 2.11 wavelengths toward the generator. The 2.00 wavelengths amount to four complete rotations on the Smith chart; so at 2.00 wavelengths from the load the impedance is the same as at the load, and such a point corresponds to the 0.280 on the wavelength-toward-the-generator scale. It is therefore necessary to go an additional 0.11 wavelength to 0.390. At this point the sending-end impedance is read at the intersection of the radius passing through 0.390 and the 2.0-standing-wave-ratio circle and is found to be

$$z_s = 0.72 - j0.53$$

This is shown in Fig. 9-6.

Many microwave impedance measurements and some acoustic impedance measurements are made by the technique described in Chap. 5, in which the standing-wave ratio is measured along with the location of the first minimum with respect to the end of the line or the reflecting surface. Previous calculations of this sort have been made using the phasor diagrams. Such calculations are quicker, however, with the Smith chart.

The Smith-chart technique in such a case is as follows:

1. Measure the standing-wave ratio.
2. Locate the first minimum with respect to the reflection point (another minimum will do if it is known which minimum it is).
3. Enter the correct standing-wave-ratio circle at its intersection with the negative horizontal axis, corresponding to the point of minimum voltage or pressure.
4. Advance *toward the load* a distance equal to the distance from the first minimum to the load.
5. Read the value of the impedance at that point.

This is exactly the same technique as was used with the phasor diagrams. There the standing-wave ratio was used to determine the relative magnitudes of incident and reflected phasors, and they were rotated from their minimum position (where the reflected wave subtracted arithmetically from the incident wave) an amount corresponding to the distance from the minimum to the load. The resulting voltage and current phasors were then computed at the load, and the impedance was obtained by their phasor ratio.

As an example of the use of the Smith chart in such problems, consider the following:

Example 9-4. *Determining Load Impedance from Standing-wave Measurements.* Given the following measured values:

VSWR = 1.50
First minimum at 0.370λ from load

Find z_R.

The 1.50-standing-wave-ratio circle intersects the horizontal axis at 1.50 and 0.667. The point corresponding to minimum voltage is 0.667. Since this is zero on the wavelength-toward-the-load scale, it is merely necessary to find the intersection of the constant-standing-wave-ratio circle with a line from the center to the 0.370 wavelength-toward-the-load point, as shown in Fig. 9-7. The result is seen to be

$$z_R = 0.95 + j0.4$$

Example 9-5. *Determining Acoustic Terminating Impedance from Standing-wave Measurements.* Given the following measured values:

Pressure standing-wave ratio = 9.0
First minimum 0.32λ from reflection point

The situation is shown in Fig. 9-8. The first minimum is at 0.11 on the real axis where the 9.0-standing-wave-ratio circle intersects the negative horizontal axis. It is merely necessary to go a distance

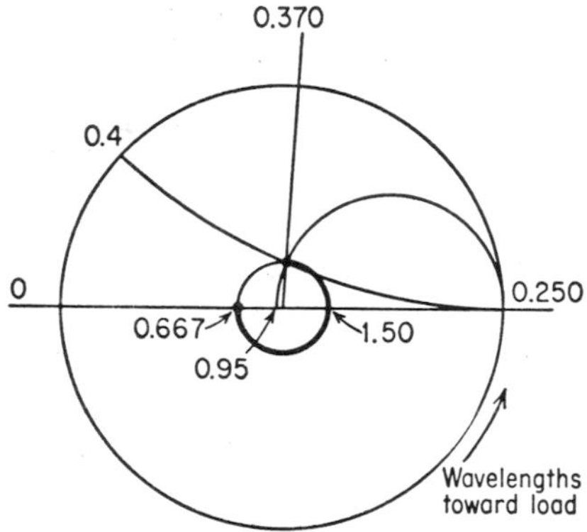

FIG. 9-7. Use of Smith chart in impedance measurement.

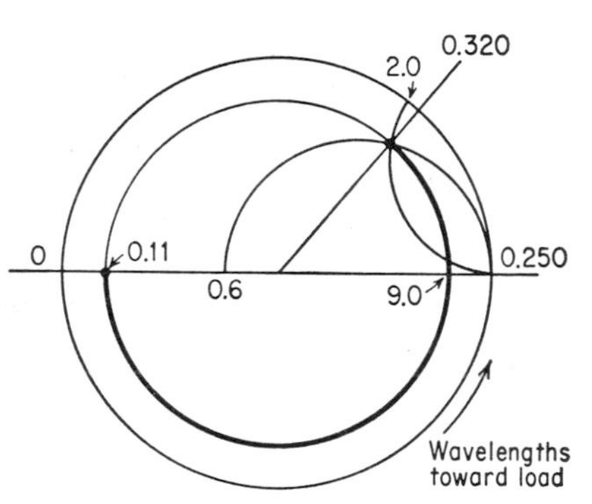

FIG. 9-8. Use of Smith chart in acoustic impedance measurement.

0.32 wavelength toward the load from there along the constant-standing-wave-ratio circle and read off the acoustic impedance of the reflecting surface. We find this to be

$$Z_R = (0.60 + j2.0)Z_0$$

The Smith chart is equally applicable for admittance and impedance. In fact, in most transmission-line matching applications it is used as an admittance chart rather than as an impedance chart, because the matching is performed by the insertion of parallel elements and is therefore the result of the addition of admittances.

To see how the chart may be used with admittances, consider the expression for the admittance at any point on the line; this is simply the reciprocal of the expression for impedance. Thus,

$$Y = Y_0 \frac{1 - \Gamma_R e^{-2\gamma d}}{1 + \Gamma_R e^{-2\gamma d}} \tag{9-7}$$

Here, of course, $Y_0 = 1/Z_0$. We may write

$$y = \frac{Y}{Y_0} = \frac{1 - Ke^{j\theta}}{1 + Ke^{j\theta}}$$

Since the Smith chart has coordinates of K in the radial direction and θ in the circumferential direction, it is desirable to ascertain the location of a given admittance on the chart in terms of location of the corresponding impedance. Consider the effect of the use of admittance on

$$\text{(9-6)} \qquad Ke^{j\theta} = \frac{z - 1}{z + 1}$$

that is,

$$Ke^{j\theta} = \frac{1/y - 1}{1/y + 1} = \frac{1 - y}{1 + y}$$

$$= -\frac{y - 1}{y + 1}$$

Hence,

$$\text{(9-8)} \qquad Ke^{j\theta} = \frac{y - 1}{y + 1} e^{j\pi}$$

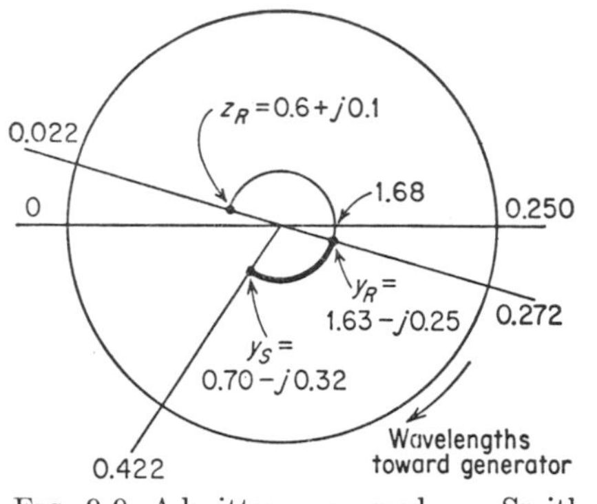

FIG. 9-9. Admittance example on Smith chart.

On the chart, therefore, the admittance is located 180° from the corresponding impedance. Thus, to get the admittance corresponding to a given impedance, it is merely necessary to rotate halfway around the chart.

This is best seen by an example. Consider Example 9-6 and the related Fig. 9-9.

Example 9-6. *Determining Admittance on the Smith Chart.* Given

$$z_R = 0.6 + j0.1$$
$$\text{Length} = 0.15\lambda$$

Find y_s.

To find y_R, we enter the Smith chart at $0.6 + j0.1$, which is at the intersection of the line corresponding to 0.022 wavelength toward the generator and a 1.68-standing-wave-ratio circle, and then rotate 180° to obtain the admittance. Thus, we come to 0.272 wavelength toward the generator and find that the admittance is

$$y_R = 1.63 - j0.25$$

We could also have discovered this by merely inverting z_R as a complex number.

To find the input admittance, it is necessary to go 0.150 wavelength farther toward the generator—from 0.272 to 0.422. At this point, the admittance is read to be $0.70 - j0.32$.

It would also have been possible to determine the sending-end impedance by going 0.15 wavelength from the receiving-end impedance and then inverting it on the diagram by the 180° rotation. The resulting point on the diagram would, of course, be the same.

Thus far this chapter has dealt with lossless transmission lines. In the case of lossy lines, the *effective* reflection coefficient is of course changed as the point of observation is moved from the load to the generator. The result is that, instead of traveling along a circle on the Smith chart in going from the load to the generator, it is necessary to travel along a spiral winding up about the center of the diagram. That is, since the effective reflection coefficient decreases, the standing-wave ratio decreases. Thus, each point is on a circle of somewhat smaller radius than the adjacent point closer to the load. This is illustrated in Example 9-7.

Example 9-7. *Finding Impedance on a Lossy Line.* Given

$$z_R = 2.0 + j2.0$$
$$\text{Length} = 2.13\lambda$$
$$\alpha = 0.2 \text{ neper/wavelength}$$

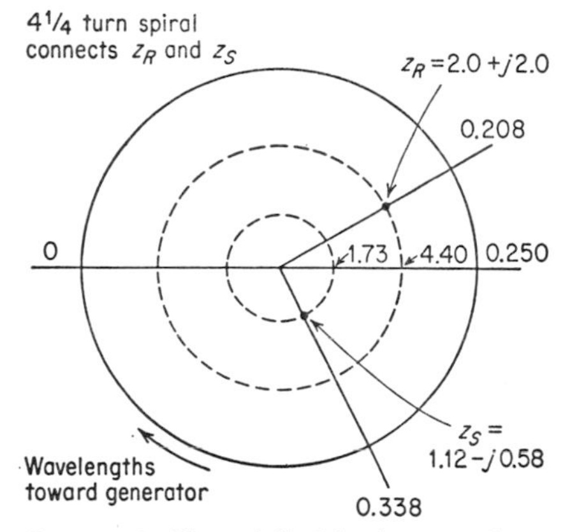

FIG. 9-10. Use of Smith chart on lossy line.

As shown in Fig. 9-10, the receiving-end impedance is on a radial line corresponding to 0.208 wavelength toward the generator, and the standing-wave ratio at the load is found by the intersection with the right-hand horizontal axis of a circle about the center of the diagram passing through the load. It is 4.40. The magnitude of the reflection coefficient at the load is therefore

$$|\Gamma_R| = \frac{4.4 - 1}{4.4 + 1} = \frac{3.4}{5.4} = 0.630$$

Hence, the effective reflection coefficient at the sending end is given by

$$\begin{aligned} |\Gamma_{R,\text{eff}}| &= 0.630e^{-2\alpha L} \\ &= 0.630e^{-2\times 2.13\times 0.2} \\ &= 0.268 \end{aligned}$$

The corresponding value of standing-wave ratio is then

$$\frac{1.268}{0.732} = 1.73$$

Hence, the sending-end impedance is found on this circle, the radius of which is considerably smaller than for the standing-wave ratio of 4.4. The sending end is on the wavelength-toward-the-generator scale at 0.208 + 0.130 = 0.338. Hence, from Fig. 9-10, it can be seen that

$$Z_s = (1.12 - j0.58)Z_0$$

For a lossless line the impedance would have been found at the intersection of the 0.338 line and the 4.4-standing-wave-ratio circle, or $(1.34 - j0.74)Z_0$.

The Smith chart is not the only impedance chart in use, but it has been treated here because it is the one most widely used today. Actually, the Smith chart was derived originally from the "rectangular impedance chart," on which the rectangular coordinates are the resistance and the reactance or the conductance and susceptance. On this chart the loci of constant standing-wave ratio and of constant distance from the end of the line are intersecting circles, whereas, of course, the loci of constant resistance and constant reactance are, respectively, vertical and horizontal lines. For small values of impedance the rectangular chart is just as easy to use as the Smith chart and in some regions more accurate. It has fallen into disuse, however, because large values of standing-wave ratio result in points off the chart. Hence, many of the readings which one must take in working practical problems fall off scale on this chart, though it is sufficiently large to solve problems for small impedances. The principal advantage of the Smith chart is that any impedance falls within the circle, and therefore nothing can be off scale. It does have the disadvantage, for a large impedance, that the scale is quite crowded and accuracy is low.

Other forms of impedance charts have also been developed, but none is so widely used as the rectangular chart, which in itself has fallen into disuse because of the advantages of the Smith chart.

9-2. Impedance Matching

With radio-frequency transmission lines (including waveguides) impedance matching is very important. Standing waves lead to increased losses and frequently result in loads that cause the transmitter to malfunction. Maximum power is transferred when the load impedance matches the transmission-line impedance (this is a standard theorem of circuit theory). For both these reasons, and for others as well, it is desirable to match radio-frequency transmission lines as closely as possible.

Sometimes matching is done with lumped-constant circuits, and sometimes short-circuited lengths of transmission line are used as reactances in the matching networks. Such sections of transmission line are also used as reactances in microwave and ultra-high-frequency filter networks.

Three or four basic techniques are applied for matching on transmission lines. Modifications of these are sometimes utilized to obtain matching over a wider band of frequencies. The problem of matching involves adjusting both resistive and reactive components of impedance or admittance. To achieve a match, therefore, it is necessary to have two quantities that may be varied.

Single-element matching is carried out with one adjustable susceptive lumped element (or short-circuited transmission line) whose position may be adjusted. Thus, the two quantities varied in this case are the position and the magnitude of the susceptance. *Double-element matching* (sometimes known as *double-stub tuning* when the susceptive elements are short-circuited transmission lines) utilizes two susceptances at fixed points on the line, both of whose magnitudes may be adjusted. This technique has the advantage that no mechanical arrangement need be made for changing the location of the susceptances; this is a very real advantage in the case of waveguide or coaxial transmission lines. The double-element tuner has the disadvantage that, with a given spacing between the susceptances, it will not match every load. Addition of a third susceptance at another point on the line makes possible matching of all loads without changing the position of the variable susceptances.[1]

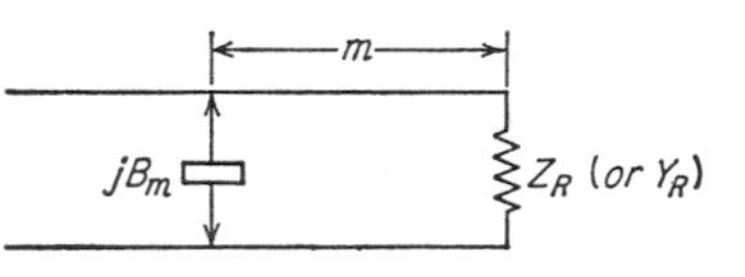

FIG. 9-11. Single-element matching.

Both the single- and the double-element matching devices are treated here. In addition, the "quarter-wavelength transformer" is described.

Figure 9-11 shows the use of single-element susceptive matching. It is assumed that the load admittance is complex. At some point on the line the *conductance* is equal to the characteristic admittance of the line (where the constant-standing-wave-ratio circle intersects the $g = 1.0$ circle on the Smith chart). At this point it is merely necessary to insert a shunt susceptance to cancel out the susceptive part of the admittance so that the line can be matched. That is, in general,

$$Y = G + jB$$

At this critical point on the line,

$$Y = Y_0 + jB$$

[1] H. J. Reich, P. F. Ordung, H. L. Krauss, and J. G. Skalnik, "Microwave Theory and Techniques," pp. 178–179, D. Van Nostrand Company, Inc., Princeton, N.J., 1953.

Thus, to make

$$Y = Y_0$$

it is merely necessary to insert an equal and opposite susceptance at this point, so that if

$$B_m = -B$$
$$Y = Y_0 + jB + jB_m = Y_0$$

Example 9-8. *Single-lumped-susceptance Matching.* It is desired to match a transmission line having the following parameters and to find the values of the inductance or capacitance which can be used to match this line. Given

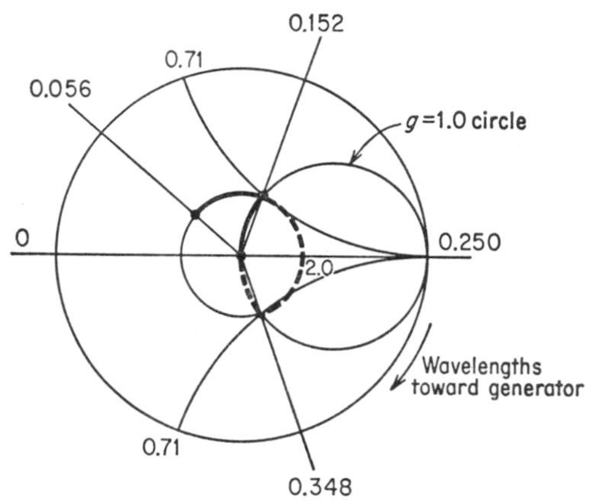

Fig. 9-12. Single-susceptance tuner.

$$y_R = 0.55 + j0.27 \text{ (VSWR} = 2.0)$$
$$\text{Frequency} = 10^9 \text{ cps}$$
$$Y_0 = \tfrac{1}{50} \text{ mho}$$

The Smith chart for this case is shown in Fig. 9-12. It is determined that the angular coordinate θ for this admittance is 0.056 on the wavelength-toward-the-generator scale. Going along the circle of constant standing-wave ratio 2.0 until it intersects the $g = 1.0$ circle, it is determined that the intersection occurs at 0.152 on the wavelength-toward-the-generator scale. Thus, the distance m between the end of the line and the susceptance is

$$m = 0.152 - 0.056 = 0.096\lambda$$

At that point the susceptance is given by

$$y = 1.0 + j0.71$$

Hence, we must add a susceptance at this point:

$$b_m = -0.71$$

A negative susceptance is an inductance, so we may write

$$B_m = -0.71Y_0 = -\frac{1}{\omega L}$$

Hence,

$$L = \frac{1}{0.71\omega Y_0} = \frac{50}{0.71 \times 2\pi \times 10^9} = 1.12 \times 10^{-8} \text{ henry}$$
$$= 0.0112\ \mu\text{h}$$

Thus, the line can be matched by inserting an inductance of this size at a point 0.096 wavelength, or about 3 cm, from the load.

It is also possible to match this line by a positive susceptance (capacitor) by going along the constant-standing-wave-ratio circle to the second interception with $g = 1.0$. This occurs in the lower part of the Smith chart at 0.348 wavelength toward the generator. Hence, the location of the susceptance is $0.348 - 0.056 = 0.292$ wavelength from the load. At that point the admittance is

$$y = 1.0 - j0.71$$

Hence, we must add a susceptance:

$$b_m = 0.71$$

Since this is a positive susceptance, we add a capacitor such that

$$\omega C = \frac{0.71}{50} \qquad \text{or} \qquad C = 2.23\ \mu\mu\text{f}$$

Although simple inductors and capacitors can be added, as indicated in Example 9-8, it is more common to utilize the susceptive properties of short-circuited sections of transmission lines. Short-circuited sections are used rather than open-circuited sections because a good short circuit is easier to achieve than a good open circuit. It would seem that an open circuit is the easiest thing in the world to obtain, but actually the capacitance between the ends of the transmission line makes it harder to achieve a good open circuit than a good short circuit.

We therefore utilize the short-circuited transmission line, whose impedance was given by

$$Z = jZ_0 \tan \beta L \tag{5-19}$$

It is perfectly reasonable to solve this equation for L. However, it is frequently somewhat easier to utilize the Smith chart, since one is likely to be used anyway in solving the problem of the matching elements.

Example 9-9. *Use of Short-circuited Transmission Line as Reactance in Single-element Matching.* Consider the use of short-circuited "stubs" of transmission line in Example 9-8. In the first case, an inductive susceptance was added whose magnitude was

$$b = -0.71$$

An inductive stub (short-circuited) is less than a quarter-wavelength long. To find the length of such a stub, enter the Smith chart at the short circuit ($b \rightarrow \infty$), where the distance scale reads 0.25 wavelength toward the generator. Then travel along the outside of the diagram to $b = -0.71$, which appears on the distance scale as 0.403. Hence, the short-circuited element is 0.153 wavelength long. The situation is shown in Fig. 9-13.

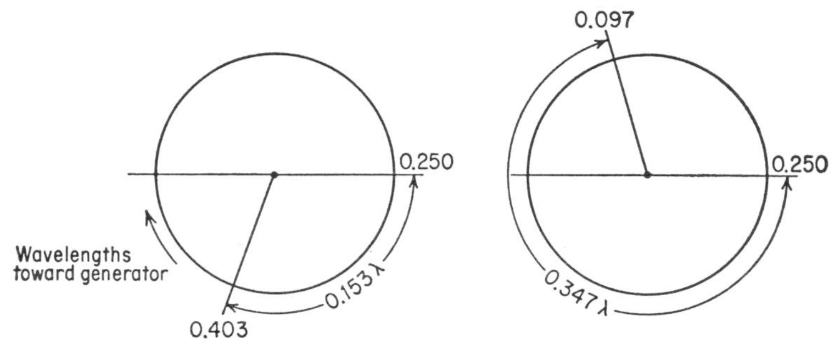

Fig. 9-13. Stub tuner.

For the capacitive matching of Example 9-8 we must add a susceptance $b = 0.71$. Again we enter the Smith chart at the short circuit, whose coordinate is 0.25. The capacitive stub is more than a quarter-wavelength long. Hence, we go more than halfway around the diagram, past 0.50, to the point on the upper semicircle where b is 0.7. This occurs where the wavelength-toward-the-generator scale reads 0.097. Actually this is the amount more than a quarter-wavelength for the stub, so the length is $0.25 + 0.097 = 0.347$ wavelength.

Thus, the admittance $0.55 - j0.27$ may be matched either by a short-circuited stub 0.153 wavelength long at 0.096 wavelength from the load or by a stub 0.347 wavelength long at a distance 0.292 wavelength from the load. For the frequency of 10^9 cps, the wavelength is 30 cm. Hence, the first arrangement is a shorted stub 4.6 cm long located at a distance 2.9 cm from the end of the line, and the second is a stub 10.4 cm long located at a distance 8.8 cm from the load.

As these examples have indicated, there are always two solutions for the single-element tuner, one using a capacitive susceptance and the other an inductive susceptance. Sometimes it is more convenient to use one and sometimes the other. In the case just discussed, for example, it may not be feasible to place a stub as close to the load as the short inductive stub would have to be. It may be much easier to use the longer capacitive stub farther from the load, or vice versa.

The double-susceptance tuner is used when it is difficult to arrange mechanically for a moving stub or reactance. In this case the two susceptances are variable in magnitude but are fixed at positions about $\frac{3}{8}$ wavelength apart. The stub nearest the load is used to adjust the susceptance so that it is located $\frac{3}{8}$ wavelength from the $g = 1$ circle, on an appropriate constant-standing-wave-ratio circle. Then the admittance at the second stub is $y = 1 + jb$, and it is merely necessary to cancel out the susceptance with that stub.

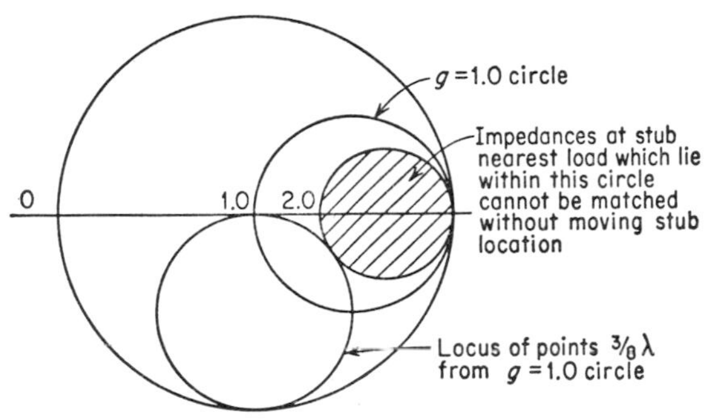

FIG. 9-14. Smith-chart geometry for double-susceptance tuner.

Sometimes a distance as short as a quarter-wavelength is used and sometimes other distances are used. Here we treat only the case of $\frac{3}{8}$-wavelength spacing, since the others may be easily handled when this one is understood.

It is not so difficult to find a point $\frac{3}{8}$ wavelength from the $y = 1 + jb$ circle as it would at first seem. On the Smith chart, $\frac{3}{8}$ wavelength represents three-quarters of a revolution. Hence, all points $\frac{3}{8}$ wavelength toward the load from the $y = 1 + jb$ circle lie three-quarters of a revolution counterclockwise from the corresponding points on this circle. This means that the locus of these points is also a circle, which has been rotated 270° from the position of the $g = 1 + jb$ circle. Thus, the locus of such points lies on a circle tangent at the top to the point $1 + j0$ and at the bottom to the bottom of the Smith chart. This is shown in Fig. 9-14.

Since it is necessary to add susceptance at the first lumped-susceptance point until the resultant admittance lies on this circle, loads for which this is impossible cannot be matched. Such loads project, when b is varied, into the exclusion circle indicated on Fig. 9-14. The admittances in this circle have conductance such that adding susceptance moves the admittance along a constant-conductance circle which does not at any point intersect the locus of points $\frac{3}{8}$ wavelength from the $g = 1$ circle. The boundary of the exclusion area is the circle for $g = 2.0$. Hence, the $\frac{3}{8}$-wavelength-spaced susceptances cannot be used to match any load which causes the conductance on the line at the point of the first tuning element to exceed twice the characteristic admittance. When such loads are encountered, it is necessary to move the stubs or to use a triple-stub tuner.

The best way to understand the operation of a double-stub tuner is to examine a specific example.

Example 9-10. *A Double-stub-tuner Problem.* Consider a double-stub tuner which is used to match the same load that was matched in Examples 9-8 and 9-9 by a single-stub tuner, that is,

$$y_R = 0.55 + j0.27$$

The standing-wave ratio at the load is 2.0. Let us assume that the fixed-position but variable-susceptance stubs are located at 0.150 and 0.525 wavelength from the load.

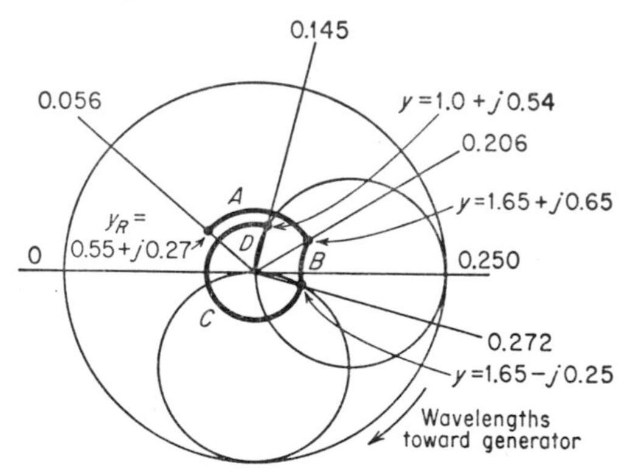

Fig. 9-15. Double-susceptance-tuner Smith chart example. A and C are on circle of constant VSWR; B and D are on circles of constant conductance. Motion along A and C shows changing position; motion along B and D shows changing susceptance at fixed position.

We enter the Smith chart as before at an angular coordinate of 0.056 wavelength from the load. In going from the end of the line to the point 0.150 wavelength, the Smith chart is traversed along the 2.0-constant-standing-wave-ratio circle, as indicated by A of Fig. 9-15. The first tuning element is located at the point on the diagram (before the effect of the element is considered) where the 2.0-standing-wave-ratio circle intersects the 0.206 wavelength-toward-the-generator line. At this point the admittance is

$$y = 1.65 + j0.65$$

The proper amount of susceptance must be added to get on the locus $\frac{3}{8}$ wavelength along a constant-standing-wave-ratio circle from $g = 1 + jb$. Thus, since only susceptance is being added, the admittance moves, with the addition of susceptance, along a constant-conductance circle for $y = 1.65 + jb$. This circle intersects the locus of points required at $y = 1.65 - j0.25$. Hence the susceptance which must be added in order to go from $y = 1.65 + j0.65$ to $y = 1.65 - j0.25$ is $b = -0.90$, an inductance. On the chart, addition of this inductance changes the susceptance along the line B. Note that B does not indicate movement from one place to another along the transmission line. It only represents the effect of adding susceptance at the fixed point 0.150 from the end of the line. The admittance $y = 1.65 - j0.25$ corresponds to a standing-wave ratio of 1.70 and an angle on the wavelength-toward-the-generator scale of 0.272. A three-fourths rotation toward the generator along the 1.70 standing-wave-ratio circle corresponds to traveling from the first tuning element to the second. At the second tuning element, where the standing-wave ratio is 1.70 and distance from the load is 0.589 wavelength (0.145 on the chart), the admittance is found to be

$$y = 1.0 + j0.54$$

To match the line, it is then necessary only to add the negative susceptance

$$b = -0.54$$

with the resulting admittance becoming

$$y = 1.0 + j0.0$$

The distance between the stubs is indicated in Fig. 9-15 by C and the result of adding this susceptance by D.

Again, note that A and C represent distances traveled along the line on a constant-standing-wave-ratio circle. B and D represent addition of susceptances at a point and hence correspond to motion along a constant g circle on the diagram, but they do not indicate travel down the line.

The lengths of the inductive short-circuited stubs necessary to give the required susceptances are 0.134 and 0.172 wavelength, respectively. These lengths are simply obtained by going from the right-hand side of the Smith chart (short circuit) toward the generator until the required susceptances are encountered and then measuring the appropriate distances.

Both double- and single-susceptive tuners involve adjustable elements which may be changed as loads are changed. For production purposes it is sometimes known that a load will be fixed, and it is then possible to perform the matching to a line by inserting a quarter-wavelength section of characteristic impedance intermediate between the line impedance and the load impedance. Such sections are known as "quarter-wavelength transformers."

Consider

$$Z_s = Z_0 \frac{1 + \Gamma_R e^{-j2\beta L}}{1 - \Gamma_R e^{-j2\beta L}} \tag{5-29}$$

For a quarter-wavelength line, $\beta L = \pi/2$. Hence,

$$Z_s = Z_0 \frac{1 + \Gamma_R e^{-j\pi}}{1 - \Gamma_R e^{-j\pi}} = Z_0 \frac{1 - \Gamma_R}{1 + \Gamma_R}$$

The impedance at the receiving end of the line is obtained by substituting $L = 0$ in Eq. (5-29), and it is found to be

$$Z_R = Z_0 \frac{1 + \Gamma_R}{1 - \Gamma_R}$$

Hence, we may substitute in the expression for Z_s for the quarter-wavelength line and find

$$Z_s = Z_0 \frac{Z_0}{Z_R} \qquad \text{or} \qquad Z_s Z_R = Z_0^2 \tag{9-9}$$

Thus, insertion of a transmission line whose characteristic impedance is the geometric mean of the characteristic impedance of the line being matched (Z_s) and the load (Z_R) results in a good match.

Sometimes these quarter-wavelength transformers are modified in various ways to improve their bandwidths. Thus a number of sections in series with smaller impedance change for each may be used to accomplish the same purpose.

Whereas stubs are usually used on coaxial and parallel-wire transmission lines for matching purposes where quarter-wavelength transformers cannot be used, lumped constants are used at both lower and higher frequencies. At lower frequencies the lumped constant takes the form of "actual" inductors and capacitors. At higher frequencies, where waveguides are used, the lumped-constant effect is obtained by some device which perturbs either the electric or magnetic field in the waveguide in such a way that the result is the same as the shunt inductance or capacitance.

In practice, once the approximate length of the stubs is known, it is usually necessary to adjust by trial and error over a narrow range. Tuning of such stubs is quite a delicate process, for perturbations are nearly always present which make the situation slightly different from that represented in the model used in the calculations.

9-3. Resonators

The impedance-transformation properties of transmission lines, pipes carrying acoustic waves, and vibrating strings and bars are utilized for resonators. Similar phenomena exist in wave mechanics. Transmission-line and waveguide resonators are used as wavemeters since they have a very sharp resonant peak on an impedance-vs.-frequency plot. They are used as tuned circuits in amplifiers, oscillators, and other devices operating above a few hundred megacycles. The electric-transmission-line and waveguide resonators are frequently used as filter elements at frequencies where lumped-constant inductances and capacitances could not be used. Both acoustic waves in air columns in pipes and acoustic waves in solid rods are used as elements of filters. A tuning fork is a form of acoustic resonator. Many musical instruments use acoustic resonators. Stringed instruments not only have resonant vibrating strings but usually have some form of resonant cavity as well. All wind instruments, both woodwind and brass, are acoustic resonators whose resonant frequency may be varied by the player. Some percussion instruments, such as xylophones and marimbas, also utilize acoustic resonators. The pipe organ is one of the best-known acoustic resonator devices. Many whistles also are based upon acoustic resonance phenomena in pipes.

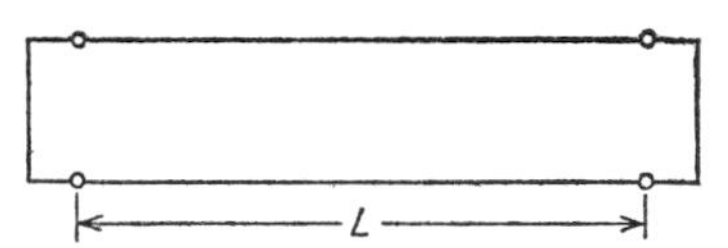

FIG. 9-16. Transmission-line resonator.

Transmission-line resonators are usually short-circuited at least at one end and frequently at both ends. Acoustic resonators are usually open to the air at least at one end (equivalent to an open circuit on the transmission line) and frequently at both ends, although one end may be used for the excitation.

Any traveling-wave resonator has a number of "natural frequencies." Consider, for example, the natural frequencies of a transmission linc of length L short-circuited at both ends. This analysis applies equally well to a pipe containing an air column which is open at both ends. The resonator is shown in Fig. 9-16. Consider that the left-hand side is the sending end and the right-hand side the receiving end. Then we may write for the sending-end voltage on this line

$$V_s = V^+(e^{j\beta L} - e^{-j\beta L}) = 2jV^+ \sin \beta L$$

Since the line is short-circuited at both ends, the sending-end voltage, as well as the receiving-end voltage, must be zero; that is,

$$0 = \sin \beta L$$

This means that

$$\beta L = \frac{\omega L}{v_p} = n\pi$$

where n is an integer, is a requirement for the standing wave to exist on the line. This occurs at frequencies where

$$\omega = \frac{n\pi v_p}{L}$$

that is, where

$$f = \frac{nv_p}{2L} \tag{9-10}$$

This may also be interpreted in terms of the wavelength. For a line to be resonant when shorted at both ends, its length should be an integral multiple n of a half-wavelength. Thus

$$L = \frac{n\lambda}{2} \tag{9-11}$$

It is customary to refer to the wave for each value of n as a *mode* of the resonator. Since all modes may be present at once (and usually are when the resonator is excited with a transient), the total voltage on the line is the sum of a number of harmonics, so that

$$v = \text{Re } 2j \sum_{n=1}^{\infty} V_n^+ \sin \frac{n\pi d}{L} \, e^{j(n\pi v_p/L)t}$$

where V_n^+ is the amplitude of the incident wave for the nth harmonic, or the nth mode. The *fundamental* frequency for such a resonance is, of course, the one where the line is a half-wavelength.

It is possible to consider this simple resonator from the standpoint of a *boundary-value problem* all the way through. The boundary-value concept was inherent in our derivation of the reflection coefficient and in the application of the sending-end short circuit above.

Consider the application of the basic wave equation for the transmission line:

$$\frac{\partial^2 V}{\partial x^2} = \gamma^2 V \tag{2-5b}$$

For a lossless line, this is

$$\frac{\partial^2 V}{\partial x^2} + \beta^2 V = 0$$

The solution of this equation is in trigonometric functions or exponentials with imaginary argument; consider the former. Then

$$V = A \cos \beta x + B \sin \beta x$$

The *boundary conditions* are

(1) $V = 0 \qquad x = 0$
(2) $V = 0 \qquad x = L$

Substituting $x = 0$ and $V = 0$, from (1),

$$0 = A$$

so

$$V = B \sin \beta x$$

Substituting $x = L$ and $V = 0$, from (2),

$$0 = B \sin \beta L$$

Hence,

$$\beta L = n\pi$$

where n is any integer. Since β may have as many values as there are integers, we designate the value associated with a particular integer n as β_n. Then

$$\beta_n = \frac{n\pi}{L}$$

Thus, the total voltage is the sum of all possible solutions, with multiplying factors B_n which cause the *initial conditions* to be matched:

$$v = \text{Re} \sum_{n=1}^{\infty} B_n \sin \beta_n x \, e^{j\omega_n t}$$

The values β_n are the *characteristic values*, or *eigenvalues*, associated with the particular problem. Characteristic values for frequency correspond to those for β_n, since

$$\beta_n = \omega_n \sqrt{LC}$$

The methods of boundary-value problems are often applied to resonance problems. Traveling-wave techniques and boundary-value techniques complement each other. In curvilinear coordinates, resonant wave phenomena are usually *determined* by boundary-value techniques but frequently *interpreted* in terms of traveling waves.

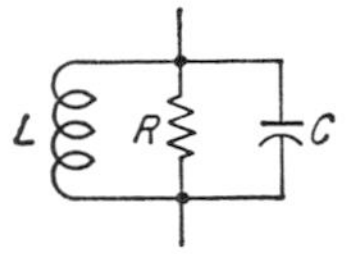

FIG. 9-17. Parallel resonant circuit.

This discussion has dealt so far with lossless lines. In practice, many of the interesting properties of traveling-wave resonators are connected with losses and their effect on impedance and bandwidth. For comparison, let us consider the analysis of a lumped-constant parallel resonant circuit of the type illustrated in Fig. 9-17. The admittance presented to the terminals of this circuit is given by

$$Y = \frac{1}{R} + j\left(\omega C - \frac{1}{\omega L}\right) = \frac{1}{R} + j\left(1 - \frac{1}{\omega^2 LC}\right)\omega C$$

This may be written in terms of the resonant frequency

$$\omega_0 = \frac{1}{\sqrt{LC}}$$

as

$$Y = \frac{1}{R} + j\omega C\left(1 - \frac{\omega_0^2}{\omega^2}\right) = \frac{1}{R} + j\omega C\left(1 + \frac{\omega_0}{\omega}\right)\left(1 - \frac{\omega_0}{\omega}\right)$$

For frequencies close to resonance, this may be approximated by

$$Y = \frac{1}{R} + j\omega C\,\frac{2(\omega - \omega_0)}{\omega} = \frac{1}{R} + j2C\,\Delta\omega$$

where $\Delta\omega$ is the deviation of the actual frequency from the resonant frequency.

The half-power bandwidth of this resonant circuit corresponds to the spacing between values of ω where conductance and susceptance are equal. For such values of ω,

$$\frac{1}{R} = 2C\,\Delta\omega_1 \qquad \text{or} \qquad \Delta\omega_1 = \frac{1}{2RC}$$

One definition of the quality factor Q for the circuit is the ratio of the resonant frequency to the half-power bandwidth. Thus, in this case,

$$Q = \frac{\omega_0}{2\,\Delta\omega_1} = \omega_0 RC \tag{9-12}$$

Now consider a resonator which consists of a transmission line a half-wavelength long with terminals at the center and short circuits at both ends. The total admittance at the center is given by

$$Y = Y_{\text{left}} + Y_{\text{right}} = 2Y_{\text{right}}$$

since symmetry indicates that the admittance looking to right and left will be the same. The admittance on the right-hand side is given by

$$Y_{\text{right}} = Y_0\,\frac{1 - \Gamma_R e^{-2\gamma\lambda_0/4}}{1 + \Gamma_R e^{-2\gamma\lambda_0/4}}$$

where λ_0 is the wavelength at resonance, but for the short circuit

$$\Gamma_R = -1$$

and

$$e^{-2\gamma L} = e^{-\alpha\lambda_0/2}e^{-2j\beta\lambda_0/4} = -e^{-\alpha\lambda_0/2}$$

Therefore the admittance for one-half of the line is

$$Y_{\text{right}} = Y_0 \frac{1 - e^{-\alpha\lambda_0/2}}{1 + e^{-\alpha\lambda_0/2}} \approx \frac{Y_0 \alpha\lambda_0/2}{2} = Y_0 \frac{\alpha\lambda_0}{4}$$

This approximation is valid where the attenuation is small enough that only the first two terms of the series expansion of the exponential need be used in the numerator and only the first term in the denominator. This is the value for the admittance of half the resonator; therefore the total admittance is

$$Y = 2Y_0 \frac{\alpha\lambda_0}{4} = Y_0 \frac{\alpha\lambda_0}{2} \tag{9-13}$$

This admittance occurs at the resonant frequency. Of course, if the line were lossless, the admittance would be zero and the impedance would be infinite.

To determine the Q for such a resonant circuit, it is desirable to expand the exponentials appearing in the admittance expression. Thus,

$$\begin{aligned} e^{-\gamma\lambda_0/2} = e^{-\alpha\lambda_0/2 - j\beta\lambda_0/2} &= e^{-\alpha\lambda_0/2} e^{-j\pi} e^{-j(\beta\lambda_0/2 - \pi)} \\ &= -e^{-\alpha\lambda_0/2} e^{-j(\beta\lambda_0/2 - \pi)} \end{aligned}$$

β may be expressed in terms of the resonant frequency ω_0 and the deviation from this frequency $\Delta\omega$ as

$$\beta = \frac{\omega_0 + \Delta\omega}{v_p} = \beta_0 \left(1 + \frac{\Delta\omega}{\omega_0}\right)$$

Hence,

$$\begin{aligned} \frac{\beta\lambda_0}{2} - \pi &= \frac{\beta_0\lambda_0}{2} + \frac{\beta_0\lambda_0}{2} \frac{\Delta\omega}{\omega_0} - \pi \\ &= \frac{\beta_0\lambda_0}{2} \frac{\Delta\omega}{\omega_0} = \frac{\pi \, \Delta\omega}{\omega_0} \end{aligned}$$

Thus,

$$e^{-\gamma\lambda_0/2} \approx -e^{-\alpha\lambda_0/2} e^{j\pi\Delta\omega/\omega_0}$$

When

$$\frac{\Delta\omega}{\omega} \ll 1$$

and also

$$\frac{\alpha\lambda_0}{2} \ll 1$$

this becomes

$$e^{-\gamma\lambda_0/2} \approx -1 + \frac{\alpha\lambda_0}{2} + \frac{j\pi \, \Delta\omega}{\omega_0}$$

Hence, we may write

$$Y_{\text{right}} = Y_0 \frac{\alpha\lambda_0/2 + j\pi \, \Delta\omega/\omega_0}{2}$$

The half-power points occur where the real and imaginary parts of the numerator of this fraction are equal; that is, where

$$\frac{\pi\,\Delta\omega_1}{\omega_0} = \frac{\alpha\lambda_0}{2}$$

This may be written as

$$\frac{\Delta\omega_1}{\omega_0} = \frac{\alpha\lambda_0}{2\pi} = \frac{\alpha}{\beta_0}$$

Thus the quality factor Q of the resonant circuit is

$$Q = \frac{\omega_0}{2\,\Delta\omega_1} = \frac{\beta_0}{2\alpha} \tag{9-14}$$

The Q is, as one would expect, proportional to the ratio of the energy stored to the energy dissipated. It may also be thought of as proportional to the power transmitted divided by the power dissipated per unit length, since the energy in the standing wave is transmitted first one way and then the other.

The values of Q obtained for transmission-line resonators are quite high compared with the values obtained ordinarily with lumped-constant circuits. An example follows.

Example 9-11. *Q of a Resonant Coaxial Cavity.* Consider a copper coaxial transmission line closed at both ends so as to form a resonant cavity. The outer diameter of this line is 4 cm and the inner diameter is 1 cm. The characteristic impedance for such a line is given by[1]

$$Z_0 = 60 \log_e \frac{\text{outer diameter}}{\text{inner diameter}} = 60 \log_e 4 = 83.2 \text{ ohms}$$

Suppose that this cavity is resonant to 500 Mc (5×10^8 cps). The wavelength here is 60 cm, so the cavity is 30 cm long.

To calculate the attenuation, it is first necessary to calculate the resistance, which may be obtained from the skin depth. The skin depth for copper at this frequency is

$$\sqrt{\frac{2}{\omega\mu\sigma}} = \sqrt{\frac{1}{\pi \times 5 \times 10^8 \times 4\pi \times 10^{-7} \times 5.8 \times 10^7}}$$
$$= 2.95 \times 10^{-6} \text{ m}$$

The resistance of the inner conductor is therefore the resistance of a circular shell whose thickness is the skin depth. Hence it is

[1] "Reference Data for Radio Engineers," 4th ed., p. 589, International Telephone and Telegraph Corporation, New York, 1956.

given by

$$\frac{1}{\sigma\pi(\text{diameter})(\text{skin depth})} = \frac{1}{5.8 \times 10^{-7}\pi \times 10^{-2}(2.95 \times 10^{-6})}$$
$$= 0.186 \text{ ohm/m}$$

Since the skin depth is the same for the outer conductor and since its diameter is four times as great, the resistance is one-fourth as great or 0.047 ohm/m. The total resistance per meter is the sum of these two:

$$R = 0.233 \text{ ohm/m}$$

According to the approximate formula for attenuation on a low-loss transmission line,

$$\alpha = \frac{R}{2Z_0} = \frac{0.233}{166.4} = 1.40 \times 10^{-3} \text{ neper/m}$$

β is of course given by

$$\beta = \frac{\omega}{v_p} = \frac{10^9\pi}{3 \times 10^8} = 10.5$$

Thus,

$$Q = \frac{\beta}{2\alpha} = \frac{10.5}{2 \times 1.40} \times 10^3 = 3{,}760$$

This should be compared with values of, at most, a few hundred for lumped-constant circuits. It can be seen that the coaxial transmission line is a much higher Q resonator than the customary lumped-constant circuit.

To determine the resonant impedance at the center of this resonator, we note that

$$Z = \frac{1}{Y} = \frac{2}{Y_0\alpha\lambda_0} = \frac{2}{(1/83.2) \times 1.40 \times 10^{-3} \times 0.60} = \frac{166.4 \times 10^3}{1.40 \times 0.60}$$
$$= 2 \times 10^5 \text{ ohms}$$

Thus, the coaxial cavity is a high-impedance, narrow-band resonator. The half-power bandwidth Δf_1 is given by

$$\Delta f_1 = \frac{f_0}{2Q} = \frac{500}{7{,}520} = 0.067 \text{ Mc}$$

The half-bandwidth of 67 kc may seem quite large, but it is really a very small percentage of the 500-Mc center frequency.

Although a straightforward shorted coaxial resonator is frequently desired, there are other applications in which it is convenient to shorten the length of the resonator considerably. This may be accomplished by the use of lumped capacitance or inductance in the circuit at some point, or it may be achieved by combining transmission lines with different characteristic impedances, letting the inductive half of the resonant circuit be a line with high characteristic impedance (small center conductor) and the capacitance end be a line with low characteristic impedance (center conductor almost as large as outer conductor). Such an application is illustrated in Example 9-12.

Example 9-12. *Foreshortened Resonator for Artificial Line.* In testing short-range radars such as altimeters, it is frequently desirable to have a delay line which will simulate the delay in the signal transmission between an aircraft and the ground and return. It is desirable to make this delay unit rather compact, and it is certainly desirable to keep its attenuation to a minimum. A bandpass filter with sections of the type illustrated in Fig. 9-18*a* was once designed for such an artificial line. The frequency involved is 450 Mc.

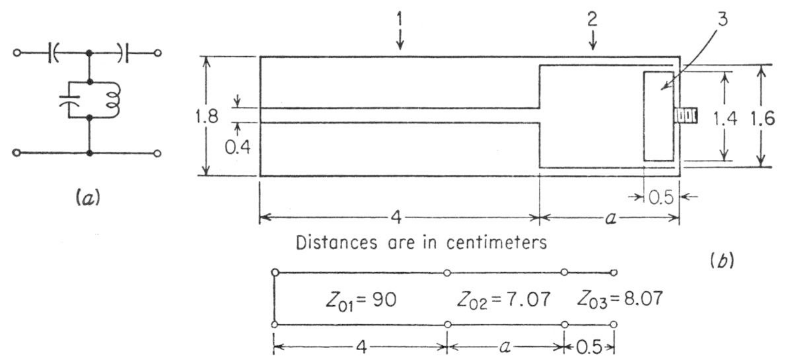

FIG. 9-18. Foreshortened resonator: (*a*) lumped equivalent; (*b*) actual resonator.

In order to make a filter element for this frequency, it is necessary, at least if low loss is a requirement, to utilize transmission-line elements. The transmission line comparable to the lumped-constant circuit of Fig. 9-18*a* is shown in Fig. 9-18*b*. The coaxial section shown was actually constructed with similar, but not identical, measurements. It is assumed here that the dimensions are all as shown, and it is desired to find how long the capacitive section *a* has to be. In the practical device section 3, the slug at the end, is a tuning element required to permit fine adjustment. The capacitance for coupling from the parallel resonant circuit to the next section is obtained by cutting a small window in the outer conductor adjacent to the capacitive part of the inner conductor.

Let us assume that the resonant maximum impedance occurs at the point where the inner conductor changes diameter. We thus have to the left an inductance (shorted transmission line less than a quarter-wavelength long) and to the right a capacitance made up of a section of line of 1.6 cm inside diameter and length a and another section of 1.4 cm inside diameter, 1.6 cm outside diameter, and 0.5 cm length. The impedances are different for all three sections. However, it is still possible to use the Smith chart.

The characteristic impedance in the inductive section is given by

$$Z_{01} = 60 \log_e \frac{1.8}{0.4} = 90 \text{ ohms} \qquad Y_{01} = 0.0111 \text{ mho}$$

Similarly, for the capacitive sections we have

$$Z_{02} = 60 \log_e \frac{1.8}{1.6} = 60 \times 0.117 \qquad Y_{02} = 0.142 \text{ mho}$$
$$= 7.068 \text{ ohms}$$

and

$$Z_{03} = 60 \log_e \frac{1.6}{1.4} = 60 \times 0.1345 \qquad Y_{03} = 0.124 \text{ mho}$$
$$= 8.070 \text{ ohms}$$

The wavelength, assuming that the velocity of propagation in the line is the same as that in air, is given by

$$\lambda = \frac{3 \times 10^8}{4.5 \times 10^8} = 0.67 \text{ m}$$

Hence, the length of the inductive section is

$$l_1 = \frac{4}{67} = 0.060\lambda$$

Using the Smith chart, we find that the admittance of a section of line this long is

$$-j2.55 Y_{01} = -j0.0284 \text{ mho}$$

It is desirable to express all the admittances in terms of the admittance in section 2. Hence,

$$-j0.0284 = -j0.200 Y_{02}$$

The length of the tuning slug is

$$l_3 = \frac{0.5}{67} = 0.0075\lambda$$

From the Smith chart, we find that an open-circuited section of line of this length has

$$Y_{s3} = j0.05 Y_{03} = j0.00625$$

In terms of Y_{02}, this is

$$j0.00625 = j0.042 Y_{02}$$

Using the Smith chart, we find that this corresponds to a distance on the wavelength scale of 0.007. In order to cancel completely the inductive susceptance of section 1, it is necessary to have a capacitive susceptance equal and opposite; that is, the capacitive susceptance must be 0.200. On the Smith chart, this corresponds to a distance toward the generator of 0.031. Hence, since the tuning slug represents 0.007 wavelength in this direction,

$$a = 0.031 - 0.007 = 0.024\lambda = 1.60 \text{ cm}$$

Thus the resonator's over-all length is 5.60 cm. If it had not been foreshortened (if it had been a quarter-wavelength long), its length would be 16.7 cm.

The choice of the particular value of inductive reactance used here was dictated by design considerations for the filter, but a discussion of these factors is beyond the scope of this presentation.

Rather than connect into the center of a half-wavelength line shorted at both ends, it is frequently desirable to use a quarter-wavelength line shorted at one end and open at the other as a resonator. A resonator of this type is frequently used as a portion of a vacuum-tube circuit, with the high-impedance end connected between two elements of the tube. Because of the effects of the interelectrode capacitance, the line must be somewhat less than a quarter-wavelength long, so that it appears as an inductive susceptance which resonates with the capacitive susceptance of the vacuum tube. Such an arrangement is shown in Fig. 9-19. Here, of course, the inductance L is given by

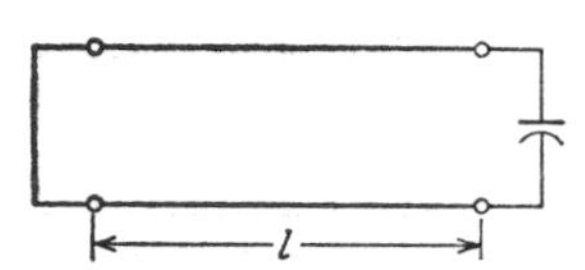

FIG. 9-19. Capacitor-tuned shorted resonator.

$$j\omega L = jZ_0 \tan \beta l$$

Acoustic resonances in pipes are used for wind musical instruments. With the trombone, the frequency is changed by changing the length of the resonator and, therefore, its resonant frequency. Most wind instruments, including all the woodwinds, control the frequency by opening holes along the resonant tube to the air. This reduces pressure at the points in the tube where the holes exist and diverts the air motion from "down the pipe" to "through the holes." Since reducing pressure is analogous to reducing current, this appears as a series impedance, the diversion of motion (velocity) representing the voltage drop. If its effect as impedance were in terms of some sort of storage of energy, either in the inertia of particles or in compression, the effect would be reactive. However, it is not, because the reduction in pressure is due to opening the inside of the tube to the outside, and any air which escapes from the tube is lost completely and does not return. Hence, the effect of the hole is the same as the effect of a series resistance on an electric transmission line.

The equivalent electric circuit for a woodwind instrument, along with the pictorial diagram of the instrument itself, is shown in Fig. 9-20. The end open to the air completely is equivalent to an open circuit. The holes along the way, when uncovered, connect resistances in series with the line. The switches of the equivalent line would be replaced on the woodwind with fingers or keys. Because of the escape of air through the holes, it is necessary to keep supplying air at the driving end, just as it is necessary to keep supplying power to the transmission line, because of dissipation in the resistors. The resonant frequency depends on which switches are closed.

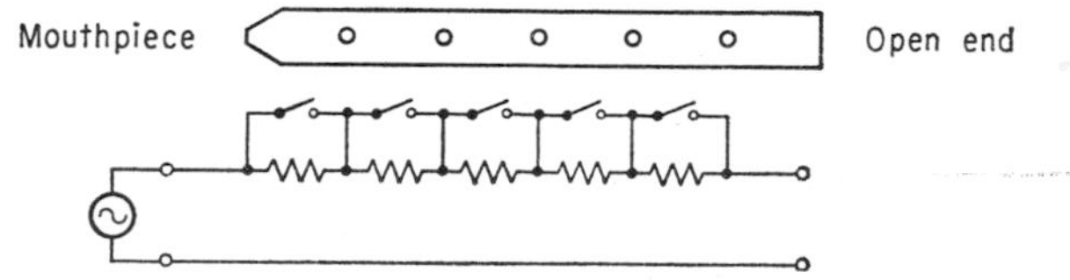

FIG. 9-20. Woodwind instrument and electrical equivalent.

Example 9-13. *Flute.* Consider a flute whose length is 0.63 m and assume that the velocity of propagation of air at this temperature is 330 m/sec. We assume that the flute represents a transmission line open at both ends and that the tone generated is that due to the half-wavelength resonance. Thus, the wavelength is 1.26 m and the resonant frequency is

$$f = \frac{330}{1.26} = 262 \text{ cps (C)}$$

First let us consider what would happen if the series resistors were open circuits. Suppose that the first opening is to resonate the flute at 294 cps (D). This shortens the line; the wavelength for 294 cps is

$$\lambda = \tfrac{330}{294} = 1.12 \text{ m}$$

Hence, half of this is 56 cm, which places the hole 7 cm from the end of the flute.

This is what would happen if the lowest hole represented a completely open circuit. Actually, of course, it must represent a resistance instead, and there will be some slight detuning effect due to the shunt impedance of the section of the flute beyond this point. In fact, there will also be higher-frequency components of lesser amplitude generated in this region, and these contribute to the tone.

Suppose that the resistance at the opening is given by

$$R = 5Z_0$$

The end section (beyond the last hole) has a length $\frac{7}{112} = 0.0625$ wavelength. Figure 9-21 shows the Smith chart involved. The reactance for the shorted section 0.0625 wavelength long appears on the chart at 0.3125 wavelength toward the generator on the bounding circle. This represents a capacitive reactance of $2.40Z_0$. Since this is in series with the resistance $R = 5Z_0$, the net impedance at this point is $5 - j2.40$. A line from the center through this point intersects the wavelength-toward-the-generator scale at 0.2625. Thus, in this case the flute is actually detuned for this frequency and will resonate at another frequency. If it were not for the effect of this section, the resistance would lie on the 0.250 line, and traveling a half-wavelength would bring us back to 0.250, which would give the proper impedance for the sending end. However, here we notice that the length is off by 0.0125/0.500, or 2.5 per cent. The effect is (approximately) that the frequency will be reduced by about 2.5 per cent. Hence, if it is desired to make the flute play this note, it is first necessary to try again with a somewhat shorter distance between the source and first hole (about 2.5 per cent shorter).

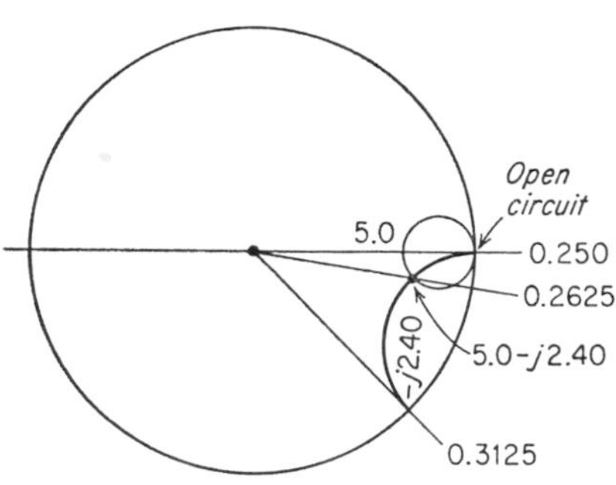

FIG. 9-21. Smith chart for flute.

This discussion has only touched on traveling-wave resonator uses. There are many other uses, but all are based upon the same principle. For example, acoustic resonances in solids are now used in filters for radio receivers. Three-dimensional resonance phenomena are present for wave potentials in atoms.

9-4. Waveguides

All transmission lines treated so far have had two conductors, although electromagnetic plane waves in air have been discussed, as have been many kinds of nonelectromagnetic plane waves. At microwave frequencies, transmission is usually in the form of a modified plane wave traveling through a pipe or guided in some other manner, rather than by a two-conductor transmission line.

As an example of the simplest type of waveguide (a pipe which guides waves), consider the situation shown in Fig. 9-22. Suppose that a transmission line consists of two parallel strips of conducting material of width w. Now place a number of quarter-wavelength short-circuited transmission lines on either side of the strips. Since these are "quarter-wavelength insulators," they do not affect the transmission line. Now suppose that these quarter-wavelength insulators are placed closer and closer together until eventually they form a solid boundary. It can be seen that the transmission line has been replaced by a rectangular cross-sectional pipe of width $w + \lambda/2$.

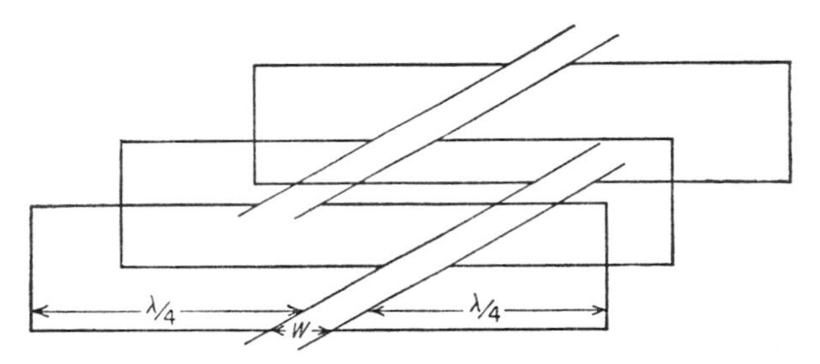

FIG. 9-22. Evolution of the waveguide.

It is instructive to consider what happens as the width of the effective transmission line changes. As w approaches zero, the inductance approaches infinity, and the capacitance approaches zero. Since the characteristic impedance is

$$Z_0 = \sqrt{\frac{L}{C}}$$

for a lossless line, Z_0 approaches infinity. Thus, at the frequency for which the pipe is just a half-wavelength wide, the impedance is infinite.

It is not so easy to see what happens to the velocity of propagation for

$$v_p = \frac{1}{\sqrt{LC}}$$

However, it can be shown that L approaches infinity approximately logarithmically and that C approaches zero more nearly linearly. Hence, their product approaches zero, and v_p approaches infinity. This happens at the frequency where the entire width of the pipe is taken up by the quarter-wavelength insulators, and there is nothing left for the transmission line. This is known as the cutoff frequency for the pipe. Obviously a lower frequency corresponds to a longer wavelength, and it is not possible to have two quarter-wavelength insulators in tandem if the over-all distance is less than a half-wavelength.

The fact that the velocity of propagation becomes infinite at cutoff frequency is interesting, since this means that the phase velocity can be higher than the velocity of light. This is indeed true for the phase velocity but not for the velocity with which energy is transmitted, for it could not exceed the velocity of light.

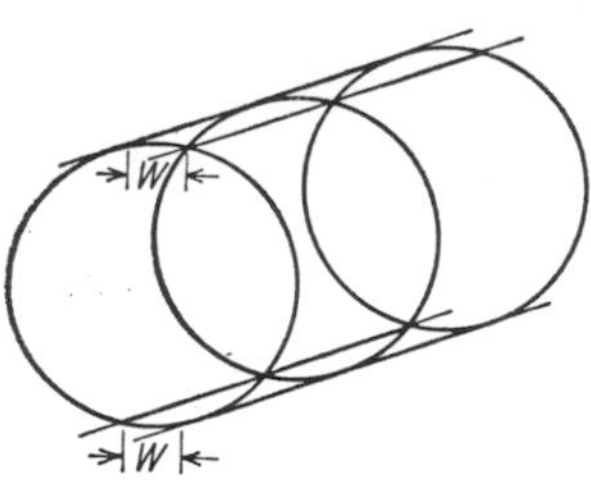

FIG. 9-23. Evolution of circular waveguides.

Figure 9-23 shows a comparable development for circular waveguides. Here the semicircular sections shown represent arcs of circles of such length that their input impedance is infinite, just as is that of the quarter-wavelength transmission line. Hence, a circular pipe may support a wave much like that of a rectangular pipe.

Only the simplest wave types have been described qualitatively above. Actually, there are many "higher modes" which do not have such a direct analogy with simple transmission lines. To see how the higher modes are established, it is necessary to make a quantitative study of the waveguides by solving the boundary-value problems. This is done here for a single simple example, more to indicate the method than to present a detailed study of the waveguide.[1]

Let us limit attention to an infinite waveguide with waves traveling in the positive z direction. Such waves are then described by time and longitudinal variations

$$e^{j\omega t - \gamma_z z}$$

Hence, the partial derivative of any one of the field components with respect to z is given by

$$\frac{\partial A}{\partial z} = -\gamma_z A$$

where A is any one of the field components.

Ampere's law may be used to express the components of the electric field in terms of the magnetic field. Thus,

$$\mathbf{E} = \frac{1}{j\omega\epsilon} \nabla \times \mathbf{H}$$

[1] See, for example, the detailed development in Simon Ramo and John R. Whinnery, "Fields and Waves in Modern Radio," chaps. 8 and 9, John Wiley & Sons, Inc., New York, 1944.

For the x and y components of the electric field, this is

$$E_x = \frac{1}{j\omega\epsilon}\left(\frac{\partial H_z}{\partial y} - \frac{\partial H_y}{\partial z}\right) = \frac{1}{j\omega\epsilon}\left(\frac{\partial H_z}{\partial y} + \gamma_z H_y\right)$$

$$E_y = \frac{1}{j\omega\epsilon}\left(-\gamma_z H_x - \frac{\partial H_z}{\partial x}\right)$$

Similarly, using Faraday's law, we have

$$\mathbf{H} = -\frac{1}{j\omega\mu}\nabla \times \mathbf{E}$$

Hence,

$$H_x = -\frac{1}{j\omega\mu}\left(\frac{\partial E_z}{\partial y} + \gamma_z E_y\right)$$

and

$$H_y = -\frac{1}{j\omega\mu}\left(-\gamma_z E_x - \frac{\partial E_z}{\partial x}\right)$$

This set of four equations may be solved simultaneously, from which we obtain the following expressions for electric and magnetic fields:

$$E_x = -\frac{1}{\gamma_z^2 + \omega^2\mu\epsilon}\left(\gamma_z \frac{\partial E_z}{\partial x} + j\omega\mu \frac{\partial H_z}{\partial y}\right) \tag{9-15}$$

$$E_y = \frac{1}{\gamma_z^2 + \omega^2\mu\epsilon}\left(-\gamma_z \frac{\partial E_z}{\partial y} + j\omega\mu \frac{\partial H_z}{\partial x}\right) \tag{9-16}$$

$$H_x = \frac{1}{\gamma_z^2 + \omega^2\mu\epsilon}\left(j\omega\epsilon \frac{\partial E_z}{\partial y} - \gamma_z \frac{\partial H_z}{\partial x}\right) \tag{9-17}$$

$$H_y = -\frac{1}{\gamma_z^2 + \omega^2\mu\epsilon}\left(j\omega\epsilon \frac{\partial E_z}{\partial x} + \gamma_z \frac{\partial H_z}{\partial y}\right) \tag{9-18}$$

Thus the transverse components of electric and magnetic fields can be expressed independently in terms of the longitudinal components of these fields. A complete set of transverse components can exist with only E_z or H_z present.

The waves are said to be transverse electric (TE) when the longitudinal component of the electric field is zero ($E_z = 0$). They are said to be transverse magnetic (TM) when the longitudinal component of the magnetic field is zero ($H_z = 0$). Both types of field can exist at the same time, but they are independent and therefore may be treated separately.

To determine the characteristics of a TE mode, it is necessary to solve for H_z. The equation for the magnetic field is comparable with that for the electric field given in (2-5*b*):

$$\nabla^2\mathbf{H} = \gamma^2\mathbf{H}$$

To solve for the longitudinal magnetic field, we take the component of this vector equation that is associated with H_z. Thus,

$$\nabla^2 H_z = \gamma^2 H_z \tag{9-19}$$

When the wave varies with distance as above, this equation becomes

$$\frac{\partial^2 H_z}{\partial x^2} + \frac{\partial^2 H_z}{\partial y^2} + \gamma_z^2 H_z = \gamma^2 H_z$$

or

$$\frac{\partial^2 H_z}{\partial x^2} + \frac{\partial^2 H_z}{\partial y^2} = (\gamma^2 - \gamma_z^2) H_z$$

Such an equation may be solved by the method of separation of variables. Let us assume, therefore, that $H_z = X(x)Y(y)$, where X is a function of x alone and Y is a function of y alone. Substituting these into the above equation and going through the usual method of solution for boundary-value problems of this sort,[1] we obtain, after some algebraic manipulation,

$$\frac{1}{X}\frac{d^2X}{dx^2} + \frac{1}{Y}\frac{d^2Y}{dy^2} = \gamma^2 - \gamma_z^2$$

The first term of this equation is a function of x alone, the second a function of y alone, and the third term a constant. If the equation is to remain true for any value of x and y, then the first two terms must also be constants. Let the first term be designated by the constant $-k_x^2$ and the second by the constant $-k_y^2$. Then the characteristic equation for the solution is

$$-k_x^2 - k_y^2 = \gamma^2 - \gamma_z^2$$

This results in second-order ordinary differential equations, as follows:

$$\frac{d^2X}{dx^2} + k_x^2 X = 0$$

$$\frac{d^2Y}{dy^2} + k_y^2 Y = 0$$

The solutions of these equations are in trigonometric functions or imaginary exponentials. Thus one form of the solution is

$$H_z = (A \cos k_x x + B \sin k_x x)(C \cos k_y y + D \sin k_y y)e^{-\gamma_z z}$$

where A, B, C, and D are constants to be determined from boundary conditions.

[1] William H. Hayt, "Engineering Electromagnetics," pp. 168–177, McGraw-Hill Book Company, Inc., 1958.

Let us now consider the waveguide shown in Fig. 9-24. To determine the values of the constants in this equation for H_z, it is necessary to consider boundary conditions. The boundary conditions which are considered, however, are for the electric field rather than for the magnetic field. The electric field may be found from Eqs. (9-15) and (9-16). Since the tangential electric field must be zero at the surface of the conducting walls of the waveguide, the boundary conditions are then as follows:

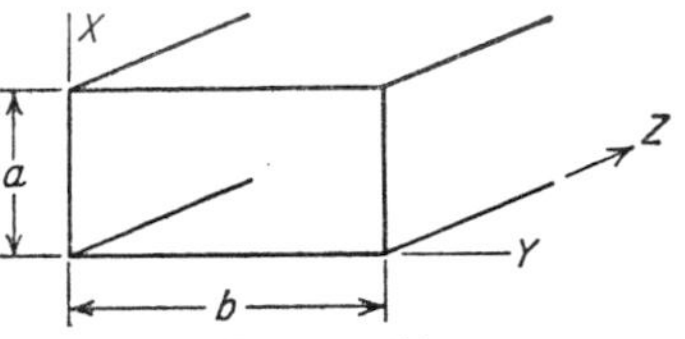

FIG. 9-24. Waveguide geometry.

$$E_y = 0 \qquad \text{where } x = 0, a$$
$$E_x = 0 \qquad \text{where } y = 0, b$$

The electric field is given by

$$E_y = \frac{j\omega\mu}{\gamma_z^2 - \gamma^2}(-Ak_x \sin k_x x + Bk_x \cos k_x x)(C \cos k_y y + D \sin k_y y)e^{-\gamma_z z}$$
$$E_x = \frac{j\omega\mu}{\gamma_z^2 - \gamma^2}(A \cos k_x x + B \sin k_x x)(-Ck_y \sin k_y y + Dk_y \cos k_y y)e^{-\gamma_z z}$$

Substituting the first boundary condition for E_y, we have

$$0 = \frac{j\omega\mu}{\gamma_z^2 - \gamma^2} Bk_x(C \cos k_y y + D \sin k_y y)e^{-\gamma_z z}$$

The right-hand side can be zero for all values of y and z only if $B = 0$, except for the trivial solution where $E = 0$ everywhere. Substituting the second boundary condition,

$$0 = \frac{j\omega\mu}{\gamma_z^2 - \gamma^2}(-Ak_x \sin k_x a)(C \cos k_y y + D \sin k_y y)e^{-\gamma_z z}$$

For this to be nontrivially zero,

$$k_x a = m\pi \qquad \text{or} \qquad k_x = \frac{m\pi}{a}$$

where m is an integer. Hence,

$$E_y = -\frac{j\omega\mu A m\pi/a}{\gamma_z^2 - \gamma^2} \sin \frac{m\pi x}{a} (C \cos k_y y + D \sin k_y y)e^{-\gamma_z z}$$

Substituting the boundary conditions for E_x, we find by the same method that

$$D = 0$$

and

$$k_y b = n\pi \qquad \text{or} \qquad k_y = \frac{n\pi}{b}$$

where n is an integer. It should be noted that k_x, k_y, and γ_z are *eigenvalues* of this problem. Hence, the general expressions for the field are

$$E_y = \frac{-j\omega\mu ACk_x}{\gamma_z^2 - \gamma^2} \sin \frac{m\pi x}{a} \cos \frac{n\pi y}{b} e^{-\gamma_z z} \tag{9-20}$$

$$E_x = \frac{j\omega\mu k_y AC}{\gamma_z^2 - \gamma^2} \cos \frac{m\pi x}{a} \sin \frac{n\pi y}{b} e^{-\gamma_z z} \tag{9-21}$$

It should be noted that these solutions are for the TE mode. Comparable solutions can also be obtained for the TM mode.

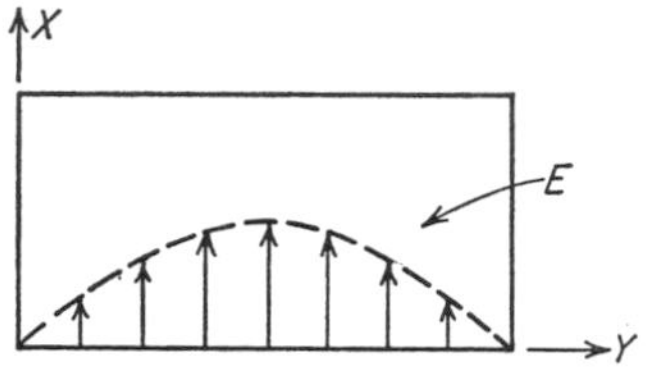

FIG. 9-25. Electric field in TE_{01} waveguide.

The particular mode associated with a given field configuration is described in terms of the number of variations across the waveguide in the x and y directions. Thus, a mode is described as the TE_{mn} mode when m and n are the values of the integers in (9-20) and (9-21).

The fundamental mode associated with the qualitative picture of Fig. 9-22 is obtained when $m = 0$, $n = 1$. In this case the electric field is

$$E_x = \frac{j\omega\mu\pi AC}{b(\gamma_z^2 - \gamma^2)} \sin \frac{\pi y}{b} e^{-\gamma_z z} \tag{9-22}$$

A comparable value for H_y could be obtained. E_y and H_x are everywhere zero in this case. The distribution of electric fields is shown in Fig. 9-25.

The value for γ_z has not yet been obtained. This comes from the characteristic equation. For this fundamental mode (TE_{01}),

$$\gamma^2 - \gamma_z^2 = -\left(\frac{\pi}{b}\right)^2$$

Thus,

$$\gamma_z^2 = \gamma^2 + \left(\frac{\pi}{b}\right)^2$$

For the lossless line,

$$\gamma = j\omega\sqrt{\mu\epsilon}$$

Hence,

$$\gamma_z = \sqrt{\left(\frac{\pi}{b}\right)^2 - \omega^2\mu\epsilon} = j\omega\sqrt{\mu\epsilon}\sqrt{1 - \left(\frac{\pi}{\omega b}\right)^2 \frac{1}{\mu\epsilon}} \tag{9-23}$$

$$= j\omega\sqrt{\mu\epsilon}\sqrt{1 - \frac{\omega_0^2}{\omega^2}}$$

Here we have defined

$$\omega_0^2 = \frac{\pi^2}{b^2\mu\epsilon}$$

The corresponding frequency is

$$f_0^2 = \frac{1}{4b^2\mu\epsilon}$$

From this and the relationship between velocity of propagation and parameters of the medium, we learn that the wavelength in air corresponding to this frequency is

$$\lambda_0 = 2b \tag{9-24}$$

These expressions indicate that the propagation constant is pure imaginary, representing phase shift and no attenuation, for frequencies higher than f_0 and that it is real, representing attenuation and no phase shift, for frequencies below f_0. Hence, the frequency f_0 is known as the cutoff frequency for the waveguide. At frequencies *above* the cutoff the waveguide *transmits* energy. At frequencies below the cutoff the waveguide *attenuates* energy. Thus, unless it is desired to use the waveguide as an attenuator, the frequency used must be above the cutoff frequency.

The wavelength associated with the cutoff frequency (λ_0) is known as the cutoff wavelength. It can be seen that the width of the waveguide is just one-half this wavelength [Eq. (9-24)]. In terms of the qualitative picture discussed at the beginning of Sec. 9-4, it can be seen that the cutoff frequency derived there heuristically is the same as that derived here quantitatively. It was shown that cutoff would occur where the two quarter-wavelength insulators of Fig. 9-22 meet at the center (the width of the strip transmission line in the center is zero).

It is thus necessary that the largest dimension of a waveguide be somewhat greater than half the wavelength in air. For this reason waveguides for frequencies of a few hundred megacycles are so large that they are not practical for most purposes (for example, at 300 Mc the waveguide would have to be more than 50 cm across). On the other hand, for frequencies of many thousands of megacycles, waveguides are quite practical. One frequency where waveguides are frequently used (and about the highest frequency where coaxial lines are used) is the S band around 10 cm wavelength or 3,000 Mc in frequency. Here the waveguide must be somewhat more than 5 cm across, a not unreasonable size. When the wavelength in air is only a few millimeters, the waveguide itself becomes difficult to construct reasonably, just as the coaxial transmission line and parallel-wire transmission lines lose their utility at about 3,000 Mc. Nevertheless, no satisfactory alternative has been found for the millimeter wavelength, and therefore extremely precise and small waveguides are used at these wavelengths.

The velocity of propagation in the waveguide may be obtained from γ_z. It can be seen that the velocity is given by

$$v_p = \frac{1}{\sqrt{\mu\epsilon}\sqrt{1 - \omega_0^2/\omega^2}} = \frac{v_{p0}}{\sqrt{1 - \omega_0^2/\omega^2}} \tag{9-25}$$

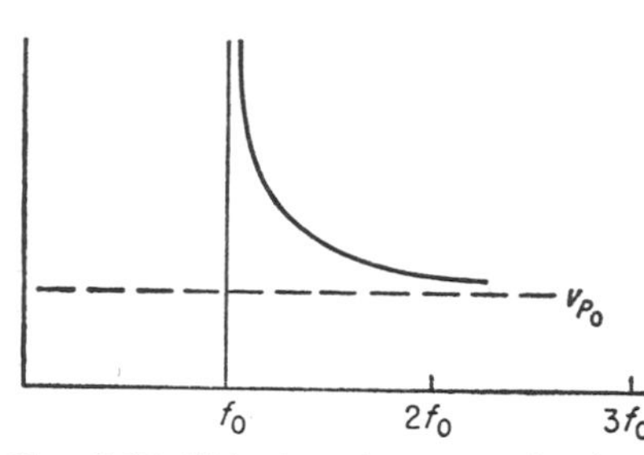

Fig. 9-26 Velocity of propagation in a waveguide.

As indicated in the preliminary qualitative discussion, this velocity is greater than the velocity of light (v_{p0}). In fact, at cutoff the velocity becomes infinite. This situation is shown in Fig. 9-26, where the velocity of propagation is plotted as a function of frequency.

When the frequency is below cutoff, the attenuation constant is given by rewriting Eq. (9-23):

$$\gamma_z = \omega\sqrt{\mu\epsilon}\sqrt{\frac{\omega_0^2}{\omega^2} - 1} \qquad \text{nepers/m} \tag{9-23a}$$

An impedance for the waveguide may be obtained as the ratio

$$\frac{E_x}{H_y}$$

It can be shown that the impedance for the transverse electric wave is given by

$$Z_{TE} = \frac{\eta}{\sqrt{1 - \omega_0^2/\omega^2}} \tag{9-26}$$

The expression for the impedance for the transverse magnetic wave goes to zero instead of infinity at the cutoff wavelength. Similar expressions may be developed for higher-order modes. With rectangular waveguides, the lowest-order mode which can be supported is the TE_{01} mode described here.

Waveguides may also be constructed with circular cross section and with various other cross sections. In general, the phase velocity for any waveguide is higher than that for the wave in air, and the impedance also is related to cutoff frequency by either (9-26) or the comparable equation (with the square-root term in the numerator) which applies for transverse magnetic waves. It should be noted that the group velocity with which energy propagates is not greater than the velocity of light.

The foregoing discussion has dealt with the special impedance and velocity characteristics of waveguides. In other respects, waveguides may be treated the same as transmission lines, and all the methods previously described for treating traveling waves are applicable to waveguides.

Example 9-14. *Waveguide Parameters.* A waveguide is operated at a frequency of 4,000 Mc (4×10^9 cps). The smallest waveguide that can pass this frequency is one for which the free-space wavelength is twice the width. Thus, in this case,

$$\lambda_0 = \frac{3 \times 10^8}{4 \times 10^9} = 7.5 \text{ cm}$$

Hence, the minimum width possible for the waveguide is 3.75 cm (the b dimension on Fig. 9-24). Assume that the width in this guide is 4.5 cm. Then, for this guide,

$$\lambda_0 = 2 \times 4.5 = 9 \text{ cm}$$

The cutoff frequency is therefore

$$f_0 = \frac{3 \times 10^8}{9 \times 10^{-2}} = 3{,}330 \text{ Mc}$$

The ratio of angular frequencies is the same as the ratio of frequencies, so

$$\frac{\omega_0}{\omega} = \frac{f_0}{f} = \frac{3.33}{4.0} = \frac{5}{6}$$

Hence,

$$\sqrt{1 - \frac{\omega_0^2}{\omega^2}} = \sqrt{1 - \tfrac{25}{36}} = 0.552$$

From this, the velocity of propagation may be calculated:

$$v_p = \frac{3 \times 10^8}{0.552} = 5.41 \times 10^8 \text{ m/sec}$$

The wavelength may be readily obtained from v_p and the frequency:

$$\lambda_g = \frac{v_{pg}}{f} = \frac{5.41 \times 10^8}{4 \times 10^9} = 13.5 \text{ cm}$$

It can be seen that the wavelength and therefore the velocity of propagation are higher than they would be if the wave were in space. It is always important to keep this in mind when dealing with waveguides, for it is possible to draw quite erroneous conclusions if one assumes that the wavelength in the guide is the same as in air.

Example 9-15. *Attenuation in Waveguides.* Consider what happens when the waveguide of Example 9-14 is operated at 2,000 Mc and at 500 Mc. Here we must use (9-23*a*). The attenuation at 2,000 Mc is

$$\alpha = \frac{\omega}{v_p}\sqrt{\frac{\omega_0^2}{\omega^2} - 1} = \frac{4\pi \times 10^9}{3 \times 10^8}\sqrt{\frac{3.33^2}{4} - 1}$$
$$= 56 \text{ nepers/m}$$
$$= 4.85 \text{ db/cm}$$

This is a very high attenuation indeed. At 500 Mc, α is approaching its limiting expression,

$$\alpha \approx \frac{\omega_0}{v_p} = 69.5 \text{ nepers/m}$$
$$= 6.0 \text{ db/cm}$$

9-5. Summary

Special techniques for transmission lines used at radio frequencies permit easier calculation of practical problems than is possible with either the formulas or the phasor diagrams discussed in Chaps. 5 and 6. The Smith chart is the most common calculation aid, and it greatly simplifies impedance and admittance calculations.

The Smith chart is a diagram whose polar coordinates are the magnitude of the reflection coefficient and an angle term containing the effects of the angle of reflection coefficient and of travel down the line. Since the maximum value of the reflection coefficient is 1, the entire diagram is contained within a unit circle. Contours of constant conductance and constant susceptance are also circles on this diagram, but they are not concentric with the center of the diagram. Any point on the diagram may be described in terms of any two of the seven coordinates associated with it, these being (1) reflection-coefficient magnitude; (2) standing-wave ratio; (3) angle factor (distance and angle of reflection); (4) resistance and (5) reactance; (6) conductance and (7) susceptance.

The Smith chart is readily adapted to solving matching problems on transmission lines, a very important type of radio-frequency line problem. Since matching involves modification of both magnitude and angle of impedance (or both real and imaginary parts), two quantities must be adjusted to obtain a match. The single-susceptance matching arrangement moves a variable susceptance to a position such that it can match the load. The double-susceptance matching arrangement uses two susceptances, whose location is fixed but whose magnitude is variable, to accomplish the same purpose. These matching arrangements are most easily handled by using the Smith chart.

Transmission lines and pipes carrying acoustic waves, as well as bars supporting acoustic waves, are frequently used as resonators. The simplest form of resonator consists of a transmission line short-circuited at both ends or a tube of air open to the free air at both ends. Such an arrangement is resonant and will support standing waves at a fundamental frequency such that the wavelength for this frequency is twice the length of the resonator. Higher-order modes are also supported where the resonator length is a multiple of a half-wavelength.

At times resonators are foreshortened by capacitors acting as tuning elements or by an equivalent capacitor made up of an open-circuited section of very low impedance transmission line.

Transmission-line resonators have a much higher Q (narrower bandwidth) than lumped-constant resonators, which makes them particularly desirable where extreme stability of frequency is required or where extremely narrow bandwidths are required.

At sufficiently high frequencies it is customary to transmit energy through pipes known as waveguides. In most respects the waveguides are quite comparable to transmission lines, the principal difference being that the phase velocity in a waveguide is higher than the velocity of the corresponding plane wave outside the waveguide.

Waveguides exhibit the phenomenon of cutoff; that is, they can operate only down to a certain minimum frequency. Below that frequency they act as attenuators and do not transmit energy. The simplest waveguide is one in which there is no longitudinal component of electric field and only a half-wavelength of variation of the electric field across the guide. The cutoff wavelength for this mode is just twice the width of the guide. Better performance is obtained for somewhat shorter wavelengths.

Various other field configurations are possible in the waveguide, as can be seen by solving the differential equations and by matching the boundary conditions. Their properties are of considerable interest in the study of microwaves, but a detailed discussion has been omitted here, since they, like the simpler modes, can be treated just like any other traveling-wave device as far as traveling and standing waves are concerned.

PROBLEMS

9-1. The impedance of an antenna is measured to be $40 + j70$ ohms. It is connected to a transmitter through a lossless line 2.10 wavelengths long. What is the load presented to the transmitter? Assume that $Z_0 = 50$ ohms.

9-2. A load is measured to have an impedance of $120 + j37$. It is connected through a lossless line of 100 ohms characteristic impedance 3.88 wavelengths long to its source. What is its input admittance?

9-3. A slotted-line measurement on a 75-ohm transmission line gives a standing-wave ratio of 2.1, with the first minimum located 0.175 wavelength from the load. What are the load impedance and admittance?

9-4. A pressure standing-wave measurement for an acoustic plane wave shows a standing-wave ratio of 7.15. The first minimum occurs 0.41 wavelengths from the reflecting surface. If the wave is traveling in air, what is the acoustic surface impedance of the reflecting surface?

9-5. Repeat Prob. 9-1 for a line with a loss of 0.1 neper/wavelength.

9-6. Repeat Prob. 9-2 for a line with a loss of 0.5 neper/wavelength.

9-7. An alternating heat wave in copper strikes an insulated surface (zero heat flow across the surface). Determine, using the Smith chart, the ratio of the impedance at a point 0.450 wavelength from the boundary to the characteristic impedance.

9-8. A slotted-line measurement on a 50-ohm transmission line gives a standing-wave ratio of 3.7, with the first minimum located 0.380 wavelength from the load. What are the load impedance and admittance?

9-9. Design a single-susceptance tuner to match the line of Prob. 9-1 to its load. Indicate both the value of the lumped element and the length of an equivalent short-circuited stub.

9-10. Repeat Prob. 9-9 for the line of Prob. 9-2.

9-11. Repeat Prob. 9-9 but use a different solution.

9-12. Assuming that the antenna impedance of Prob. 9-1 is unchanged, determine the input impedance using the tuner of Prob. 9-9, but with a frequency 5 per cent higher than the design frequency.

9-13. Design a double-stub tuner to match the line of Prob. 9-1.

9-14. Design a different double-stub tuner to match the line of Prob. 9-1.

9-15. Repeat Prob. 9-12 using the double-stub tuner of Prob. 9-13.

9-16. Use the Smith chart to determine the input impedance in Prob. 5-8.

9-17. Use the Smith chart to determine the input impedance in Prob. 5-9.

9-18. A quarter-wavelength transformer is used to connect a 25-ohm resistive load to a 100-ohm line. Sketch the variation of reactance as a function of frequency between a frequency 25 per cent below the design frequency and one 25 per cent above the design frequency.

9-19. The quarter-wavelength transformer of Prob. 9-18 is replaced by two quarter-wavelength transformers in tandem. Determine the characteristic impedance for the two sections.

9-20. Repeat the sketch of Prob. 9-18 for the line of Prob. 9-19.

9-21. Determine the fundamental resonant frequency of longitudinal waves in a steel rod 3 cm long.

9-22. Determine the fundamental resonant frequency for longitudinal waves in an aluminum rod 10 cm long.

9-23. A short-circuited 50-ohm transmission line is used as the plate tuning circuit of an amplifier operating at 300 Mc. If the plate-to-cathode capacitance is 1.5 $\mu\mu$f, how long is the line? Assume $v_p = 3 \times 10^8$ m/sec.

9-24. An open-circuited 300-ohm line is used as the tuned circuit for an oscillator whose tubes introduce a capacitance of 2.5 $\mu\mu$f. If the oscillator is tuned to 415 Mc, how long is the line? Assume $v_p = 3 \times 10^8$ m/sec.

9-25. A resonator a half-wavelength long is made with a line having attenuation of 1 db/100 m. What is the Q of this resonator if the center frequency is 1,000 Mc? What is its resonant impedance, if Z_0 is 80 ohms? What is its bandwidth?

9-26. Repeat Prob. 9-25 assuming that $Z_0 = 50$ ohms and $\alpha = 0.1$ neper/m.

9-27. How long should an organ pipe be for a resonant frequency of 440 cps?

9-28. For the flute of Example 9-13, determine the nonharmonic frequencies present when the note E is played.

9-29. For the flute of Example 9-13, determine the location of the second hole, corresponding to E (330 cps), assuming that both the D and the E holes are uncovered at the same time to play E.

9-30. RG-52/U waveguide has a cutoff frequency of 6,590 Mc. Its operating range is stated to be from 8,200 to 12,400 Mc. At the ends of its usable range, find v_p, λ_g, and β_z. Compare the length of a quarter-wavelength short-circuited stub using the waveguide at these two frequencies with a stub using an air-filled coaxial line.

9-31. Compute and sketch the attenuation-vs.-frequency curve for this waveguide for frequencies below cutoff.

9-32. Develop the expression for cutoff frequency for a waveguide operating in the TE_{mn} mode. Compare the resulting expression for v_p with that for the TE_{01} mode.

9-33. RG-49/U waveguide has a stated operating range of 3,950 to 5,850 Mc. It is 2.000×1.000 in. in cross section. Find its cutoff frequency. At the ends of its usable range, find v_p, λ_g, and β_z.

10. Oblique-incidence Plane-wave Reflections and Spherical Waves

Plane waves of various kinds—electromagnetic, acoustic, thermal, and diffusion—have been discussed in the previous chapters. Heretofore, attention has been confined to situations where the plane-wave reflections are obviously analogous to transmission-line reflections. Plane waves have always been incident on plane boundaries perpendicular to the direction of wave travel. This chapter treats the complications which arise when the plane waves strike boundaries at other angles.

The concept of a plane wave is a very general one, and any other kind of wave may be built up out of plane waves, just as any time function may be constructed out of sine waves. Thus any wave in space may be Fourier-analyzed into its plane-wave components.[1]

With time functions it is frequently inconvenient to deal with the Fourier series. For example, a square wave may frequently be dealt with as just that—not as a series of harmonics. In the same way, it is frequently desirable to think of certain kinds of waves in space separately, rather than in terms of their plane-wave spectrum. The two most common such sets of waves are spherical and cylindrical waves. The latter are of particular importance in solving boundary-value problems associated with waves on a wire or in a circular waveguide, but they will not be treated in this chapter. The former are nearly always of importance with waves in space, for most waves become essentially spherical far removed from their source; hence they are treated here.

Electromagnetic waves (which are discussed first) are more compli-

[1] J. A. Stratton, "Electromagnetic Theory," p. 362, McGraw-Hill Book Company, Inc., New York, 1941.

cated than acoustic and thermal waves because they are characterized by two vectors perpendicular to the direction of wave travel, whereas the acoustic and thermal waves are characterized by only one vector, in the direction of wave travel, and a scalar. The general techniques applied, however, are the same for all waves.

10-1. Rotation of Coordinates

The expressions for electromagnetic plane waves making an oblique angle with a reflection plane are easily obtained by rotating coordinates. We first consider an electromagnetic wave traveling in the X' direction whose electric field is in the Y' direction. Rotation of the coordinates about the Y' axis is then assumed, and the field expressions are modified accordingly.

The electric field of a plane wave traveling in the X' direction with **E** along the Y' axis is

$$\mathbf{E} = \mathbf{1}_{y'}E_{y'} = \mathbf{1}_{y'}E_0^{-\gamma x'} \tag{10-1}$$

Here $\mathbf{1}_{y'}$ is a unit vector in the Y' direction. Since this is a plane wave, the partial derivatives in the Z' and Y' directions of all the fields are zero, and only X' partial derivatives exist.

The magnetic field could be determined independently, but it is just as easy to get it from the electric field, using Faraday's law. Thus,

$$\nabla \times \mathbf{E} = \begin{vmatrix} \mathbf{1}_{x'} & \mathbf{1}_{y'} & \mathbf{1}_{z'} \\ \dfrac{\partial}{\partial x'} & 0 & 0 \\ 0 & E_{y'} & 0 \end{vmatrix} = \mathbf{1}_{z'} \frac{\partial E_{y'}}{\partial x'} = -j\omega\mu\mathbf{H}$$

with $\mathbf{1}_{x'}$, $\mathbf{1}_{y'}$, and $\mathbf{1}_{z'}$ unit vectors in the X', Y', and Z' directions, whence

$$\mathbf{H} = \mathbf{1}_{z'} \frac{j}{\omega\mu} \frac{\partial E_{y'}}{\partial x'} = \mathbf{1}_{z'} \frac{\gamma}{j\omega\mu} E_0 e^{-\gamma x'} = \mathbf{1}_{z'} \frac{E_0}{\eta} e^{-\gamma x'} \tag{10-2}$$

Here η is the intrinsic impedance.

The power flowing per unit area in the electromagnetic plane wave is given by the Poynting vector. The average value of power flow in watts per square meter is given by

$$\mathbf{P} = \frac{1}{2} \operatorname{Re} \mathbf{E} \times \mathbf{H}^* = \mathbf{1}_{x'} \frac{1}{2} \frac{E_0^2}{\eta} e^{-2\alpha x'} \tag{10-3}$$

where the asterisk represents the complex conjugate.

The wave is traveling in the X' direction of the coordinate system (X',Y',Z') shown in Fig. 10-1. It is desired to find the expressions for the fields in the unprimed coordinate system (X,Y,Z) obtained by rotating the primed coordinate system about the Y' axis (also the Y axis) by an angle θ.

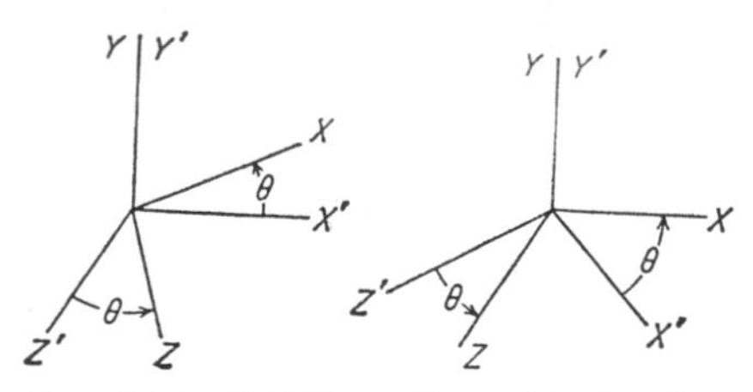

FIG. 10-1. Rotation of coordinates for plane waves.

The Poynting vector (in the X' direction) lies in the ZX plane but not along either unprimed axis; that is, the wave is traveling in a direction intermediate between the directions of the coordinate axes. Likewise, the magnetic field $\mathbf{H}$ is not along an axis but, rather, in the ZX plane normal to the Poynting vector.

In the unprimed system (after rotation), fields may be described by

$$\begin{aligned}
\mathbf{E} &= \mathbf{1}_y E_0 e^{-\gamma x'} \\
\mathbf{H} &= \frac{E_0}{\eta} e^{-\gamma x'}[\mathbf{1}_z(\mathbf{1}_{z'} \cdot \mathbf{1}_z) + \mathbf{1}_x(\mathbf{1}_{z'} \cdot \mathbf{1}_x)] \\
&= \frac{E_0}{\eta} e^{-\gamma x'}[\mathbf{1}_z \cos\theta + \mathbf{1}_x(-\sin\theta)] \\
\mathbf{P} &= \frac{1}{2}\frac{E_0^2}{\eta} e^{-2\alpha x'}[\mathbf{1}_z(\mathbf{1}_z \cdot \mathbf{1}_{x'}) + \mathbf{1}_x(\mathbf{1}_x \cdot \mathbf{1}_{x'})] \\
&= \frac{1}{2}\frac{E_0^2}{\eta} e^{-2\alpha x'}(\mathbf{1}_z \sin\theta + \mathbf{1}_x \cos\theta)
\end{aligned}$$

The remainder of the discussion here deals with lossless media, where α is zero. A complete discussion is possible with lossy media, but it is better to understand the lossless case before tackling the more difficult lossy one.

Note that the results above are for rotation about the direction of the $\mathbf{E}$ vector and that this vector is normal to the plane containing $\mathbf{P}$ and $\mathbf{H}$. Such a wave is said to be *normally polarized* or polarized normal to the *plane of incidence*. Rotation about the $\mathbf{H}$ vector would give polarization "in the plane of incidence." The polarization of an electromagnetic wave is described by the direction of the $\mathbf{E}$ vector. Polarization is defined in terms of the direction the $\mathbf{E}$ vector takes with respect to the plane of incidence. With acoustic waves, the pressure is a scalar, so no polarization need be considered.

In the preceding equations, x' was not changed to the new coordinate system. The modification of x' is important, and applicable to any polarization. Thus,

$$\begin{aligned}
x' &= (\mathbf{1}_{x'} \cdot \mathbf{1}_{x'})x' = \mathbf{1}_{x'} \cdot [\mathbf{1}_z(\mathbf{1}_{x'} \cdot \mathbf{1}_z) + \mathbf{1}_x(\mathbf{1}_{x'} \cdot \mathbf{1}_x)]x' \\
&= \mathbf{1}_{x'} \cdot (\mathbf{1}_z \sin\theta + \mathbf{1}_x \cos\theta)x'
\end{aligned}$$

but

$$x' \sin \theta = z \qquad \text{and} \qquad x' \cos \theta = x$$

so

$$x' = \mathbf{1}_{x'} \cdot (\mathbf{1}_z z + \mathbf{1}_x x) = \mathbf{1}_{x'} \cdot \mathbf{R}$$

where **R** is the radius vector from the origin to a point in the ZX plane with coordinates (z,x). The unit vector $\mathbf{1}_{x'}$ is in the direction of wave travel and is labeled henceforth **N**, because it is *normal* to the equiphase plane.

Hence,

$$x' = \mathbf{N} \cdot \mathbf{R} \tag{10-4}$$

and the fields for lossless media are

$$\mathbf{E} = \mathbf{1}_y E_0 e^{-j\beta \mathbf{N}\cdot\mathbf{R}} \tag{10-5}$$

$$\mathbf{H} = (\mathbf{1}_z \cos \theta - \mathbf{1}_x \sin \theta) \frac{E_0}{\eta} e^{-j\beta \mathbf{N}\cdot\mathbf{R}} \tag{10-6}$$

This derivation applies to waves traveling in the YZ plane, but it can be shown that the form of the exponential in (10-5) and (10-6) is applicable for waves in any direction in space. Thus **N** and **R** may have three components each.

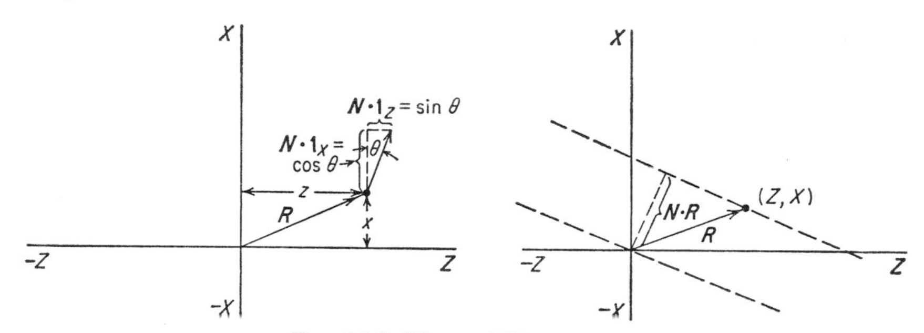

FIG. 10-2. Phase-shift geometry.

The full meaning of the term $\mathbf{N} \cdot \mathbf{R}$ is better understood by reference to Fig. 10-2. Here a wave is assumed to be coming from the $-Z$, $-X$ direction at an incidence angle of θ. Although **R** is the distance from the origin to the point at which the field is determined, it can be seen from the figure that $\mathbf{N} \cdot \mathbf{R}$ does indeed describe the distance between the wavefront through the point (0,0) and that through the point (z,x).

It is convenient to describe the phase shift in terms of the unprimed coordinates Z, X. The radial vector is

$$\mathbf{R} = \mathbf{1}_z z + \mathbf{1}_x x$$

so the exponential appearing in Eqs. (10-5) and (10-6) may be written as

$$e^{-j\beta(\mathbf{1}_z z + \mathbf{1}_x x)\cdot\mathbf{N}}$$

Performing the indicated scalar multiplication, this is

$$e^{-j\beta(z \sin \theta + x \cos \theta)}$$

Hence the field of (10-5) may be written as

$$\mathbf{E} = \mathbf{1}_y E_0 e^{-j\beta z \sin \theta} e^{-j\beta x \cos \theta} \tag{10-7}$$

If either of the exponential factors is considered along with the magnitude, the other describes a wave traveling in the direction of one of the axes, but with a different velocity. This wave is not a *uniform* plane wave, however, for its fields possess derivatives in a direction other than the direction of its travel.

Thus, considering the wave as traveling in the X direction, the total phase shift is given by $\beta x \cos \theta = \beta_x x$, where β_x is defined by

$$\beta_x = \beta \cos \theta \tag{10-8}$$

This is a phase-shift constant for a particular direction, and it can be seen that such a phase-shift constant is less than that in the direction of

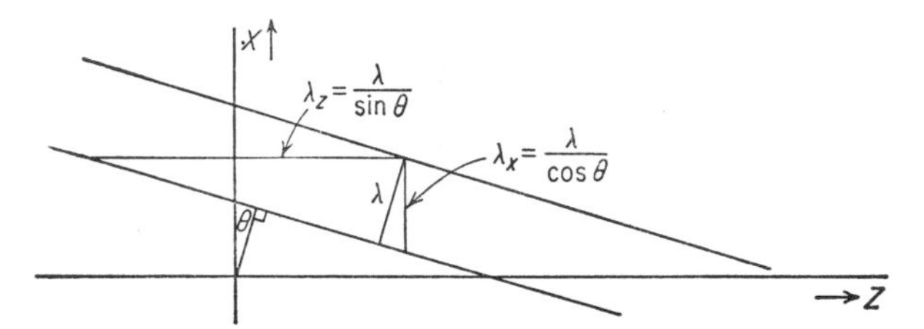

Fig. 10-3. Directional wavelengths.

wave travel (β). Phase velocity in the X direction may be obtained from the relation

$$\beta = \frac{\omega}{v_p}$$

whence

$$v_{px} = \frac{v_p}{\cos \theta} \tag{10-9}$$

It is interesting to note that the phase velocity in a direction other than that of wave travel is greater than in the direction of wave travel. This is due to the fact that it is necessary to travel a greater distance between successive constant-phase surfaces when the direction of travel is other than a direction normal to the surfaces. Phase velocities greater than the velocity of light do not violate the relativity principle, for no object moves with the phase velocity. Actually, the energy in the wave moves with a velocity less than the velocity of light.

This situation is more readily seen in Fig. 10-3. Here the wavelength in the Z and X directions is indicated. Note that the definition of wavelength here is the same as that originally used for transmission lines: the distance between two points separated in phase by 360°. The wavelength in any direction but that of the wave advance is longer than the wavelength in that direction. Since the wavelength and velocity are related by

$$\frac{\omega\lambda}{v_p} = 2\pi$$

$$\lambda_x = \frac{\lambda}{\cos\theta} = \frac{v_{px}}{f} \tag{10-10}$$

Similarly, for the Z direction,

$$\beta_z = \beta \sin\theta \tag{10-11}$$

$$v_{pz} = \frac{v_p}{\sin\theta} \tag{10-12}$$

$$\lambda_z = \frac{\lambda}{\sin\theta} \tag{10-13}$$

10-2. Directional Impedance and Snell's Law

The wave impedance for plane waves is the ratio of magnitudes of **E** to **H**. **E** and **H** are vectors normal to the direction of wave travel and to each other. Thus, a plane wave traveling in the X direction, with $\mathbf{E} = \mathbf{1}_y E_y$ and $\mathbf{H} = \mathbf{1}_z H_z$, has an impedance

$$\eta = \frac{E_y}{H_z} \tag{10-14}$$

If this wave travels in the ZX plane, but not along the X axis, η also describes the ratio of E to H. This ratio is not dependent on the choice of coordinates, since a choice is purely a geometric consideration, and the intrinsic impedance is a parameter of the medium. If, as indicated in connection with (10-7), we wish to consider such a wave as moving in the direction of one coordinate but with phase shift in the direction of the other (so that it is not a plane wave in the usual sense), we may also wish to consider only those field components which contribute to the power flow in the direction of wave travel. Figure 10-4 illustrates the necessary components.

Consider, for example, the fields of (10-5) to (10-7). In terms of a wave in the X direction,

$$\mathbf{E} = \mathbf{1}_y E_0 e^{-j\beta z \sin\theta} e^{-j\beta x \cos\theta} \tag{10-15}$$

$$\mathbf{H} = (\mathbf{1}_z \cos\theta - \mathbf{1}_x \sin\theta)\,\frac{E_0}{\eta}\, e^{-j\beta x \sin\theta} e^{-j\beta z \cos\theta} \tag{10-16}$$

$$\mathbf{P} = \tfrac{1}{2}\,\mathrm{Re}\left(\mathbf{1}_x \cos\theta\,\frac{E_0^2}{\eta} + \mathbf{1}_z \sin\theta\,\frac{E_0^2}{\eta}\right) \tag{10-17}$$

The Z component of **H** contributes to the X component of **P**. Thus, we can write an impedance for that part of the wave contributing to travel in the X direction:

$$Z_x = \frac{E_y}{H_z} = \frac{E}{H \cos \theta} = \frac{\eta}{\cos \theta} \tag{10-18}$$

Similarly, the impedance in the Z direction is

$$Z_z = \frac{E_y}{-H_x} = \frac{\eta}{\sin \theta} \tag{10-19}$$

Z_x and Z_z are known as *directional impedances*. It should be noted that, for this type of polarization (**E** vector normal to plane of incidence), the directional impedances are greater than the intrinsic impedance.

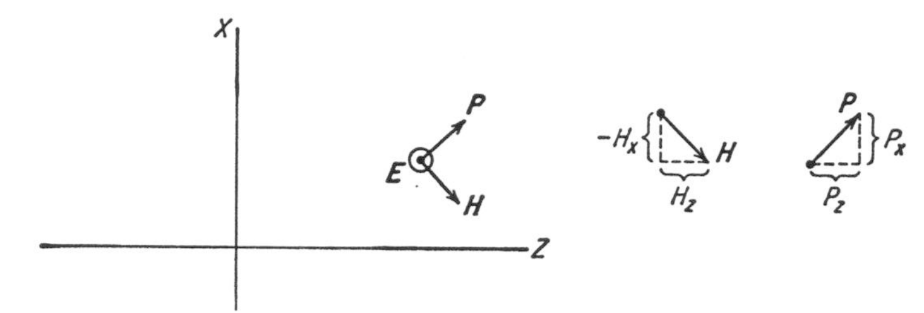

FIG. 10-4. Field components.

When the **E** vector is in the plane of incidence, directional impedances are less than η. These directional impedances may be employed in the same way that characteristic impedances are employed for transmission lines or that intrinsic impedances are employed for complete plane waves traveling through space.

Thus, it is possible to apply the impedances defined above to problems of plane waves incident upon plane boundaries at angles other than normal to the boundaries. With them we can calculate reflected and transmitted waves for this *oblique incidence* just as we did for *normal incidence* in Chaps. 4 to 6. Before we can do this, however, we must know the angles made by the waves with the normal to the reflecting surface, for these angles determine the impedances.

At a boundary between two media there is usually bending of waves. That is, the direction in which the transmitted plane wave leaves the boundary is not the same as that from which the incident wave arrives at the boundary. The relationship between these angles is known as Snell's law. Figure 10-5 shows the necessary geometry.

The basis for the derivation of Snell's law is that the wave traveling along the boundary must travel at the same speed in each of the media. Otherwise the relative phase of these two waves would be continually changing, and boundary conditions could not be satisfied. In Fig. 10-5 a wave whose normal lies in the ZX plane is assumed to strike a boundary between two media identified by subscripts 1 and 2. The boundary is the YZ plane—which appears in our sketch as the Z axis. For the above condition to be satisfied, the velocity of propagation in the Z direction must be the same for both media. Thus, from (10-12),

$$v_{pz1} = v_{pz2}$$

$$\frac{v_{p1}}{\sin \theta_i} = \frac{v_{p2}}{\sin \theta_t}$$

so

$$\sin \theta_t = \frac{v_{p2}}{v_{p1}} \sin \theta_i \tag{10-20}$$

Here θ_i is called the *angle of incidence*, and θ_t is called the *angle of transmission*, or *angle of refraction*. Note that these angles are measured

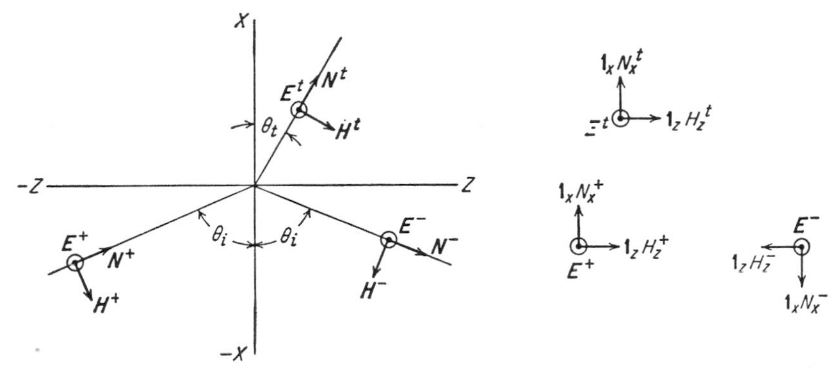

FIG. 10-5. Obliquely incident plane wave showing reflection and transmission. Components shown are those that are pertinent in reflection.

with respect to the *normal* to the interface between the two media. Their complements, measured with respect to the surface itself, are known as *grazing angles.*

In the above derivation, it was assumed that the media were lossless dielectrics. The basic principle of Snell's law is applicable to lossy media as well, for the attenuation along the Z axis must also be the same in both media. To account for both attenuation and phase shift at the same time, we use the propagation constant. Thus,

$$\gamma_1 z \sin \theta_i = \gamma_2 z \sin \theta_t \tag{10-21}$$

is the form Snell's law takes for complex γ. The effect of the complex value for $\sin \theta_t$ is to make both attenuation and phase shift occur in the reflected and transmitted waves. Usually the planes of equal amplitude and the planes of equal phase are not parallel.[1]

[1] J. A. Stratton, "Electromagnetic Theory," pp. 500–505, McGraw-Hill Book Company, Inc., New York, 1941.

In the special case of waves in lossless media the velocity of propagation is given by

$$v_p = \frac{1}{\sqrt{\mu\epsilon}}$$

so for media with equal permeability, (10-20) becomes

$$\sin \theta_t = \sqrt{\frac{\epsilon_1}{\epsilon_2}} \sin \theta_i \tag{10-22}$$

This is a very familiar form of Snell's law.

10-3. Reflection of Plane Electromagnetic Waves at Oblique Incidence

The treatment of the reflection of plane electromagnetic waves when the electric vector is normal to the plane of incidence is different from the treatment when the vector is parallel to that plane. The case of normal polarization is treated first, followed by that for parallel polarization, or polarization in the plane of incidence.

Figure 10-5 shows the fields for a normally polarized plane wave incident upon the boundary between medium 1 and medium 2. The wave originates in the $-Z$, $-X$ quadrant. It is reflected from the YZ plane; therefore, as far as reflection is concerned, we must consider the wave as traveling in the X direction. The impedance in the X direction is given by (10-18), with the appropriate angles in the two media.

The angles involved are described as θ_i, the incident-wave angle, and θ_t, the transmitted-wave angle. The resulting impedances are

$$Z_{x1} = \frac{\eta}{\cos \theta_i} \tag{10-23}$$

$$Z_{x2} = \frac{\eta}{\cos \theta_t} \tag{10-24}$$

Applying Snell's law (10-20), the denominator of the impedance expression in medium 2 is given by

$$\cos \theta_t = \sqrt{1 - \sin^2 \theta_t} = \sqrt{1 - \frac{v_{p2}^2}{v_{p1}^2} \sin^2 \theta_i} \tag{10-25}$$

Since θ_i is already known, it is merely necessary to evaluate for a particular case the directional impedances of (10-23) and (10-24) and obtain a reflection coefficient from the usual expression

$$\Gamma_{Rx} = \frac{Z_{x2} - Z_{x1}}{Z_{x2} + Z_{x1}} \tag{10-26}$$

The reflection coefficient may be used in the usual manner to get the electric field in the reflected wave,

$$E_y^- = \Gamma_R E_y^+ \tag{10-27}$$

and the magnetic field,

$$H_z^- = -\Gamma_R H_z^+ \tag{10-28}$$

Note that the magnetic field here is the Z component, which is the component associated with travel in the X direction and with Z_x. The other component of the magnetic field may be obtained from

$$H_x^- = \frac{E_y^-}{Z_{z1}} = \frac{\Gamma_R E_y^+}{Z_{z1}} = \Gamma_R H_x^+ \tag{10-29}$$

Thus, the electric field of the reflected wave is reduced and phase-shifted by Γ_R, and the component of the magnetic field associated with the wave being reflected (H_z) is reduced, phase-shifted, and reversed by $-\Gamma_R$. The other component of the magnetic field is not reversed but is reduced and phase-shifted.

Heretofore in this chapter the X coordinate has been described in the usual manner for a wave starting at $x = 0$. In accordance with the policy used for other traveling waves (Chaps. 5 and 6), the coordinate x is replaced by the coordinate d, whose positive value increases in the direction of the X component of travel of the reflected wave. Using this coordinate, the net field in the medium of incidence is given by

$$E_{y1} = E_y^+(e^{j\beta_{1x}d} + \Gamma_R e^{-j\beta_{1x}d})e^{-j\beta_{1z}z} \tag{10-30}$$

$$H_{z1} = \frac{E_y^+}{Z_{x1}}(e^{j\beta_{1x}d} - \Gamma_R e^{-j\beta_{1x}d})e^{-j\beta_{1z}z} \tag{10-31}$$

$$H_{x1} - \frac{E_{y1}}{Z_{z1}} \tag{10-32}$$

Here the directional phase-shift constants are as indicated before, except for the additional subscript 1, which refers to the medium of incidence. Thus,

$$\beta_{1x} = \beta_1 \cos \theta_i \tag{10-8a}$$

$$\beta_{1z} = \beta_1 \sin \theta_i \tag{10-11a}$$

The subscript 2 is reserved for the medium into which the wave is transmitted.

Figure 10-5 shows the resulting field components. Note that, as a result of the reversal of H_z without reversal of the other vectors, the direction of the wave normal is changed from a direction making an angle of θ_i with the normal and pointed toward the interface to a direction making the same angle with the normal but pointed away from the interface. The direction of the reflected wave is therefore given by the familiar relationship

$$\text{Angle of reflection} = \text{angle of incidence}$$

which is well-known from optics. The normal unit vectors are shown instead of the Poynting vectors, because the Poynting vectors, representing power as they do, cannot be directly superimposed but, rather, must be calculated with total fields. These normal vectors are in the direction that a Poynting vector associated with only the incident or only the reflected wave would go if the other wave were not present.

With normal incidence (as in the case of tandem transmission lines), a transmission coefficient relating the transmitted electric field to the incident field may be calculated. This procedure was originally introduced in Chap. 5 [Eq. (5-30)]. Using the directional impedances here, a comparable expression may be calculated and used to determine the field in medium 2. Thus,

$$\Gamma_t = \frac{2Z_{x2}}{Z_{x1} + Z_{x2}} \tag{10-33}$$

from which the transmitted fields are

$$E_y^t = E_y^+ \Gamma_t e^{j\beta_{2x}d} e^{-j\beta_{2z}z} \tag{10-34}$$

$$H_z^t = \frac{E_y^+}{Z_{x2}} \Gamma_t e^{j\beta_{2x}d} e^{-j\beta_{2z}z} \tag{10-35}$$

$$H_x^t = \frac{E_y^t}{Z_{z2}} \tag{10-36}$$

The field expressions in medium 1 indicate the presence of a standing wave there, with its surfaces of constant amplitude parallel to the interface (at constant d or constant x). This standing wave travels in the Z direction, along the interface. Standing-wave calculations may be made for this wave with the Smith chart, just as for a wave on a transmission line. In medium 2, of course, there is only one wave if the medium is infinite. If it is finite, the situation may be treated as a tandem-transmission-line problem.

Example 10-1. *Wavelength in a Waveguide in Terms of a Directional Wavelength.* It is possible to describe the performance of a waveguide in terms of directional wavelengths and impedances, obtaining the same result as in Chap. 9. Consider a normally polarized wave incident at angle θ_i on a perfect conductor. For a perfect conductor, $\Gamma_R = -1$. Hence the fields in the incident medium become, by Eq. (10-30),

$$E_y = E_y^+(e^{j\beta_x d} - e^{-j\beta_x d})e^{-j\beta_z z} = 2jE_y^+ \sin \beta_x d \; e^{-j\beta_z z}$$

The standing-wave pattern associated with this field and having no reflections in the Z direction has zeros wherever $\beta_x d = n\pi$, as shown in Fig. 10-6. At any point where the electric field is zero, it is possible to insert a conductor in the ZY plane, since the boundary conditions will be satisfied whether or not the conductor is inserted. Insert such a conducting plane and two other conducting planes parallel to the XZ plane. These latter may be inserted, since there is no electric field parallel to them and charges on them can terminate the field already indicated. Now let the value of d associated with

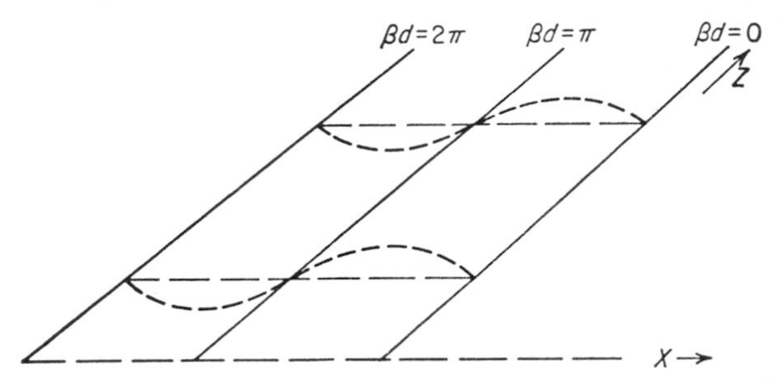

FIG. 10-6. Standing-wave pattern for reflection from perfect conductor.

the zero field point in the standing wave be $d = b$. Using this to define β_x, the electric field is

$$E_y = 2jE_y^+ \sin \frac{n\pi d}{b} e^{-j\beta_z z}$$

This equation is essentially the same as Eq. (9-21), except that here $\gamma_z = j\beta_y$. The multiplying factor is $2jE_y^+$; we are using the coordinate d instead of z.

When $n = 1$, this corresponds to the field for the dominant mode. It is given by

$$E_y = 2jE_y^+ \sin \frac{\pi d}{b} e^{-j\beta_z z}$$

Note here that $\beta_z = \beta \sin \theta$, from Eq. (10-11), and that $\beta_x = \beta \cos \theta$, from Eq. (10-8). Writing

$$\cos \theta = \frac{\beta_x}{\beta}$$

we find that

$$\beta_z = \beta \sqrt{1 - \left(\frac{\beta_x}{\beta}\right)^2}$$

The ratio appearing here is given by

$$\frac{\beta_x}{\beta} = \frac{n\pi}{b\beta} = \frac{n\lambda}{2b}$$

If

$$\lambda_0 = \frac{2b}{n}$$

then

$$\frac{\beta_x}{\beta} = \frac{\lambda}{\lambda_0} = \frac{\omega_0}{\omega}$$

so

$$\beta_z = \beta \sqrt{1 - \left(\frac{\omega_0}{\omega}\right)^2}$$

as indicated in Chap. 9. Thus, the parameters of a waveguide may be derived from oblique-incidence reflection just as they are derived from the boundary-value problem. With nonrectangular cross sections for the waveguides, it is necessary to solve the boundary-value problem as such.

Example 10-2. *Reflection of a Normally Polarized Wave.* Suppose the following wave parameters apply:

$E^+ = 1$ volt/m
$\theta_i = 30°$
$\epsilon_1 = \epsilon_0$ (air)
$\epsilon_2 = 4\epsilon_0$ (glass)
$f = 3 \times 10^9$ cps

It is desired to find θ_t, $\mathbf{E}^-$, $\mathbf{E}^t$, $\mathbf{H}^+$, $\mathbf{H}^-$, and $\mathbf{H}^t$.

To find θ_t, we apply Snell's law. Thus,

$$\sin \theta_t = \frac{\beta_1}{\beta_2} \sin \theta_i = \frac{\omega \sqrt{\mu_0 \epsilon_0}}{\omega \sqrt{\mu_0 4 \epsilon_0}} \sin \theta_i$$
$$= \tfrac{1}{2} \sin \theta_i = 0.25$$

Hence,

$$\theta_t = 14.4°$$

To find the reflection coefficient, it is necessary to obtain the impedances in the X direction. Thus,

$$Z_{x1} = \frac{\eta_0}{\cos \theta_i} = 1.16\eta_0$$

and

$$Z_{x2} = \frac{\eta_2}{\cos \theta_t} = \frac{\sqrt{\mu_0/4\epsilon_0}}{\cos \theta_t} = \frac{\eta_0}{2 \cos 14.4°}$$
$$= 0.52\eta_0$$

Hence, the reflection coefficient for the X direction is

$$\Gamma_{Rx} = \frac{0.52 - 1.16}{0.52 + 1.16} = -\frac{0.64}{1.68} = -0.381$$

Hence,

$$\mathbf{E}^- = -\mathbf{1}_y 0.38 \text{ volt/m}$$

$$\mathbf{H}^+ = \frac{E^+}{\eta_0}(\mathbf{1}_z \cos 30° - \mathbf{1}_x \sin 30°) = \tfrac{1}{377}(\mathbf{1}_z 0.866 - \mathbf{1}_x 0.50) \text{ amp/m}$$

Similarly, the reflected magnetic field is given by

$$\mathbf{H}^- = -\frac{0.38}{377}(-\mathbf{1}_z 0.866 - \mathbf{1}_x 0.50) \text{ amp/m}$$

The transmission coefficient, from (10-33), is

$$\Gamma_t = \frac{2Z_{x2}}{Z_{x1} + Z_{x2}} = \frac{2 \times 0.518}{1.68} = \frac{1.036}{1.68} = 0.657$$

Hence, the transmitted electric field is

$$\mathbf{E}^t = \mathbf{1}_y 0.657 \text{ volt/m}$$

and the transmitted magnetic field is

$$\begin{aligned}\mathbf{H}^t &= \frac{0.657}{\frac{1}{2} \times 377}(\mathbf{1}_z \cos 14.4° - \mathbf{1}_x \sin 14.4°) \\ &= \frac{1.314}{377}(\mathbf{1}_z 0.975 - \mathbf{1}_x 0.25) \text{ amp/m}\end{aligned}$$

Thus, the technique used here is the same as that used for reflections on a transmission line, the only additional complication being due to the extra components of the field vectors.

Total Reflections. When a wave goes from a medium of high dielectric constant (low velocity of propagation) to one of lower dielectric constant (higher velocity of propagation), there is a fixed range of angles for which power may be transmitted through the interface. There is another range of angles where "total reflection" takes place; that is, there is no net power flow into the second medium. The reason for this can be seen from examination of Snell's law in the form

$$\text{(10-22)} \qquad \sin \theta_t = \sqrt{\frac{\epsilon_1}{\epsilon_2}} \sin \theta_i$$

When $\epsilon_1 > \epsilon_2$, there is a range of values of θ_i where $\sin \theta_t > 1$.

To one accustomed to thinking only of real numbers for angles, this appears an impossible state of affairs. However, if complex angles are allowed, it is quite easy to understand how this can hold. Consider the sine of an angle α which has real and imaginary parts:

$$\begin{aligned}\sin \alpha &= \sin (\alpha_r + j\alpha_i) \\ &= \sin \alpha_r \cosh \alpha_i + j \sinh \alpha_i \cos \alpha_r\end{aligned}$$

The value of sin α is a real number greater than 1 when

$$\cos \alpha_r = 0$$

and consequently

$$\sin \alpha_r = 1$$

In this case,

$$\sin \alpha = \cosh \alpha_i$$

whose minimum value is 1. It is frequently necessary to use the cosine of an angle whose sine is greater than 1. The normal trigonometric relation is used; that is,

$$\cos \alpha = \sqrt{1 - \sin^2 \alpha} = \sqrt{1 - \cosh^2 \alpha_i} = \pm j \sqrt{\cosh^2 \alpha_i - 1}$$

Thus, when the sine is greater than 1, the cosine is pure imaginary. This relationship is used in what follows.

The impedance for wave travel away from the interface in medium 2 is given by (10-24) and is dependent on the cosine, so it becomes

$$Z_{x2} = \frac{\eta_2}{\sqrt{1 - \sin^2 \theta_t}}$$

but in this case the denominator is imaginary, so

$$Z_{x2} = \frac{\eta_2}{j \sqrt{\sin^2 \theta_t - 1}} \tag{10-37}$$

Since this represents a purely reactive load to the wave in medium 1, the reflection coefficient is of unity magnitude, and no energy is transmitted into medium 2.

The angle at which this phenomenon first occurs is called the angle of total reflection for θ_i, and it occurs where sin θ_t is unity. Thus, at

$$\theta_{\text{tot refl}} = \arcsin \sqrt{\frac{\epsilon_2}{\epsilon_1}} \tag{10-38}$$

Z_{x2} is infinite, so the reflected electric field is in phase with the incident field. For larger incident angles, the reactive load is finite, and the reflection coefficient is complex.

Although there is no average energy flow into medium 2, energy is stored there, and **E** and **H** fields exist. They are in time quadrature, as indicated by (10-37). To see what happens to them with increasing distance from the interface, consider β_{x2}:

$$j\beta_{x2} = j\beta_2 \cos \theta_t = \beta_2 \sqrt{\sin^2 \theta_t - 1} = \alpha_{x2} \tag{10-39}$$

Thus, the electric field is given by

(10-40) $$E_{y2} = \Gamma_t E_y^+ e^{\alpha_{x2}d} e^{-j\beta_{z2}z} = \Gamma_t E_y^+ e^{-\alpha_{x2}x} e^{-j\beta_{z2}z}$$

with a similar expression for each H component. The important characteristic of this expression is the exponential attenuation in medium 2 in the X direction, away from the interface. Waves of this type are known as *evanescent waves.*

Energy storage is associated with evanescent waves, as there must be whenever fields exist, but there is no average flow of power into the second medium once the fields have been established.

Example 10-3. *Angles of Total Reflection.* Consider two examples at opposite extremes for dielectric constant: a wave emerging from water into air and a wave emerging from dry snow into air.

The dielectric constant of water is $81\epsilon_0$. Hence,

$$\theta_{\text{tot refl}} = \arcsin \sqrt{\tfrac{1}{81}} = 6.4°$$

Thus, there is only a small range of angles for which total reflection does not occur with light waves traveling from water into air. It is easy to observe this effect in a swimming pool by looking up from under water. Looking up at any angle very far from the vertical, the surface appears as a mirror, which indicates that total reflection is taking place.

With dry snow, the dielectric constant may be as low as $1.5\epsilon_0$. In this case, substituting in (10-38) gives total reflection occurring at a 55° angle of incidence. Thus, there is a very wide range of angles for which waves may leave dry snow.

Parallel Polarization. When the **E** vector lies in the plane of incidence, that is, when the **E** vector and the Poynting vector form a plane perpendicular to the plane from which reflection takes place, we say that the wave is *parallel-polarized.* The fields for this situation can be obtained by rotating the primed coordinate system of Fig. 10-1 about the **H** field instead of about the **E** field.

Instead of carrying out the rotation indicated, we here assume that the plane of incidence is the ZX plane, as before, and that the magnetic field is normal to this plane (in the Y direction). The situation is shown in Fig. 10-7. Here it can be seen that

(10-41) $$\mathbf{H}^+ = \mathbf{1}_y H_0^+ e^{-j\beta \mathbf{N}_i \cdot \mathbf{R}}$$

(10-42) $$\mathbf{E}^+ = (-\mathbf{1}_z \cos\theta_i + \mathbf{1}_x \sin\theta_i)\eta_1 H_0^+ e^{-j\beta \mathbf{N}_i \cdot \mathbf{R}}$$

(10-43) $$\mathbf{N}_i = \mathbf{1}_z \sin\theta_i + \mathbf{1}_x \cos\theta_i$$

(10-44) $$\mathbf{N}_i \cdot \mathbf{R} = z \sin\theta_i + x \cos\theta_i$$

The impedance expressions for parallel polarization turn out to be different from those for normal polarization. They are

$$Z_{x1} = -\frac{E_z^+}{H_y^+} = -\frac{\eta_1 H_0^+}{H_0^+}(-\cos\theta_i) = \eta_1 \cos\theta_i \tag{10-45}$$

$$Z_{x2} = \eta_2 \cos\theta_t \tag{10-46}$$

$$Z_{z1} = \frac{E_x^+}{H_y^+} = \eta_1 \sin\theta_1 \tag{10-47}$$

$$Z_{z2} = \eta_2 \sin\theta_t \tag{10-48}$$

θ_t may be obtained from θ_i by Snell's law, which is independent of polarization [see Eqs. (10-20) to (10-22)]. Note that the angular functions appear in the numerator for parallel polarization, whereas they are in the denominator for normal polarization. Thus, the wave impedances

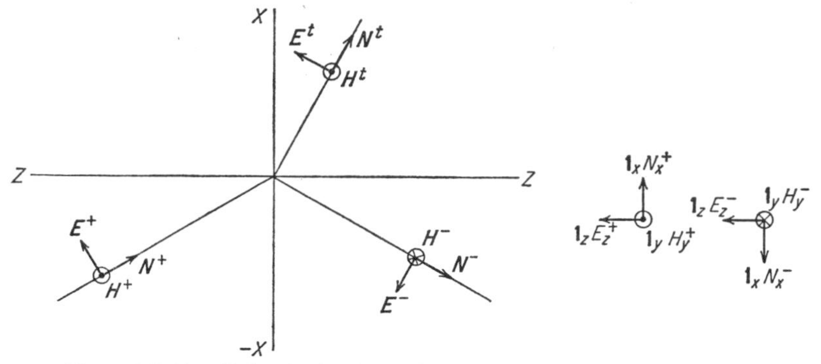

FIG. 10-7. Parallel-polarization reflection of electromagnetic wave.

in the Z direction are less than (or equal to) the intrinsic impedance for parallel polarization and greater than (or equal to) the intrinsic impedance for normal polarization.

Brewster Angle. One may calculate reflected waves for the parallel-polarization case just as for the normal-polarization case—and again it turns out that the reflected wave travels away from the interface at an angle of reflection equal to the angle of incidence. One new phenomenon is possible, however. This is the presence of an angle of *total transmission*, or *zero reflection*. This angle is known as the *Brewster angle*. In optics it is sometimes called the *polarizing angle*, for it is an angle at which normally polarized waves are reflected and parallel-polarized waves are not; hence a wave striking the surface with both polarizations present is reflected with only one polarization, and it is said to be polarized.

The Brewster angle is easily obtained from the standard formula for the reflection coefficient. The condition for zero reflection is that the wave in medium 1 be properly terminated; that is, the impedances (in the X direction) must be the same in medium 1 and medium 2. This condition occurs when

$$\eta_2 \cos \theta_t = \eta_1 \cos \theta_i \tag{10-49}$$

but

$$\cos \theta_t = \sqrt{1 - \sin^2 \theta_t} = \sqrt{1 - \frac{\beta_1^2}{\beta_2^2} \sin^2 \theta_i}$$
$$= \sqrt{1 - \frac{\epsilon_1}{\epsilon_2} \sin^2 \theta_i}$$

so

$$\cos^2 \theta_i = \frac{\epsilon_1}{\epsilon_2}\left[1 - \frac{\epsilon_1}{\epsilon_2}(1 - \cos^2 \theta_i)\right]$$

which reduces to

$$\cos^2 \theta_i = \frac{1}{1 + \epsilon_2/\epsilon_1} \quad \text{(Brewster angle)} \tag{10-50}$$

When one of the media is a conductor (not necessarily a good one), the match is usually impossible. The angle at which the closest approach to a match (closest approach to total transmission) occurs is called the *pseudo Brewster angle.*

Example 10-4. *Fields with Parallel Polarization.* Let us suppose that the magnetic field is given and that the other parameters are the same as in Example 10-2. That is, we are given

$\mathbf{H}^+ = \mathbf{1}_y 10$ amp/m
$\theta_i = 30°$
$\epsilon_1 = \epsilon_0$ (air)
$\epsilon_2 = 4\epsilon_0$ (glass)
$f = 3 \times 10^9$ cps

We are to find θ_t, $\mathbf{E}^-$, $\mathbf{E}^t$, $\mathbf{E}^+$, $\mathbf{H}^-$, and $\mathbf{H}^t$.

Since Snell's law is the same for both normal and parallel polarization, the angle of transmission is the same; that is,

$$\theta_t = 14.4°$$

The incident value of the electric field can be seen to be

$$\mathbf{E}^+ = 3{,}770(-\mathbf{1}_z 0.866 + \mathbf{1}_x 0.5)e^{-j\beta_{1z}z}e^{j\beta_{1x}d}$$

The characteristic impedance in medium 1 for wave travel in the X direction is

$$Z_{x1} = \eta_0 \cos \theta_i = 0.866\eta_0$$

In medium 2, this is

$$Z_{x2} = \eta_2 \cos \theta_t = \frac{\eta_0}{2} 0.97 = 0.485\eta_0$$

Hence, the reflection coefficient is

$$\frac{0.485 - 0.866}{0.485 + 0.866} = \frac{-0.381}{1.356} = -0.272$$

This should be compared with the normal-polarization value, found in Example 10-2, of -0.381. Thus, with parallel polarization in this particular case, the reflection coefficient is less in magnitude than with normal polarization. This is ordinarily the case.

The reflected field is, of course, given by

$$\mathbf{E}^- = 3{,}770(\mathbf{1}_z 0.866 \times 0.278 - \mathbf{1}_x 0.5 \times 0.278)e^{-j\beta_{1z}z}e^{-j\beta_{1x}d}$$

The transmission coefficient is obtained as usual from

$$\Gamma_t = \frac{2Z_{z2}}{Z_{z1} + Z_{z2}} = \frac{2 \times 0.485}{1.356} = \frac{0.97}{1.356} = 0.74$$

This should be compared with the value for normal polarization: 0.657. Hence, the electric field in the second medium is greater in the case of parallel polarization. This too is a customary state of affairs. The values of $\mathbf{E}^t$ and $\mathbf{H}^t$ are given by

$$\mathbf{E}^t = 3{,}770 \times 0.74(-\mathbf{1}_z 0.97 + \mathbf{1}_x 0.25)e^{-j\beta_{2z}z}e^{j\beta_{2x}d}$$
$$\mathbf{H}^t = 10 \times \mathbf{1}_y 0.74 e^{-j\beta_{2z}z}e^{j\beta_{2x}d}$$

It can be seen from this that the methods for approaching parallel and normal polarization are very similar.

Example 10-5. *Brewster Angles.* The Brewster angle corresponds to a match between the wave in the one medium and that in the other. In this example we consider waves going between air and water. Brewster angles can occur for transmission in either direction.

For the wave going from air into water, we find, using (10-50),

$$\cos^2 \theta_i = \frac{1}{1 + 81}$$

In this case the Brewster angle comes out to be 83.67°. Thus, a wave in air must be going almost *horizontally* before it is matched to the water. When the wave in the water is coming out (in which case it can also be totally reflected), we have

$$\cos^2 \theta_i = \frac{1}{1 + \frac{1}{81}}$$

so that the Brewster angle is 6.33°. Thus, a wave coming from water into air is matched at almost normal incidence.

Most dielectrics have a relative dielectric constant under 10, in which case the Brewster angles are less extreme than they are with water.

10-4. Reflection of Plane Acoustic Waves at Oblique Incidence

The techniques of directional impedance and Snell's law used for electromagnetic waves may frequently be used with acoustic, thermal, and other waves. The differences in application are due to the different vector and scalar nature of the waves and to the different boundary conditions. Whereas electromagnetic waves are characterized entirely by vectors, acoustic, thermal, and chemical-concentration waves are each characterized by a scalar and a vector. Longitudinal waves in solids involve *two* longitudinal vectors. Waves on vibrating strings and drums are entirely described by scalars.

We normally think of acoustic and thermal waves as scalar, but in Chap. 3 we found that, in each case, the wave is described by the familiar scalar quantity plus a vector in the direction of wave travel. For the acoustic wave, this vector is the differential particle velocity (whether the wave is in a fluid or a solid), and for the thermal wave it is the rate of heat flow per unit area.

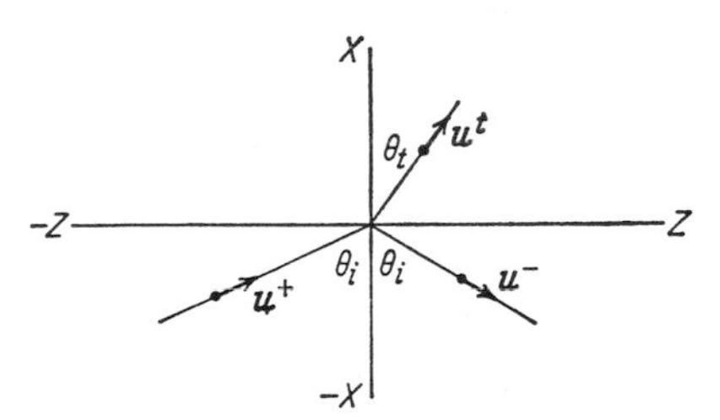

FIG. 10-8. Reflection of acoustic plane wave.

Figure 10-8 shows the coordinates and the pertinent quantities for acoustic waves. It will be recalled that the impedance was defined as the ratio of particle velocity to pressure. To define a directional impedance, it is necessary to use the component of velocity contributing to wave travel in the direction considered. This is somewhat more obvious than in the electromagnetic case, because the wave velocity and the wave normal are in the same direction. Thus,

$$Z_1 = \frac{u^+}{p^+} \tag{10-51}$$

$$Z_{1x} = \frac{u_x^+}{p^+} = \frac{u^+ \cos \theta_i}{p^+} = Z_1 \cos \theta_i \tag{10-52}$$

Of course, by a similar argument,

$$Z_{2x} = Z_2 \cos \theta_t \tag{10-53}$$

with Snell's law being used to find θ_t, as with electromagnetic waves. Since Snell's law is based upon identical velocities of propagation along the boundary in the two media, it must apply for acoustic waves as well as for electromagnetic waves.

The form of the directional impedance for acoustic waves is the same as that for parallel-polarized electromagnetic waves, so both total reflection and total transmission may take place. The reflection coefficient is, of course,

$$\Gamma_{Rx} = \frac{u_x^-}{u_x^+} = \frac{Z_2 \cos \theta_t - Z_1 \cos \theta_i}{Z_2 \cos \theta_t + Z_1 \cos \theta_i} \tag{10-54}$$

and the transmission coefficient is

$$\Gamma_{tx} = \frac{u_x^t}{u_x^+} = \frac{2Z_2 \cos \theta_t}{Z_2 \cos \theta_t + Z_1 \cos \theta_i} \tag{10-55}$$

The intrinsic acoustic impedance, regardless of the type of medium, is given by

$$Z_0 = \frac{1}{\rho_v v_p} \tag{3-22}$$

where ρ_v is the density and v_p the velocity of propagation. Since the velocity of propagation is, for acoustics, one of the "handbook parameters," the most convenient form of Snell's law, for acoustics, is

$$\sin \theta_t = \frac{v_{p2}}{v_{p1}} \sin \theta_i \tag{10-20}$$

Since total reflection is possible with the parallel-polarized electromagnetic wave, it is also possible with the acoustic wave—and, in fact, frequently occurs.

It is important to note that this analogy holds only when the boundary conditions involve pressure and the *normal* component of particle velocity. This means that it must be possible for tangential *slip* to occur along the boundary, without viscous shear stresses.

Example 10-6. *Acoustic Wave across an Air-Water Boundary.* In Example 3-3 we found that the characteristic impedance for air is 0.00235 m³/newton-sec and that the velocity of propagation is 330 m/sec. Example 3-5 shows that the characteristic impedance for water is 6.90×10^{-7} m³/newton-sec and that the velocity of propagation in water is 1,450 m/sec. The angle of total reflection occurs where

$$\beta_1 \sin \theta_i = \beta_2 \qquad \text{or} \qquad \sin \theta_i = \frac{\beta_2}{\beta_1} = \frac{\omega}{v_{p2}} \frac{v_{p1}}{\omega} = \frac{v_{p1}}{v_{p2}}$$

In this case, therefore,

$$\sin \theta_i = \frac{330}{1{,}450} = 0.227$$

$$\theta_i = 13.15°$$

is therefore the angle of total reflection for an acoustic wave going from air to water. Since this is nearly vertical, it probably helps to explain why conversation along the boundary of a swimming pool is not readily heard within the water. The small transmission coefficient also helps to explain this.

Let us now consider a wave in air incident upon the water at an angle of incidence of 10°. Applying Snell's law, we find

$$v_{p2} \sin \theta_i = v_{p1} \sin \theta_t$$

Hence,

$$\sin \theta_t = \frac{1{,}450 \sin 10°}{330} = 0.765$$

$$\theta_t = 49.9°$$

When $\theta_i = 10°$, the following conditions prevail:

$$Z_{x1} = Z_1 \cos \theta_i = 0.00235 \cos 10° = 0.00231 \text{ m}^3/\text{newton-sec}$$

$$Z_{x2} = Z_2 \cos \theta_t = 6.90 \times 10^{-7} \cos 49.9° = 4.45 \times 10^{-7} \text{ m}^3/\text{newton-sec}$$

Hence, the reflection coefficient is given by

$$\Gamma_{Rx} = \frac{4.45 \times 10^{-7} - 2.31 \times 10^{-3}}{4.45 \times 10^{-7} + 2.31 \times 10^{-3}} \approx -1$$

The transmission coefficient is, of course,

$$\Gamma_{tx} = \frac{8.90 \times 10^{-7}}{2.31 \times 10^{-3}} = 3.85 \times 10^{-4}$$

Hence, the pressures in the water are very small compared with those in the air.

10-5. Oblique-incidence Reflection of Waves in Solids

Waves in solids obey reflection laws similar to those for acoustic waves, but the situation in solids is more complicated. Both longitudinal (acoustic) waves and shear waves (which have polarization) exist in solids. When a longitudinal wave strikes the boundary between two solids at oblique incidence, reflected and transmitted longitudinal waves are set up, just as for acoustic waves in gases. With the waves in solids, however, shear (transverse) waves are also set up at the interface; both reflected and transmitted shear waves are present. Thus, any treatment of reflection of waves in solids at oblique incidence must consider the transfer of energy to these different types of waves.

With acoustic waves, the components of particle velocity normal to the boundary were assumed to be the same in both media. The pressure, a scalar, was also assumed to be continuous across the boundary. The first was treated as analogous to the presence of the same voltage on both lines at the junction of two transmission lines, and the latter was considered analogous to continuity of current in such a junction. With these analogies, the directional impedances could be applied to determine reflection and transmission. It was assumed that slip could exist along the boundary without causing stresses—that the particle velocities parallel to the boundary could be different in the two media.

With waves in solids, the pressure of the acoustic wave is replaced by a stress vector (frequently, in fact, by a stress tensor). It is not reasonable to assume that slip can occur in solids without setting up shear stresses. Thus, the boundary conditions require that[1] (1) the stress *vector* be continuous across the boundary (both components) and (2) the particle velocity *vector* be continuous across the boundary. A good electrical analogy is somewhat difficult to devise for this situation. Four equations are necessary—one each for the tangential and normal components of stress and of velocity.

Both reflection and transmission coefficients may be calculated for both longitudinal and shear waves by solving simultaneously the four boundary-condition equations. The results are complicated and are not given here.[2] Actually, three separate cases must be considered, depending on the type of incident wave. Incident waves may be longitudinal, transverse-polarized in the plane of incidence, or transverse-polarized normal to the plane of incidence.

In order for the boundary conditions to be matched at all points along the boundary, it is necessary that all waves travel along the boundary at the same speed. Since this is just the condition by which Snell's law was derived, the angles of reflection and transmission may be determined for both longitudinal and shear waves by application of Snell's law. Because of the difference in velocity of propagation for longitudinal and shear waves, the angle of reflection for a type of wave different from the incident wave is not the same as the angle of incidence. These principles are illustrated in Example 10-7.

Example 10-7. *Acoustic Wave between Rock Layers.* In seismic prospecting, the bending of waves between layers of different kinds of rock is important. Consider here a longitudinal wave in a sandstone overburden incident upon a granite layer at an angle of 20°. It is desired to find the angles of reflection and refraction for the

[1] C. F. Richter, "Elementary Seismology," pp. 670–672, W. H. Freeman & Co., San Francisco, 1958.

[2] See, for example, W. Muskat and M. W. Meres, Reflection and Transmission Coefficients for Plane Waves in Elastic Media, *Geophysics*, vol. 5, pp. 115–148, 1940.

longitudinal wave and for the shear wave which is set up at the boundary.

The values used for the parameters of sandstone and granite appear to be representative, although they are not quoted for a particular sandstone and a particular granite.[1] These representative values are shown in Table 10-1.

TABLE 10-1

Parameter	Sandstone	Granite
Y_0	1.5×10^{10} newtons/m^2	4×10^{10} newtons/m^2
v_p (longitudinal)	2,000 m/sec	5,000 m/sec
v_p (transverse)	1,200 m/sec	2,700 m/sec

For plane waves it is necessary to use the bulk modulus rather than Y_0. It was stated in Chap. 3 that the bulk modulus $Y_B \approx \frac{4}{3}Y_0$, and this approximation is used here.

The angle of reflection for the longitudinal wave is, of course, equal to the angle of incidence (20°). The angle of transmission is given by Snell's law as

$$\sin \theta_t = \frac{v_{p2}}{v_{p1}} \sin \theta_i = \tfrac{5}{2} \sin 20° = 0.855$$

so

$$\theta_t = 58.9°$$

Thus, the longitudinal incident wave is bent from 20° to almost 59° away from the vertical.

For the transverse wave,

$$\sin \theta_{rt} = \frac{v_{t1}}{v_{p1}} \sin \theta_i = \frac{1.2}{2.0} \sin 20° = 0.205$$

where v_t denotes the transverse velocity. Thus, the angle of reflection for the transverse wave is 11.8°. The transmitted angle for the transverse wave is given by

$$\sin \theta_{tt} = \frac{v_{t2}}{v_{p1}} \sin \theta_i = \frac{2.7}{2.0} \sin 20° = 0.462$$

so the angle of transmission is 27.5°.

[1] C. A. Heiland, "Geophysical Exploration," pp. 467–472, Prentice-Hall, Inc., Englewood Cliffs, N.J., 1940.

The complete assembly of incident, reflected, and transmitted waves is shown in Fig. 10-9.

10-6. Thermal-plane-wave Reflection

For thermal waves the plane-wave impedance was defined as

$$Z = \frac{T}{Q} \tag{10-56}$$

where T is the temperature and **Q** the rate of heat flow per unit area. The latter variable is a vector, pointing in the direction of heat flow, so that the heat wave is, in a sense, a longitudinal wave like the acoustic wave. The directional impedance for acoustics uses the component of the longitudinal flow vector (in acoustics the *velocity*) in the desired direction, and a directional thermal impedance is formed in the same way. Thus,

$$Z_{1x} = \frac{T}{Q_x} \tag{10-57}$$

This time it is the x component of the *flow* rate that is used. Similar expressions may be developed for the directional thermal impedances in other directions. When values are known for the angle of transmission and the angle of incidence, it is possible to calculate reflection for the thermal wave in the same way that it is calculated for acoustic and electromagnetic waves. Of course, reflection is less important in the case of the thermal wave, because of the high attenuation of the reflected wave. However, the amount of heat transmitted across the boundary is quite important, and such calculations are also a part of this kind of development.

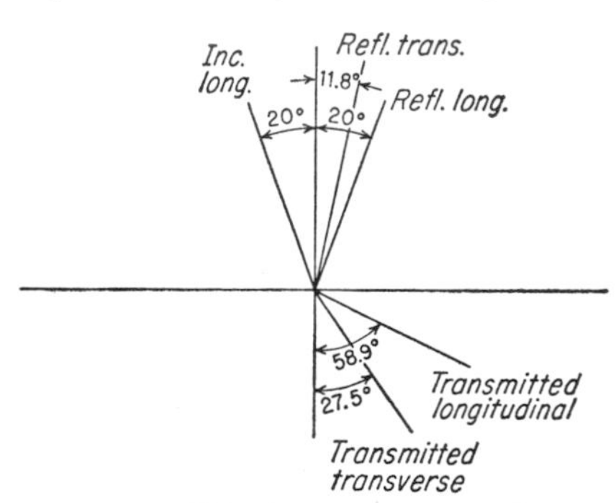

FIG. 10-9. Seismic waves at sandstone-granite boundary.

Snell's law was established for lossless media. For the lossy thermal waves and diffusion waves, the general expression given by Eq. (10-21) must be used rather than the velocity form of Snell's law. Since the propagation constant for the thermal wave is complex (with angle 45°), one would expect the angle of transmission to be complex. Since both propagation constants have the same phase angle, the complex angle of transmission only occurs for total reflection. Since a discontinuity in heat flow tangent to the boundary does not set up any other kind of wave, the complications present for waves in solids do not occur here.

The treatment for the thermal wave is exactly like that for the acoustic wave, and it is therefore not repeated here. The effect of the complex propagation constant is to complicate calculations. Planes of constant amplitude are in the same direction as planes of constant phase in the refracted waves.

10-7. Spherical Acoustic Waves

As indicated in the introduction to this chapter, it is often inconvenient to resolve waves by Fourier technique into plane-wave components. Waves symmetrical in cylindrical or spherical coordinates occur in many practical problems. At this point we consider spherical waves. These are waves for which the surfaces of constant phase are spheres. Frequently this is not true of the surfaces of constant amplitude. Acoustic waves are treated, for complications arise with spherical electromagnetic waves because of their vector nature.

The wave equation for pressure in an acoustic wave is, from (3-26),

$$\nabla^2 p = \frac{1}{v_p^2}\frac{\partial^2 p}{\partial t^2} = \left(\frac{j\omega}{v_p}\right)^2 Pe^{j\omega t} \tag{10-58}$$

In spherical coordinates, this is

$$\frac{1}{R^2}\frac{\partial}{\partial R}\left(R^2\frac{\partial P}{\partial R}\right) + \frac{1}{R^2 \sin\theta}\frac{\partial}{\partial\theta}\left(\sin\theta\frac{\partial P}{\partial\theta}\right) + \frac{1}{R^2\sin^2\theta}\frac{\partial^2 P}{\partial\phi^2} = \left(\frac{j\omega}{v_p}\right)^2 P \tag{10-59}$$

The types of spherical wave considered here correspond directly to plane waves in that they are *uniform* spherical waves and have derivatives only in the R direction. That is, it is assumed that there are no variations in the θ and ϕ directions; just as with the plane wave traveling in the X direction, it is assumed there are no variations in the Z and Y directions. Because of this, only the first term of the left-hand side of Eq. (10-59) exists, and the wave equation which must be solved for this type of acoustic wave is

$$\frac{1}{R^2}\frac{\partial}{\partial R}\left(R^2\frac{\partial P}{\partial R}\right) = -\left(\frac{\omega}{v_p}\right)^2 P \tag{10-60}$$

This equation may be integrated to obtain a solution. The result of such integration is

$$P = \frac{P_0^- e^{j\omega R/v_p}}{R} + \frac{P_0^+ e^{-j\omega R/v_p}}{R} \tag{10-61}$$

The first term of this equation represents a wave traveling in the negative R direction. Since such a wave travels from infinity toward the center, it is neglected at present, although it exists when there is reflection. The second term represents a wave traveling from the source outward. It is the only one considered here. It corresponds to the solution for a wave on the infinite transmission line. It can be readily verified by substitution that this is indeed a solution of (10-60)—or (10-60) may be integrated twice and the appropriate constants inserted.

Although the two waves of (10-61) represent inward and outward traveling waves similar to waves on a transmission line, the difference between the spherical wave and the transmission-line wave or plane wave is clearly evident when the inverse distance factor is considered. Thus, with no attenuation at all, the spherical pressure wave decreases as the distance from the source is increased, whereas a plane pressure wave or a wave on a transmission line would remain constant.

The sound intensity, corresponding to the Poynting vector for electromagnetic waves, expresses the power carried by the wave per unit area. Since p has units of force per square meter and u is expressed in meters per second, their product has units of energy per square meter-second, or power per square meter. It is easy to show that the intensity is, indeed, given by

$$\text{(10-62)} \qquad \boldsymbol{\mathcal{I}} = \text{power/unit area} = p\mathbf{u} \qquad \text{(newton-m/sec)/m}^2$$
$$= p\mathbf{u} \qquad \text{watts/m}^2$$

With sinusoidally varying pressure waves, as with sinusoidally varying electromagnetic waves, the average power is obtained from

$$\text{(10-63)} \qquad \boldsymbol{\mathcal{I}}_{\text{av}} = \tfrac{1}{2} \operatorname{Re} P^{+}(\mathbf{U}^{+})^{*} = \tfrac{1}{2} \operatorname{Re} |P^{+}|^{2} Z \mathbf{1}_{u} = \frac{1}{2} \frac{(P_0^{+})^{2} Z}{R^{2}} \mathbf{1}_{u}$$

This is an expression for the power per unit area across a spherical surface of radius R.

The value of P_0^+ is best obtained in terms of the total power radiated (transmitted through the surface of the sphere). To calibrate P_0^+ in terms of the total power, it is merely necessary to sum up the total power by integrating $\boldsymbol{\mathcal{I}}$ over the area of a sphere surrounding the source. Thus,

$$W_{\text{tot}} = \oiint \boldsymbol{\mathcal{I}} \cdot \mathbf{n}\, dA = \frac{Z}{2} \oiint \frac{(P_0^{+})^{2} R^{2} \sin\theta \, d\theta \, d\phi}{R^{2}}$$

where $\mathbf{n}$ is a unit vector normal to an area element dA on the surface of the sphere. Since P_0^+ is independent of angle and since the distance term cancels out, all factors may be removed from the integral except the differential solid angle ($\sin\theta\, d\theta\, d\phi$). Integrating over the complete sphere, this is just 4π, so that

$$W_{tot} = \frac{Z}{2} (P_0^+)^2 4\pi$$

Hence P_0^+ may be described by

$$P_0^+ = \sqrt{\frac{2W_{tot}}{4\pi Z}} = \sqrt{\frac{W_{tot}}{2\pi Z}} \qquad \text{newtons/m} \tag{10-64}$$

which may be substituted in Eq. (10-61) to obtain the sound pressure in terms of the radiated power. This same technique is used to relate electromagnetic fields of an antenna to the radiated power.

Example 10-8. *Sound Pressure in Terms of Radiated Power.* Assume that a spherical radiator emits 10 watts into free air of characteristic impedance 0.00235 m³/newton-sec. Then,

$$P_0^+ = \sqrt{\frac{10}{2\pi \times 0.00235}} = 26.0 \text{ newtons/m}$$

Hence, at a distance of 2 m, by (10-61),

$$P = 13 \text{ newtons/m}^2$$

or at a distance of 2 km,

$$P = 0.013 \text{ newton/m}^2$$

These are really quite small pressures! At a distance of 2 m, the intensity is

$$\mathcal{I} = \frac{1}{2} \frac{26^2 \times 0.00235}{4} = 0.199 \text{ watt/m}^2$$

Of course, at a distance of 2 km, this is reduced by a factor of 10^6, so it becomes 0.199 μw/m².

10-8. Arrays

One of the principal uses of the type of spherical radiator described above, or its near analog for electromagnetic waves, is as an element of some radiating system. The single isotropic acoustic source radiates equally in all directions, but combinations of sources radiate more energy in one direction than in another. This principle is utilized in acoustics for directional microphones, loudspeakers, and sonar transducers, as well as for transducers in ultrasonic cleaning and inspection devices. In electromagnetics, it is the basis for directional antennas. The directivity depends on the principle of phase interference, which is shown below for a simple example.

Consider the situation shown in Fig. 10-10, where two acoustic sources of spherical, uniform radiation are located along the axis of a spherical coordinate system. At any point in the surrounding space, the total pressure is the sum of the partial pressures due to individual sources. That is,

$$p = p_1 + p_2 \tag{10-65}$$

Now, consider that the two sources are fed with the same phase. Then, if the amplitudes are also the same,

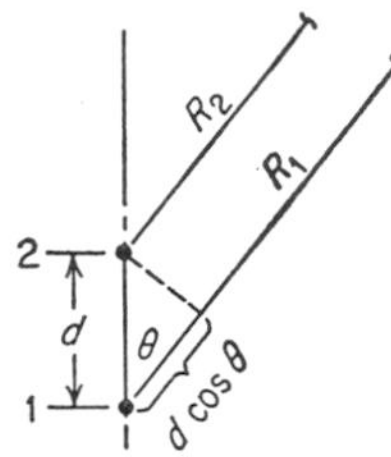

FIG. 10-10. Acoustic two-source radiator.

$$P_{10} = P_{20} = P_0 \tag{10-66}$$

and the total pressure is the *phasor* sum of the components. If the distance to the point of observation is sufficiently large compared with the distance between the two sources, the lines connecting the sources with the point are essentially parallel. This argument is the same as that used for finding the field of a static electric dipole, except that phase must be considered here, and our example assumes like radiators (as contrasted with the unlike charges of a dipole). This assumption is made in finding the distances shown in Fig. 10-10 and in the calculations which follow.

The remote field is

$$P = P_0 \left(\frac{e^{-j\omega R_1/v_p}}{R_1} + \frac{e^{-j\omega R_2/v_p}}{R_2} \right) \tag{10-67}$$

but

$$R_2 \approx R_1 - d \cos \theta \tag{10-68}$$

so

$$P = \frac{P_0}{R} e^{-j\beta R_1}(1 + e^{j\beta d \cos \theta}) \tag{10-69}$$

Here we have assumed that the small difference in the magnitude of the denominators for the two terms may be neglected in comparison with the phase difference, so that

$$R_2 \approx R_1 = R$$

as far as denominators are concerned.

In (10-69), we may take out exp $(j\frac{1}{2}\beta d \cos \theta)$ as a common factor, leaving

$$P = \frac{P_0}{R} e^{-j\beta[R-(d/2)\cos\theta]}[e^{j\beta(d/2)\cos\theta} + e^{-j\beta(d/2)\cos\theta}]$$

which can be seen to be

$$P = \frac{2P_0}{R} e^{-j\beta[R-(d/2)\cos\theta]} \cos\frac{\beta d \cos\theta}{2} \tag{10-70}$$

The cosine factor at the end of (10-70) modifies the amplitude, and the exponential represents the phase. Thus, as far as the amplitude of the pressure is concerned, we may write

$$|P| = \frac{2P_0}{R} \cos\frac{\omega d \cos\theta}{2v_p} = \frac{2P_0}{R} \cos\frac{\pi d \cos\theta}{\lambda} \tag{10-71}$$

The angular factor is called the *radiation pattern* of the *array* of sources. More complicated radiation patterns may be built up for more complicated arrays, by this same technique. If the sources are arranged in a line, they are called a *linear array*. When they are all fed in phase, the maximum always comes out normal to the line of the array, and they are called a *broadside array*. Other combinations of phases may cause the maximum radiation to be oriented in different directions; some antennas change the direction "in which they point" by adjusting phase while the sources remain fixed in space.

Sometimes it is convenient to consider a continuous source. In this case, the excitation of the continuous source is broken up into differential elements, and the summation carried out above becomes an integral.

With electromagnetic waves, a purely spherical wave does not satisfy Maxwell's equations. The elementary form used there is the radiation from a current element, for which the fields at great distance (E_θ and H_ϕ) contain a sin θ factor multiplying the expressions for the acoustic radiator discussed above. Of course, since **E** and **H** are vectors, the fields must be added vectorially as well as phasorially. Nevertheless, the principles enunciated here for acoustic radiation are applicable, with only slight modification, to electromagnetic radiation, and a good part of antenna theory is based on them.

10-9. Summary

If directional phase shift, wavelength, and impedances are used, various types of plane waves incident upon plane surfaces at oblique angles may be analyzed by the methods used for plane waves at normal incidence and for transmission-line waves. Since any wave may be resolved into a summation of plane waves at various angles of incidence (real and complex), this type of treatment is very powerful for the solution of many problems.

Electromagnetic waves are characterized by their polarization. The polarization is related to the direction of the electric-field vector. If the electric field is perpendicular to the plane of incidence (a plane passing through the wave normal and perpendicular to the reflecting surface), the wave is said to be normally polarized. If the electric vector lies in this plane, the wave is said to be parallel-polarized or polarized in the plane of incidence.

The phase shift may be expressed for a wave in three dimensions, as indicated in Eq. (10-5), the electric-field equation for normal polarization:

$$\mathbf{E} = \mathbf{1}_y E_0 e^{-j\beta \mathbf{N}\cdot\mathbf{R}} \tag{10-5}$$

Here **N** is a unit vector normal to the wavefront of the plane wave, and **R** is a radius vector from the origin out to the point at which the field is evaluated.

This expression may be broken down into Z and X components, so that we have

$$\mathbf{E} = \mathbf{1}_y E_0 e^{-j\beta z \sin\theta} e^{-j\beta x \cos\theta} = \mathbf{1}_y E_0 e^{-j\beta_z z} e^{-j\beta_x x} \tag{10-7a}$$

It is convenient to think of one of the exponentials as being part of the magnitude and phase factor for a nonuniform plane wave. Then the other exponential can be used to represent the travel of this nonuniform plane wave in either the X or the Z direction, depending on which is included in the magnitude factor and which is left separate. When the phase-shift factors associated with travel in certain directions are broken down into their components, expressions for the velocity and wavelength in those directions are developed:

$$v_{px} = \frac{v_p}{\cos\theta} \tag{10-9}$$

$$\lambda_x = \frac{\lambda}{\cos\theta} \tag{10-10}$$

$$v_{pz} = \frac{v_p}{\sin\theta} \tag{10-12}$$

$$\lambda_z = \frac{\lambda}{\sin\theta} \tag{10-13}$$

The velocities of propagation are higher and the wavelengths larger at angles with the wave normal because planes of constant phase are farther apart in any direction other than normal to the phase fronts. Since a wavelength is the distance between two points 360° apart in phase, this wavelength is longer in any direction other than normal to the phase fronts. Similarly, the phase velocity must be larger, because a greater distance must be traveled in the same time (1 wavelength per period).

When the wave is considered to be a nonuniform plane wave traveling in the direction of one of the coordinate axes, it is possible to write the directional impedances associated with the components of **E** and **H** involved in the component of the Poynting vector going in the direction of assumed wave travel. These expressions are different for normal and parallel polarization. For normal polarization, they are

$$Z_x = \frac{E_y}{H_z} = \frac{\eta}{\cos\theta} \tag{10-18}$$

and

$$Z_z = \frac{E_y}{-H_x} = \frac{\eta}{\sin\theta} \tag{10-19}$$

For polarization in the plane of incidence, they are

$$Z_x = -\frac{E_z}{H_y} = \eta\cos\theta \tag{10-45a}$$

$$Z_z = \frac{E_x}{H_y} = \eta\sin\theta \tag{10-47a}$$

Thus it can be seen that for normal polarization the directional impedances are greater than the intrinsic impedance and that for parallel polarization the directional impedances are less than the intrinsic impedance.

To determine the angles to use, we must describe an angle of incidence (the angle of the wave normal, before reflection, with the normal to the reflector) and an angle of transmission (the angle of the wave normal in the second medium). These are related by Snell's law, which is based upon the idea that attenuation and phase shift *along* the reflecting surface must be the same on both sides if the boundary conditions are to be met. Mathematically, Snell's law is given by

$$\gamma_1 \sin\theta_i = \gamma_2 \sin\theta_t \tag{10-21}$$

For lossless media the angle θ_t is real over a range up to the point where $\sin\theta_t$ reaches unity. When $\sin\theta_t$ becomes greater than 1, in accordance with (10-21), total reflection takes place. The impedances in the second medium become pure imaginary, and no power is transferred from the first to the second medium. This can happen, of course, for both normal and parallel polarization, since it is dependent on Snell's law, which applies to both. When lossy media are considered, the situation is more complicated.

Only with parallel polarization is it possible to obtain a match for the wave going through the boundary by a suitable choice of incident angle. In this case transmission through the boundary is complete, and there is no reflection. The condition for this is that the reflection coefficient be zero in the normal traveling-wave reflection-coefficient equation. This condition occurs when

$$\cos^2\theta_i = \frac{1}{1 + \epsilon_2/\epsilon_1} \tag{10-50}$$

and the angle at which it occurs is called the Brewster angle.

Nonelectric waves may also be described by directional impedances, directional wavelengths, etc. The impedances for an acoustic wave turn out to have the same form as those for the normally polarized electromagnetic wave. Thus, for the incident acoustic wave we have

$$Z_{1x} = \frac{u_x^+}{p^+} = \frac{u^+ \cos \theta_i}{p^+} = Z_1 \cos \theta_i \tag{10-52}$$

The other impedances and the directional wavelengths and phase constants can be developed in the same manner as for the parallel-polarized electromagnetic wave, and reflection is governed by the same reflection coefficient. Thermal waves have the same type of vector properties as acoustic waves, and therefore they also may be described in this manner; but for the A analogies they compare with *normally* polarized electromagnetic waves. However, the situation is more complicated with the thermal waves, because they must always be considered the same as electromagnetic waves in lossy media. Waves in solids are still more complicated, since incident longitudinal waves establish both longitudinal and shear reflected and transmitted waves.

Radiation from small sources always gives spherical waves at large distances and, in the case of an acoustic spherical radiator, even at small distances. Solving the acoustic-wave equation in spherical coordinates, we find incident and reflected waves in the radial direction. Considering only the outgoing wave, the expression for pressure may be written as

$$P = \sqrt{\frac{W_{\text{tot}}}{4\pi Z}} \frac{e^{-j\beta R}}{R} \tag{10-72}$$

The difference between this and the plane-wave case is due to the inverse distance factor, which is present even though the medium does not attenuate. This inverse distance factor comes about because the waves are spreading out, and therefore the intensity must be reduced as $1/R^2$. The intensity of an acoustic wave can be shown to be

$$\mathcal{I} = p\mathbf{u} \qquad \text{watts/m}^2 \tag{10-62}$$

Electromagnetic waves may be treated in a similar manner, but because of the complications due to their vector nature this treatment has been omitted here. The $1/R$ factor does occur in the expression for electromagnetic waves at large distances, but at smaller distances reciprocal distance variations as $1/R^2$ and $1/R^3$ also occur.

PROBLEMS

10-1. For a plane wave polarized in the plane of incidence, calculate the components of **E** and **P** by rotating coordinates about the axis in which the **H** vector lies.

10-2. By rotating coordinates, calculate the components of q for a thermal wave.

10-3. Consider the normally polarized plane wave of Sec. 10-1. Suppose that this wave is also rotated by an angle ϕ about the Z axis. Determine its fields. Separate them into waves polarized normal to, and in, the plane of incidence.

10-4. A horizontally polarized plane wave from a transmitter in an airplane is reflected from a smooth fresh-water lake (dielectric constant 81, conductivity negligible). Calculate the reflected wave in the air if the incident field is 1 mv/m and the angle of incidence is 85°. Also calculate the wave transmitted into the water. Determine both **E** and **H**.

10-5. When a reflected wave like that of Prob. 10-4 arrives at an aircraft in phase, it adds to the direct signal. When it arrives out of phase, it causes a reduction in the direct signal. For the wave of Prob. 10-4, assume that the aircraft is at a height of 10 km. Assuming that the waves travel with the velocity of light, determine from the geometry the phase difference between the direct and reflected waves for two such aircraft. At what height (for the 85° incident angle) would the aircraft have to fly for the signals to cancel? Assume frequency of 300 Mc.

10-6. A microwave-relay system is operated over the fresh-water lake of Prob. 10-4. If the relay stations are 40 km apart and atmospheric effects are neglected, what height (identical at both ends) would you recommend for the antennas? Wavelength is 4 cm. *Hint:* Addition of incident and reflected signals is desirable; cancellation is undesirable. Neglect curvature of the earth.

10-7. The velocity of propagation in the atmosphere is dependent on pressure, temperature, and humidity. Changes of 50 parts per million can occur. Would this cause trouble on the link of Prob. 10-6?

10-8. A plane wave strikes a plane dielectric sheet at an angle of incidence of 75°. The relative dielectric constant of the sheet is 2. The wavelength in air is 10 cm, and the sheet is 2 cm thick. Calculate the fields beyond the sheet if the incident parallel-polarized wave is 1 mv/m. *Hint:* Use Snell's law to determine the direction of the wave in each medium. Then treat this as a tandem-transmission-line problem, using the Smith chart.

10-9. A circularly polarized wave has two equal components at right angles to each other which are also 90° out of time phase. If such a wave is incident at an angle of 30° on an infinite plane dielectric with relative dielectric constant 4, describe the reflected components. Assume initially that $E_{y'} = jE_{z'}$, using the nomenclature of Sec. 10-1.

10-10. For an interface between media with $\epsilon = 4\epsilon_0$ and $\epsilon = \epsilon_0$, determine the Brewster angle (when possible) and the angle of total reflection for waves going in both directions.

10-11. Repeat Prob. 10-10 with $\epsilon = 9\epsilon_0$ and $\epsilon = \epsilon_0$.

10-12. An acoustic plane wave in water strikes a boundary with aluminum at an angle of 25°. The incident pressure is 10^3 newtons/m^2. Find the reflected and transmitted pressures and velocities.

10-13. Repeat Prob. 10-12, assuming the aluminum is replaced by rubber.

10-14. What are the angles of total reflection for the situations of Probs. 10-12 and 10-13?

10-15. Discuss the standing-wave patterns for waves in a square cross section of steel bar surrounded by air.

10-16. Discuss the standing-wave patterns for electromagnetic waves in a square cross section of dielectric rod in air. Consider in particular what happens if the linear dimensions in the transverse direction are $2\lambda_{die}/3$.

10-17. A longitudinal wave in steel strikes a boundary with aluminum at an angle of 15°. Determine the angles of reflection and transmission for both longitudinal and transverse waves.

10-18. Repeat Prob. 10-17 with the wave originating in the aluminum.

10-19. A thermal sine wave with a period of 1 sec strikes a boundary between aluminum and glass, from within the aluminum, at an angle of 10°. If the amplitude in aluminum is 10°C along the wavefront passing through the origin, describe the reflected and transmitted temperature waves and the heat flux in incident, transmitted, and reflected waves.

10-20. Repeat Prob. 10-19 for a wave in brick entering asbestos, for which the period is 1 day.

10-21. An acoustic spherical radiator emits 1 watt. Find the pressure and intensity at a distance of 100 ft.

10-22. An acoustic array consists of three radiators in a line spaced a half-wavelength apart. Find the intensity pattern and plot it, assuming the radiators are fed in phase with equal amplitudes.

10-23. Repeat Prob. 10-22 with the two end radiators in phase and the center one 180° out of phase.

10-24. Develop an expression for the pattern of N in-phase radiators spaced a distance a, all fed with the same amplitude.

Appendixes

A. Vector Analysis

Vector analysis has been used throughout "Traveling-wave Engineering" wherever it is appropriate. This Appendix is intended to aid the reader by recalling for him the pertinent points in vector analysis, but it is not intended as a text in the subject. It emphasizes in particular the close analogies between many vector operations and the comparable operations in ordinary algebra and calculus.

A-1. Vector Algebra

A vector has both *direction* and *magnitude*. We may separate these by writing each vector as the product of a *unit vector*, which has the direction specified but which always has magnitude *one*, with a nonunity magnitude. Thus,

$$\mathbf{A} = \mathbf{1}_A|\mathbf{A}| = \mathbf{1}_A A \tag{A-1}$$

where $\mathbf{1}_A$ = the unit vector in the same direction as $\mathbf{A}$

$|\mathbf{A}|$ = the magnitude of $\mathbf{A}$, sometimes also written as A

Here we use boldface to designate vectors and italic type for *scalars*. A scalar may have magnitude and sign, but not direction.

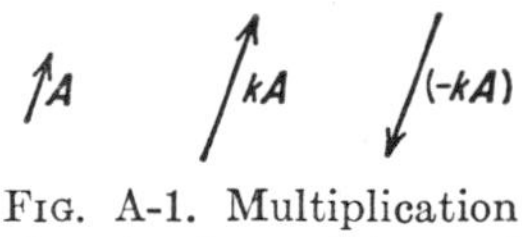

Multiplying a vector by a scalar may change both the magnitude and the sign of the vector. This is illustrated in Fig. A-1, where the result of multiplying the vector $\mathbf{A}$ by the scalar k and also by $-k$ is shown. It is assumed here that k is a positive real number, although the notion of multiplication by a scalar may be extended to include the case where the scalar is a complex number. Note that

Fig. A-1. Multiplication of vector by positive and negative scalars.

$$k\mathbf{A} = \mathbf{1}_A(kA) \qquad -k\mathbf{A} = (-\mathbf{1}_A)(kA) \tag{A-2}$$

where $-\mathbf{1}_A$ is understood to be a vector parallel to $\mathbf{1}_A$ but oriented in the opposite direction. In general, the insertion of a minus sign before a vector leaves its magnitude the same but reverses its direction.

The sum of two vectors is said to obey the "parallelogram rule"; that is, if the two vectors to be added are placed as adjacent sides of a parallelogram, the sum is along the diagonal. As indicated in Fig. A-2, this process may also be depicted by placing the *tail* of the second vector at the *head* of the first and drawing the sum as a vector whose tail is at the tail of the first and whose head is at the head of the second. If

(A-3) $$\mathbf{A} + \mathbf{B} = \mathbf{C}$$

it is not ordinarily true that

$$A + B = C$$

In fact,

(A-4) $$A + B \geqslant C$$

and an equal sign applies only when the vectors are parallel.

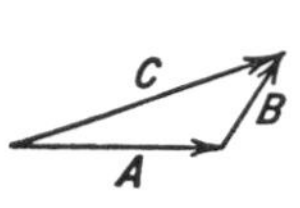

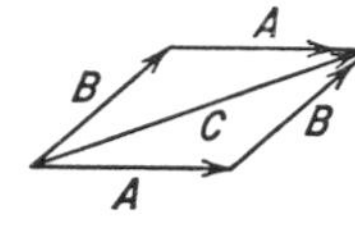

FIG. A-2. Sum of two vectors.

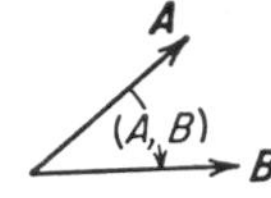

FIG. A-3. Vectors involved in scalar product.

Vectors obey the same commutative, associative, and distributive laws for addition as do scalars. Thus,

		Scalar equivalent
(A-5)	$\mathbf{A} + \mathbf{B} = \mathbf{B} + \mathbf{A}$	$A + B = B + A$
(A-6)	$(\mathbf{A} + \mathbf{B}) + \mathbf{C} = \mathbf{A} + (\mathbf{B} + \mathbf{C})$	$(A + B) + C = A + (B + C)$
	$= \mathbf{A} + \mathbf{B} + \mathbf{C}$	$= A + B + C$
(A-7)	$k(\mathbf{A} + \mathbf{B}) = k\mathbf{A} + k\mathbf{B}$	$k(A + B) = kA + kB$

Two kinds of products between vectors are defined, the *scalar* product and the *vector* product. The scalar product of two vectors is defined by

		Scalar equivalent
(A-8)	$\mathbf{A} \cdot \mathbf{B} = AB \cos (A,B)$	$AB = AB$

where (A,B) is the angle from **A** to **B**, as shown in Fig. A-3. The scalar, or *dot*, product is interpreted as

$$\mathbf{A} \cdot \mathbf{B} = \mathbf{A} \text{ (component of } \mathbf{B} \text{ in direction of } \mathbf{A}) \\ = \mathbf{B} \text{ (component of } \mathbf{A} \text{ in direction of } \mathbf{B})$$

When **A** and **B** are expressed in terms of unit vectors,

$$\mathbf{A} \cdot \mathbf{B} = AB\mathbf{1}_A \cdot \mathbf{1}_B$$

Thus,

$$\mathbf{1}_A \cdot \mathbf{1}_B = \cos(A,B) \tag{A-9}$$

Both commutative and associative laws apply to scalar products. The former means that

Scalar equivalent

$$\mathbf{A} \cdot \mathbf{B} = \mathbf{B} \cdot \mathbf{A} \qquad\qquad AB = BA \tag{A-10}$$

Note also that

$$\mathbf{A} \cdot \mathbf{A} = A^2 \tag{A-11}$$

and that a zero scalar product of two nonzero vectors, **A** and **B**, implies that the vectors are perpendicular, since it means that

$$\cos(A,B) = 0 \qquad \text{or} \qquad (A,B) = \frac{\pi}{2}$$

The vector product of two vectors is itself a vector oriented at right angles to both of them. It is defined by

Scalar equivalent

$$\mathbf{A} \times \mathbf{B} = \mathbf{1}_C AB \sin(A,B) \qquad\qquad AB = AB \tag{A-12}$$

where $\mathbf{1}_C$ is perpendicular to the plane formed by **A** and **B** and is positive in the direction a right-hand screw would move if turned through the smaller angle from **A** to **B**. This is illustrated in Fig. A-4. The vector, or *cross*, product is interpreted as

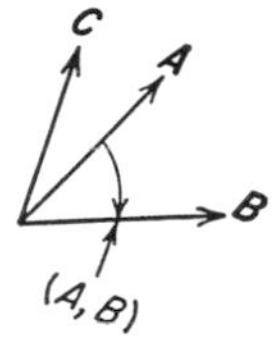

$$\begin{aligned} \mathbf{A} \times \mathbf{B} &= \mathbf{1}_C A \text{ (component of } \mathbf{B} \perp \mathbf{A}) \\ &= \mathbf{1}_C B \text{ (component of } \mathbf{A} \perp \mathbf{B}) \end{aligned}$$

FIG. A-4. Vectors involved in vector product.

It can also be seen from the figure that

$$|\mathbf{A} \times \mathbf{B}| = 2 \text{ (area of triangle with } \mathbf{A} \text{ and } \mathbf{B} \text{ as two sides)}$$

The associative law applies to vector products as to scalars. The commutative law is modified, however, to

Scalar equivalent

$$\mathbf{A} \times \mathbf{B} = -\mathbf{B} \times \mathbf{A} \qquad\qquad AB = BA \tag{A-13}$$

The minus sign arises because a right-hand screw turned from **A** to **B** moves in a different direction from the direction of such a screw turned from **B** to **A**.

Note also that

$$\mathbf{A} \times \mathbf{A} = 0 \tag{A-14}$$

and that, if the cross product of two nonzero vectors is zero, the vectors are parallel.

A number of *triple products*, involving both scalar and vector multiplication, should be considered:

Scalar equivalent

(A-15) $\mathbf{A} \cdot (\mathbf{B} \times \mathbf{C}) = (\mathbf{A} \times \mathbf{B}) \cdot \mathbf{C} = \mathbf{C} \cdot (\mathbf{A} \times \mathbf{B}) = \mathbf{B} \cdot (\mathbf{C} \times \mathbf{A})$ $\qquad ABC = BAC = CAB = CBA$

(A-16) $\mathbf{A} \times (\mathbf{B} \times \mathbf{C}) = (\mathbf{A} \cdot \mathbf{C})\mathbf{B} - (\mathbf{A} \cdot \mathbf{B})\mathbf{C}$

(A-17) $(\mathbf{A} \times \mathbf{B}) \times \mathbf{C} = (\mathbf{A} \cdot \mathbf{C})\mathbf{B} - (\mathbf{C} \cdot \mathbf{B})\mathbf{A}$

A-2. Coordinate Systems

To describe the location of a point in space, and the direction of a vector, it is customary to use *coordinate systems.* The location of a point in space is specified by three numbers, known as coordinates, each of which belongs to a set of such numbers. A point is also describable as the intersection of three surfaces, and the coordinates of the point are "tags" used to distinguish the particular surfaces from others in their sets.

In this book only two coordinate systems are used: the *rectangular,* or *cartesian,* system and the *spherical* system. These are both *orthogonal;* that is, at every point, the three surfaces describing the point are mutually perpendicular.

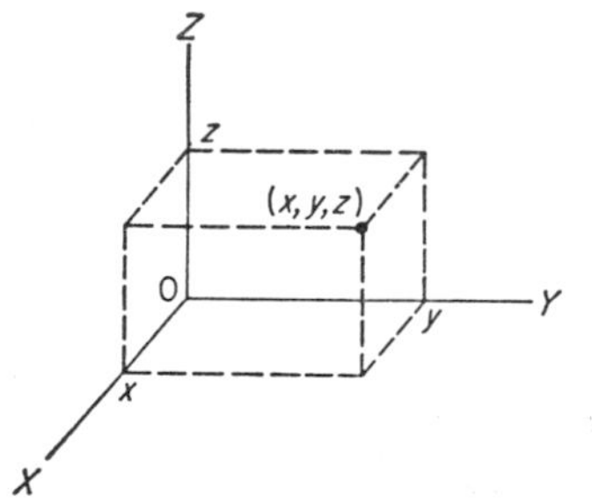

FIG. A-5. Rectangular coordinate surfaces describing a point.

Figure A-5 shows the surfaces describing the location of a point in rectangular coordinates, along with the *axes* of the coordinate system. The three surfaces are planes, each of which is perpendicular to one of the axes. The "tag" for each surface is its distance from the center of the system, 0. Thus the coordinate surface parallel to the plane formed by $0X$ and $0Y$ is a distance z from 0 (in the $0Z$ direction). It should be noted that this is also the projection of the point onto the line $0Z$, called the Z axis. Hence the point is described by the three numbers (x,y,z) indicated alongside it.

Figure A-6 shows the direction of three unique unit *base* vectors associated with the coordinate system and the point. These unit vectors are perpendicular to the coordinate surfaces (and parallel to the axes, for the rectangular system). Thus $\mathbf{1}_x$ is parallel to the X axis, $\mathbf{1}_y$ to the Y axis, and $\mathbf{1}_z$ to the Z axis. These unit vectors, with appropriate multipliers, are used in describing the direction of a vector.

A vector may be expressed as the sum of vectors in the directions of the three base vectors of the coordinate system. Such *resolution* is shown in Fig. A-7. The vector **A** is shown to be the sum of vectors in the X, Y, and Z directions. The scalar multiplying the unit vector $\mathbf{1}_x$ is defined as the *component* of **A** in the X direction and is designated A_x.

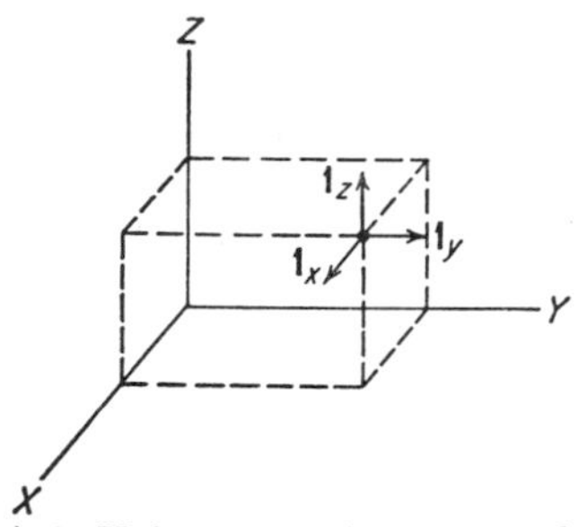

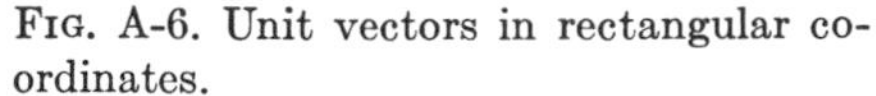
FIG. A-6. Unit vectors in rectangular coordinates.

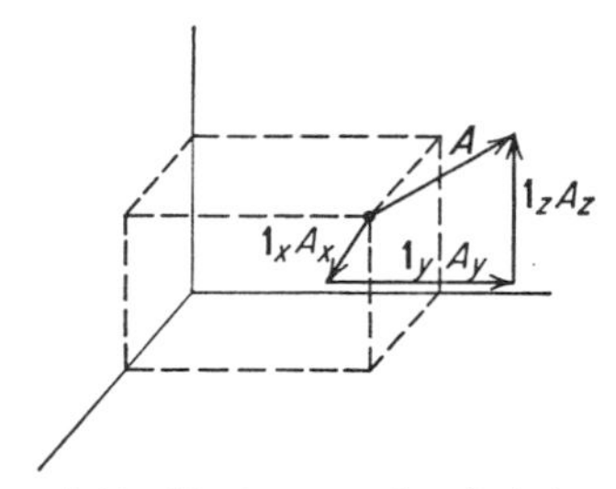

FIG. A-7. Vector resolved into rectangular components.

Similarly for the Y and Z components. Thus the vector sum pictured in Fig. A-7 is described analytically by

$$\text{(A-18)} \qquad \begin{aligned} \mathbf{A} &= \mathbf{1}_x A_x + \mathbf{1}_y A_y + \mathbf{1}_z A_z \\ &= \mathbf{1}_x(\mathbf{A} \cdot \mathbf{1}_x) + \mathbf{1}_y(\mathbf{A} \cdot \mathbf{1}_y) + \mathbf{1}_z(\mathbf{A} \cdot \mathbf{1}_z) \end{aligned}$$

It is extremely important to realize that a vector can exist at a point in space without the "crutch" of a coordinate system. The coordinate system helps us describe the vector, but the same vector may be described in *any* coordinate system, or it may be described by a picture without reference to any coordinate system at all. In vector analysis, as many operations as possible are carried out prior to resorting to coordinate systems, since the work is simpler.

The products of two vectors may be expressed readily in terms of their components. Thus,

$$\mathbf{A} \cdot \mathbf{B} = (\mathbf{1}_x A_x + \mathbf{1}_y A_y + \mathbf{1}_z A_z) \cdot (\mathbf{1}_x B_x + \mathbf{1}_y B_y + \mathbf{1}_z B_z)$$

But it is readily seen that

$$\text{(A-19)} \qquad \begin{aligned} \mathbf{1}_x \cdot \mathbf{1}_x &= \cos 0^\circ = 1 = \mathbf{1}_y \cdot \mathbf{1}_y = \mathbf{1}_z \cdot \mathbf{1}_z \\ \mathbf{1}_x \cdot \mathbf{1}_y &= \cos 90^\circ = 0 = \mathbf{1}_x \cdot \mathbf{1}_z = \mathbf{1}_y \cdot \mathbf{1}_z \end{aligned}$$

It is true for all orthogonal coordinate systems that the unit-vector scalar "self-products" are unity and scalar "mixed products" are zero.

Utilizing this relation,

$$\text{(A-20)} \qquad \mathbf{A} \cdot \mathbf{B} = A_x B_x + A_y B_y + A_z B_z$$

The vector product of two vectors $\mathbf{A} \times \mathbf{B}$ is based on the following relations between unit vectors:

$$\begin{aligned} \mathbf{1}_x \times \mathbf{1}_x &= \sin 0° = 0 = \mathbf{1}_y \times \mathbf{1}_y = \mathbf{1}_z \times \mathbf{1}_z \\ \mathbf{1}_x \times \mathbf{1}_y &= \mathbf{1}_z \sin 90° = \mathbf{1}_z = -\mathbf{1}_y \times \mathbf{1}_x \\ \mathbf{1}_y \times \mathbf{1}_z &= \mathbf{1}_x = -\mathbf{1}_z \times \mathbf{1}_y \\ \mathbf{1}_z \times \mathbf{1}_x &= \mathbf{1}_y = -\mathbf{1}_x \times \mathbf{1}_z \end{aligned} \tag{A-21}$$

It should be noted that the (positive) products at the left-hand side of the equations have directions in accordance with the rotational scheme *xyzxyz*. That is, entering this sequence at any point, a positive result is obtained by taking the next three letters in order.

Combining the results of (A-21) and expressing them as a determinant,

$$\mathbf{A} \times \mathbf{B} = \begin{vmatrix} \mathbf{1}_x & \mathbf{1}_y & \mathbf{1}_z \\ A_x & A_y & A_z \\ B_x & B_y & B_z \end{vmatrix} \tag{A-22}$$

Spherical coordinates for a point are the tags on a sphere, a cone, and a plane intersecting at the point. These surfaces are shown in Fig. A-8

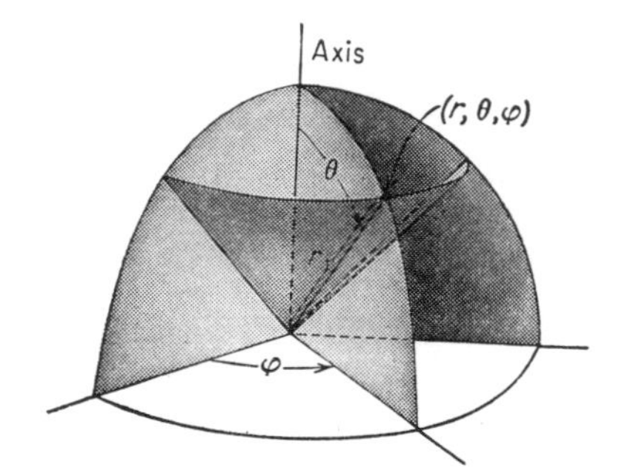

FIG. A-8. Spherical coordinate surfaces.

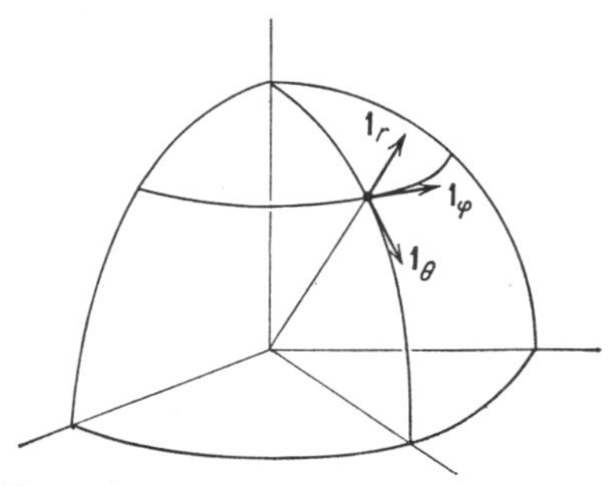

FIG. A-9. Unit vectors in spherical coordinates.

as they appear when superimposed on the axes of a rectangular system. It is customary that the *axis* of the spherical system be the Z axis of the corresponding rectangular system. The sphere is described by its radius, the cone by its apex semiangle (measured with respect to the axis of the coordinate system), and the plane (which passes through the axis) by its angle with an arbitrary reference in the plane normal to the axis. This latter reference is frequently the X axis of a related rectangular system. The angles are θ for the cone and φ for the plane.

The corresponding unit vectors are shown in Fig. A-9. The radial unit vector is pointed outward. The angle of the cone is called the *colatitude*. Its unit vector is tangent to a great circle on the sphere passing through the intersection with the axis and through the point described. The unit vector for the third coordinate, the longitude, is normal to the coordinate plane and tangent at the point described to the circle of constant latitude through that point.

A vector may be resolved in any coordinate system into three components. Thus the same vector that was resolved in rectangular coordinates may be resolved into three *different* components in the spherical system:

$$\mathbf{A} = \mathbf{1}_r A_r + \mathbf{1}_\theta A_\theta + \mathbf{1}_\varphi A_\varphi \tag{A-23}$$

The expressions for the scalar and vector products look much the same as for rectangular coordinates:

$$\mathbf{A} \cdot \mathbf{B} = A_r B_r + A_\theta B_\theta + A_\varphi B_\varphi \tag{A-24}$$

$$\mathbf{A} \times \mathbf{B} = \begin{vmatrix} \mathbf{1}_r & \mathbf{1}_\theta & \mathbf{1}_\varphi \\ A_r & A_\theta & A_\varphi \\ B_r & B_\theta & B_\varphi \end{vmatrix} \tag{A-25}$$

A-3. Integrals

Three types of integrals occur often in vector analysis: the line integral, the surface integral, and the volume integral. The first two have different interpretations in vector analysis from what they do in ordinary scalar analysis.

A line integral between two points in space is defined along a specific curve between the two points. It is the integral along that curve of the component of some vector parallel to the path. Analytically, the line integral of **A** between two points P and Q along the path C is defined by

$$\int_{C\; P(x,y,z)}^{Q(x,y,z)} \mathbf{A} \cdot d\mathbf{L} = \lim_{\substack{\Delta L_i \to 0 \\ N \to \infty \\ \text{in such a} \\ \text{way that the} \\ \Delta L\text{'s cover } C \\ \text{from } P \text{ to } Q}} \sum_{i=1}^{N} \mathbf{A}_i \cdot \mathbf{\Delta L}_i \tag{A-26}$$

$$= \lim \Sigma A_i \text{ (component of } \mathbf{\Delta L}_i \text{ in direction of } \mathbf{A}_i)$$

$$= \lim \Sigma \text{ (component of } \mathbf{A}_i \text{ in direction of } \mathbf{\Delta L}_i) \, \Delta L_i$$

Scalar comparison

$$\int_{C\; P}^{Q} A \, dL = \lim_{\substack{\Delta L_i \to 0 \\ N \to \infty \\ \text{in such a} \\ \text{way that the} \\ \Delta L\text{'s cover } C \\ \text{from } P \text{ to } Q}} \sum_{i=1}^{N} A_i \, \Delta L_i$$

The difference between the vector and scalar integrals is the use of the *component* in the vector integral. Note that the result of a line integra of this type is a *scalar*. Figure A-10 shows the quantities in the sum of Eq. (A-26).

In rectangular coordinates, we may write

$$\mathbf{dL} = \mathbf{1}_x\, dx + \mathbf{1}_y\, dy + \mathbf{1}_z\, dz$$

so that

$$\int_{P\atop C}^{Q} \mathbf{A} \cdot \mathbf{dL} = \int_{x_P}^{x_Q} A_x\, dx + \int_{y_P}^{y_Q} A_y\, dy + \int_{z_P}^{z_Q} A_z\, dz \qquad \text{(A-27)}$$

Along the path only one of the differentials is independent. If this one is dx, then dy/dx and dz/dx are both specified.

Closed line integrals occur frequently in physical problems. Such an integral is one along some path that is closed upon itself. Thus, if one starts along the path at point P, he returns to P to complete the integration. A circle would be an example of a closed path. Integrals

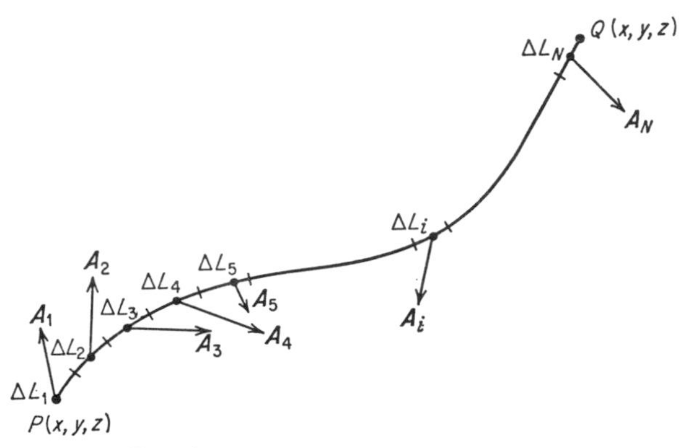

FIG. A-10. Definition of line integral.

along a closed path, since they have no limiting points, must be described by a special symbolism:

$$\oint \mathbf{A} \cdot \mathbf{dL}$$

To describe a surface integral, it is first necessary to describe the vector nature of a surface element. An incremental surface may be considered, to a first approximation, plane. The only vector which may be ascribed uniquely to a plane is one normal to the plane. Thus, we may express the differential surface element as

$$\mathbf{dS} = \mathbf{1}_n\, dS \qquad \mathbf{\Delta S} = \mathbf{1}_n\, \Delta S \qquad \text{(A-28)}$$

where $\mathbf{1}_n$ is a unit vector normal to the surface element. The positive direction of $\mathbf{1}_n$ must be defined for each problem, except when $\mathbf{dS}$ is an element of a *closed* surface, in which case convention dictates that $\mathbf{1}_n$ is always *outward.*

The surface integral of **A** over a surface S is defined as

$$\int_S \mathbf{A} \cdot d\mathbf{S} = \lim_{\substack{\Delta S \to 0 \\ \text{at a point} \\ N \to \infty \\ \text{in such a way} \\ \text{that the } \Delta S\text{'s} \\ \text{cover } S}} \sum_{i=1}^{N} \mathbf{A}_i \cdot \Delta\mathbf{S}_i = \lim \sum A_{ni} \, \Delta S_i \tag{A-29}$$

The difference between the surface integral of a vector and the surface integral of a scalar is, of course, that only the component of the vector normal to the surface (A_{ni}) multiplies the area element for the vector surface integral.

A surface integral may be evaluated numerically using (A-29). When it is evaluated analytically, it is necessary to express ΔS_i in terms of its components. Thus, in some coordinate systems, $\Delta S = \Delta u \, \Delta v$, so the surface integral may be evaluated as

$$\int_S \mathbf{A} \cdot d\mathbf{S} = \iint A_n \, du \, dv \tag{A-30}$$

In rectangular coordinates,

$$d\mathbf{S} = \mathbf{1}_x \, dy \, dz + \mathbf{1}_y \, dx \, dz + \mathbf{1}_z \, dx \, dy$$

so

$$\int_S \mathbf{A} \cdot d\mathbf{S} = \iint A_x \, dy \, dz + \iint A_y \, dx \, dz + \iint A_z \, dx \, dy$$

For the surface, two of the differentials are independent, but the third is specified by the equation of the surface. Thus, if dz is dependent, $\partial z/\partial x$ and $\partial z/\partial y$ are specified.

In many physical problems, a vector describes a rate of flow of some quantity in a particular direction per unit area. Such a flow rate is known as a *flux density*, and the total flow across some surface per unit time as the *flux*. The concept is also extended to other vectors. Thus, in general, we state that

$$\int_S \mathbf{A} \cdot d\mathbf{S} = \text{flux of } \mathbf{A} \text{ through } S$$

Closed surface integrals occur when the surface itself is closed. In accord with the convention for sign of $d\mathbf{S}$, the net flux is positive when it is *outward*. A closed surface integral is designated by

$$\oint \mathbf{A} \cdot d\mathbf{S}$$

Volume integrals occur in vector analysis, but only for scalar integrands. Thus, the normal definition applies:

(A-31)
$$\int_V A\,dv = \lim_{\substack{\Delta v\to 0 \\ \text{at a point} \\ N\to\infty \\ \text{in such a way} \\ \text{that } \Delta v\text{'s} \\ \text{completely fill} \\ V}} \sum_{i=1}^{N} A_i\,\Delta v_i$$

In rectangular coordinates this is

$$\int_V A\,dv = \iiint A\,dx\,dy\,dz$$

A-4. Derivatives

The derivative of a vector is defined the same as the derivative of a scalar, except that the changes in both magnitude and direction must be considered. Thus,

(A-32)
$$\frac{d\mathbf{A}}{du} = \lim_{\Delta u\to 0} \frac{\mathbf{A}(u+\Delta u) - \mathbf{A}(u)}{\Delta u}$$

Here u is a distance measure that may vary in any direction. Note that the derivative of a vector is the vector sum of the derivatives of its components. Thus,

(A-33)
$$\frac{d\mathbf{A}}{du} = \mathbf{1}_x \frac{dA_x}{du} + \mathbf{1}_y \frac{dA_y}{du} + \mathbf{1}_z \frac{dA_z}{du}$$

There are three special types of derivatives of great importance in vector analysis: the *gradient* of a scalar and the *divergence* and *curl* of a vector. Mathematically, each of them is defined in a manner similar to, but somewhat different from, the ordinary definition of a derivative. The similarities are important here because they help show the analogies between the one-dimensional problems treated here and the corresponding three-dimensional problems.

The gradient of a scalar field G is a vector. Figure A-11 shows two surfaces of constant G; surface S_Δ corresponds to a value of G, which we shall call G_Δ, that is greater than G on S by a small amount, ΔG. At some point (x,y,z) on S, a normal to S is drawn to S_Δ (it is also approximately normal to S_Δ). This normal is given by $\mathbf{1}_w\,\Delta w$. Now define

(A-34)
$$\text{grad}\,G = \lim_{\Delta w\to 0} \frac{G_\Delta - G}{\Delta w}\mathbf{1}_w = \frac{dG}{dw}\mathbf{1}_w$$

Note that the limiting process involved in shrinking Δw also involves taking surfaces S_Δ closer and closer to S and values of G_Δ closer and closer to G.

Gradient G is a *directional* derivative, and it is a vector. These properties distinguish it from derivatives in one-dimensional analysis. It is, however, a very particular directional derivative, since it is in a direction normal to the surface of constant G. Any other directional derivative would involve a distance between S and S_Δ greater than

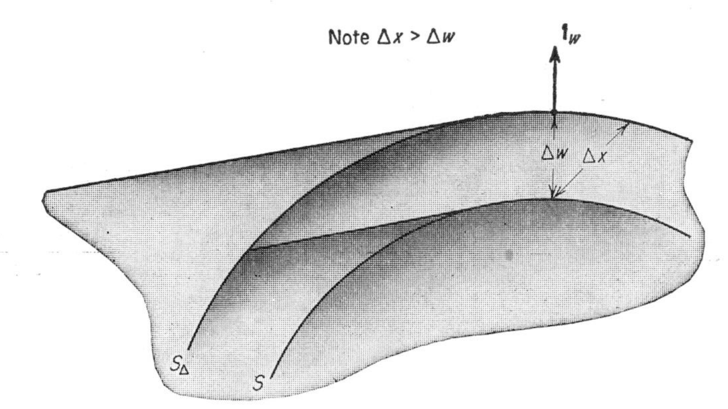

FIG. A-11. Adjacent surfaces of constant G.

Δw, so any other directional derivative would be less than the gradient. Thus

(A-35) $|\text{grad } G| =$ maximum positive directional derivative of G

It is easy to show, in rectangular coordinates, that

$$\text{grad } G = \mathbf{1}_x \frac{\partial G}{\partial x} + \mathbf{1}_y \frac{\partial G}{\partial y} + \mathbf{1}_z \frac{\partial G}{\partial z} = \nabla G \tag{A-36}$$

where we have defined the operator

$$\nabla = \mathbf{1}_x \frac{\partial}{\partial x} + \mathbf{1}_y \frac{\partial}{\partial y} + \mathbf{1}_z \frac{\partial}{\partial z} \tag{A-37}$$

In spherical coordinates the gradient is given by

$$\text{grad } G = \mathbf{1}_r \frac{\partial G}{\partial r} + \frac{\mathbf{1}_\theta}{r} \frac{\partial G}{\partial \theta} + \frac{\mathbf{1}_\varphi}{r \sin \theta} \frac{\partial G}{\partial \varphi} \tag{A-38}$$

Although ∇ is defined only for rectangular coordinates, it is customary to refer to gradient G as ∇G regardless of the coordinate system used.

When the gradient is integrated along a line, we find

$$\int_P^Q \nabla G \cdot d\mathbf{L} = \int \frac{dG}{dw} \mathbf{1}_w \cdot d\mathbf{L} = \int \frac{dG}{dw} dw$$

Hence

$$G_Q - G_P = \int_P^Q \nabla G \cdot d\mathbf{L} \tag{A-39}$$

Note that the fact that this can be done illustrates that the line integral of a gradient is independent of the path taken, as long as the end points remain the same for different paths.

The second type of special derivative in vector analysis is the *divergence.* This is a derivative of a vector with a scalar result. We define the divergence as

$$\text{(A-40)} \qquad \text{div } \mathbf{B} = \text{efflux of } \mathbf{B} \text{ per unit volume}$$

Mathematically, this is expressed as

Scalar comparison

$$\text{(A-41)} \quad \text{div } \mathbf{B} = \lim_{\substack{\Delta v \to 0 \\ \text{at a} \\ \text{point}}} \frac{\oint_{\text{surf bounding } \Delta v} \mathbf{B} \cdot \mathbf{dS}}{\Delta v} \qquad \frac{dB}{dx} = \lim_{\Delta x \to 0} \frac{B(x + \Delta x) - B(x)}{\Delta x}$$

The comparison with the ordinary derivative is easier to see in terms of a single rectangular coordinate. Consider the small volume shown in

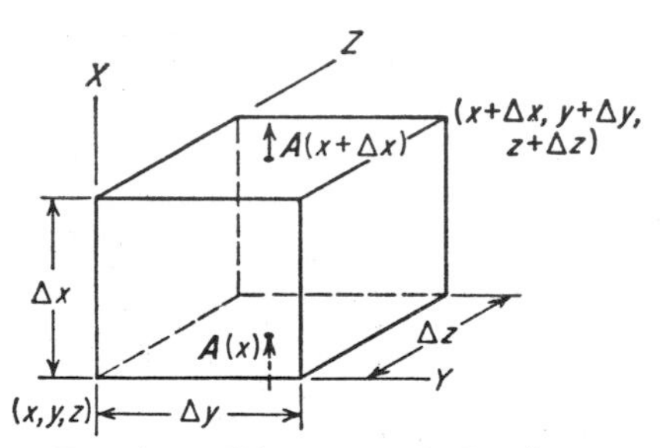

FIG. A-12. Divergence derivation.

Fig. A-12. Here a small rectangular solid has one corner at the point (x,y,z) and dimensions Δx by Δy by Δz. Consider a field **B** that is only in the X direction:

$$\mathbf{B} = \mathbf{1}_x B_x$$

Then

$$\oint \mathbf{B} \cdot \mathbf{dS} = \oint_{\text{bottom}} \mathbf{B} \cdot \mathbf{dS} + \oint_{\text{top}} \mathbf{B} \cdot \mathbf{dS}$$

On the bottom,

$$\mathbf{dS} = -\mathbf{1}_x \, dS$$

and on the top,

$$\mathbf{dS} = +\mathbf{1}_x \, dS$$

Thus

$$\oint \mathbf{B} \cdot \mathbf{dS} = -\int_{\text{bottom}} B_x \, dS + \int_{\text{top}} B_x \, dS$$

so

$$\operatorname{div} \mathbf{B} = \lim_{\Delta x \to 0} \frac{-B_x(x)\,\Delta y\,\Delta z + B_x(x + \Delta x)\Delta y\,\Delta z}{\Delta x\,\Delta y\,\Delta z}$$
$$= \lim_{\Delta x \to 0} \frac{B_x(x + \Delta x) - B_x(x)}{\Delta x}$$
$$= \frac{\partial B_x}{\partial x}$$

Thus, after cancellation, the form of the divergence in this case is exactly the same as that for an ordinary derivative.

When all components are present for **B**,

(A-42) $$\operatorname{div} \mathbf{B} = \frac{\partial B_x}{\partial x} + \frac{\partial B_y}{\partial y} + \frac{\partial B_z}{\partial z} = \nabla \cdot \mathbf{B}$$

In spherical coordinates,

(A-43) $$\operatorname{div} \mathbf{B} = \frac{1}{r^2}\frac{\partial(r^2 B_r)}{\partial r} + \frac{1}{r \sin\theta}\frac{\partial(\sin\theta B_\theta)}{\partial\theta} + \frac{1}{r\sin\theta}\frac{\partial B_\varphi}{\partial\varphi}$$

Transposing the Δv in Eq. (A-41),

Scalar comparison

$$(\operatorname{div} \mathbf{B})\Delta v = \oint_{\text{surf bounding } \Delta v} \mathbf{B} \cdot d\mathbf{S} \qquad \frac{dB}{dx}\Delta x = \Delta B$$

It can be shown that summing this equation over various Δv's results in the *divergence theorem:*

Scalar comparison

(A-44) $$\int_V (\nabla \cdot \mathbf{B})dv = \oint_{\text{surf bounding } V} \mathbf{B} \cdot d\mathbf{S} \qquad \int \frac{dB}{dx}\,dx = \int dB$$

The last special derivative of vector analysis is the *curl.* This is a derivative of a vector which is itself a vector. It is necessary to define the *circulation* of a vector before defining curl. The circulation of the vector **A** is simply its line integral around a closed path. Thus

(A-45) $$\text{circulation of } \mathbf{A} = \oint \mathbf{A} \cdot d\mathbf{L}$$

The curl of the vector **A** is defined as

(A-46) curl **A** = circulation of **A** per unit area

Analytically, this may be stated as

(A-47) $$\operatorname{curl} \mathbf{A} = \lim_{\substack{\Delta S \to 0 \\ \text{at a} \\ \text{point}}} \frac{\oint_{\text{path bounding } \Delta S} \mathbf{A} \cdot d\mathbf{L}}{\Delta S} \mathbf{1}_{\Delta S}$$

The analogy of the curl to an ordinary derivative, like that for the divergence, is more apparent when the special case of a vector in the

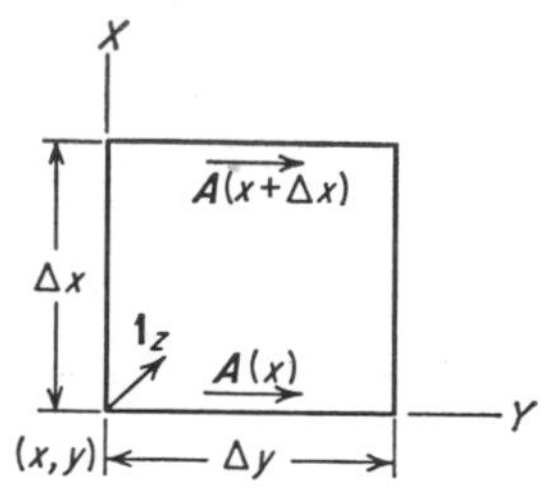

FIG. A-13. Curl derivation.

direction of one of the coordinate axes is considered. Thus, if

$$\mathbf{A} = \mathbf{1}_y A_y$$

and we consider for the path of integration an incremental rectangle in the XY plane with sides parallel to the axes, as shown in Fig. A-13,

$$\mathbf{\Delta S} = \mathbf{1}_z \, \Delta x \, \Delta y$$

and

$$\oint \mathbf{A} \cdot \mathbf{dL} = \int_{\text{top}} \mathbf{A} \cdot \mathbf{dL} + \int_{\text{bottom}} \mathbf{A} \cdot \mathbf{dL}$$

Along the top,

$$\mathbf{dL} = \mathbf{1}_y \, dy$$

and along the bottom

$$\mathbf{dL} = -\mathbf{1}_y \, dy$$

so that

$$\oint \mathbf{A} \cdot \mathbf{dL} = A_y(x + \Delta x) \, \Delta y - A_y(x) \, \Delta y$$

Consequently,

$$\text{curl } \mathbf{A} = \lim_{\substack{\Delta y \to 0 \\ \Delta x \to 0}} \frac{[A_y(x+\Delta x) - A_y(x)]\Delta y}{\Delta y \, \Delta x} \mathbf{1}_z = \frac{\partial A_y}{\partial x} \mathbf{1}_z$$

Scalar comparison

$$\frac{dA}{dx} = \lim_{\Delta x \to 0} \frac{A(x+\Delta x) - A(x)}{\Delta x}$$

When **A** has all components, it can be shown that the curl is given in rectangular coordinates by

$$\text{(A-48)} \qquad \text{curl } \mathbf{A} = \begin{vmatrix} \mathbf{1}_x & \mathbf{1}_y & \mathbf{1}_z \\ \dfrac{\partial}{\partial x} & \dfrac{\partial}{\partial y} & \dfrac{\partial}{\partial z} \\ A_x & A_y & A_z \end{vmatrix} = \nabla \times \mathbf{A}$$

In spherical coordinates, it is

$$\text{(A-49)} \qquad \text{curl } \mathbf{A} = \frac{1}{r^2 \sin \theta} \begin{vmatrix} \mathbf{1}_r & r\mathbf{1}_\theta & r \sin \theta \mathbf{1}_\varphi \\ \dfrac{\partial}{\partial r} & \dfrac{\partial}{\partial \theta} & \dfrac{\partial}{\partial \varphi} \\ A_r & rA_\theta & r \sin \theta A_\varphi \end{vmatrix}$$

When both sides of (A-47) are scalarly multiplied by $\mathbf{\Delta S}$,

Scalar comparison

$$(\nabla \times \mathbf{A}) \cdot \mathbf{\Delta S} = \oint_{\text{path bounding } \Delta S} \mathbf{A} \cdot d\mathbf{L} \qquad \frac{dA}{dx} \Delta x = \Delta A$$

This can be shown, upon summing over the ΔS's involved in surface S, to result in

(A-50)
$$\int_S (\nabla \times \mathbf{A}) \cdot d\mathbf{S} = \oint_{\text{path bounding } S} \mathbf{A} \cdot d\mathbf{L}$$

This relation is known as *Stokes' theorem.*

One important special second derivative occurs in vector analysis. This is the *Laplacian,* which is defined as

(A-51) Laplacian G = divergence of gradient of G

Thus

(A-52)
$$\text{Laplacian } G = \nabla^2 G = \nabla \cdot \nabla G = \frac{\partial^2 G}{\partial x^2} + \frac{\partial^2 G}{\partial y^2} + \frac{\partial^2 G}{\partial z^2}$$

In spherical coordinates, this is

(A-53)
$$\nabla^2 G = \frac{1}{r^2} \frac{\partial}{\partial r}\left(r^2 \frac{\partial G}{\partial r}\right) + \frac{1}{r^2 \sin\theta} \frac{\partial}{\partial \theta}\left(\sin\theta \frac{\partial G}{\partial \theta}\right) + \frac{1}{r^2 \sin^2\theta} \frac{\partial^2 G}{\partial \varphi^2}$$

The Laplacian of a vector is defined as the sum of the Laplacians of its *rectangular* components, with appropriate unit vectors:

(A-54)
$$\nabla^2 \mathbf{A} = \mathbf{1}_x \nabla^2 A_x + \mathbf{1}_y \nabla^2 A_y + \mathbf{1}_z \nabla^2 A_z$$

Note that $\nabla^2 \mathbf{A}$ is *not* equal to a comparable sum in any nonrectangular system.

Vector identities:

(A-55) $\nabla \times (\nabla \times \mathbf{A}) = \nabla(\nabla \cdot \mathbf{A}) - \nabla^2 \mathbf{A}$

Scalar comparison

(A-56) $\nabla \cdot (G\mathbf{A}) = \mathbf{A} \cdot \nabla G + G\nabla \cdot \mathbf{A}$

(A-57) $\nabla \times (G\mathbf{A}) = \nabla G \times \mathbf{A} + G\nabla \times \mathbf{A}$

(A-58) $\nabla \cdot (\mathbf{A} \times \mathbf{B}) = \mathbf{B} \cdot (\nabla \times \mathbf{A}) - \mathbf{A} \cdot (\nabla \times \mathbf{B})$

$$\frac{d(yz)}{dx} = y\frac{dz}{dx} + z\frac{dy}{dx}$$

(A-59) $\nabla \times (\nabla G) \equiv 0$

(A-60) $\nabla \cdot (\nabla \times \mathbf{A}) \equiv 0$

B. Units Conversion Table

Electricity and magnetism

Quantity	Symbol	Dimensions	To obtain the value in rationalized mks units:	Multiply the value in cgs electrostatic units:	By:	Multiply the value in cgs electromagnetic units:	By:
Charge	Q	Q	coulombs	statcoulombs	$10^{-9}/3$	abcoulombs	10
Electric field	**E**	$MLT^{-2}Q^{-1}$	volts/m or newtons/coulomb	statvolts/cm	3×10^{4}	abvolts/cm	10^{-6}
Electric displacement	**D**	QL^{-2}	coulombs/m²		$10^{-5}/12\pi$		$10^{5}/4\pi$
Electric potential	V, v	$ML^{2}T^{-2}Q^{-1}$	volts	statvolts	300	abvolts	10^{-8}
Current	I, i	QT^{-1}	amp	statamp	$10^{-9}/3$	abamp	10
Current density	**J**, **j**	$QL^{-2}T^{-1}$	amp/m²	statamp/cm²	$10^{-5}/3$	abamp/cm²	10^{5}
Conductivity	σ	$M^{-1}L^{-3}TQ^{2}$	mhos/m	statmhos/cm	$10^{-9}/9$	abmhos/cm	10^{11}
Conductance	G	$M^{-1}L^{-2}TQ^{2}$	mhos	statmhos	$10^{-11}/9$	abmhos	10^{9}
Resistance	R	$ML^{2}T^{-1}Q^{-2}$	ohms	statohms	9×10^{11}	abohms	10^{-9}
Permittivity	ϵ	$M^{-1}L^{-3}T^{2}Q^{2}$	farads/m		$10^{-9}/36\pi$		$10^{11}/4\pi$
Capacitance	C	$M^{-1}L^{-2}T^{2}Q^{2}$	farads	statfarads	$10^{-11}/9$	abfarads	10^{9}
Magnetic field	**H**	$L^{-1}T^{-1}Q$	amp/m		$10^{-7}/12\pi$	oersteds	$10^{3}/4\pi$
Magnetic flux density	**B**	$MT^{-1}Q^{-1}$	webers/m²		3×10^{6}	gausses	10^{-4}
Permeability	μ	MLQ^{-2}	henrys/m		$36\pi \times 10^{13}$		$4\pi \times 10^{-7}$
Inductance	L	$ML^{2}Q^{-2}$	henrys	stathenrys	9×10^{11}	abhenrys	10^{-9}

Diffusion

Quantity	Symbol	Dimensions	To obtain the value in mks units:	Multiply the value in metric units:	By:	Multiply the value in English units:	By:
Mass concentration	c, C	ML^{-3}	kg/m³	g/cm³	10^{3}	lb/ft³	17.00
Mass flux	$\mathbf{m}_v$	$ML^{-2}T^{-1}$	kg/m²-sec	g/cm²-sec	10	lb/ft²-sec	4.880
Diffusivity	D	$L^{2}T^{-1}$	m²/sec	cm²/sec	10^{-4}	ft²/sec	0.0929
				cm²/day	1.157×10^{-9}	ft²/day	1.075×10^{-6}
Mobility of charged particle	μ	$ML^{4}T^{-3}Q^{-1}$	m²/volt-sec	cm²/volt-sec	10^{-4}		

Mechanics

Quantity	Symbol	Dimensions	To obtain the value in mks units:	Multiply the value in cgs units:	By:	Multiply the value in English units:	By:
Mass	m	M	kg	g	10^{-3}	lb	0.4536
						slugs	14.60
Length	Various	L	m	cm	10^{-2}	in.	0.02540
						ft	0.3048
Time	t	T	sec	sec	1	sec	1
Velocity	$\mathbf{u}, \mathbf{U}$	LT^{-1}	m/sec	cm/sec	10^{-2}	ft/sec	0.3048
						miles/hr	0.6214
Force	$\mathbf{f}, \mathbf{F}$	MLT^{-2}	newtons	dynes	10^{-5}	lb	4.448
				g	9.80×10^{-3}	poundals	0.1383
Pressure	p, P	$ML^{-1}T^{-2}$	newtons/m²	dynes/cm²	10^{-1}	lb/in.²	0.6894×10^{4}
				g/cm²	98.0		
Density	ρ_v	ML^{-3}	kg/m³	g/cm³	10^{3}	lb/in.³	2.768×10^{4}
						lb/ft³	16.02
						slugs/ft³	515.4
Work or energy	Various	ML^2T^{-2}	joules	ergs	10^{-7}	ft-lb	1.356
				g-cm	9.80×10^{-5}	hp-hr	2.686×10^{6}
Power	Various	ML^2T^{-3}	watts	ergs/sec	10^{-7}	ft-lb/sec	1.356
				g-cm/sec	9.80×10^{-5}	hp	745.7
Power flux	Various	MT^{-3}	watts/m²	ergs/sec-cm²	10^{-3}	ft-lb/sec-ft²	1.954
Compressibility	K	$M^{-1}LT^2$	m-sec²/kg	cm-sec²/kg	10	in.-sec²/slug	1.450×10^{-4}
						unit vol/atm	9.869×10^{-6}
Modulus of elasticity	Y_0, Y_B, μ	$ML^{-1}T^{-2}$	newtons/m²	dynes/cm²	10^{-1}	lb/in.²	6.894×10^{3}

Heat

Quantity	Symbol	Dimensions	To obtain the value in mks mechanical units:	Multiply the value in metric heat units:	By:	Multiply the value in English heat units:	By:
Temperature	τ, T	τ		°C		°F − 32	$\frac{5}{9}$
Quantity of heat	q, Q	ML^2T^{-2}	joules	kg-cal	4,185	Btu	1,055
				g-cal	4.185		
Heat flux	$\mathbf{q}, \mathbf{Q}$	MT^{-3}	watts/m²	$\frac{\text{g-cal}}{\text{sec-cm}^2}$	4.185×10^{4}	Btu/hr-ft²	3.130
Thermal conductivity	k	$MLT^{-3}\tau^{-1}$	watts/m-°C	$\frac{\text{g-cal}}{\text{cm-°C-sec}}$	418.5	$\frac{\text{Btu}}{\text{ft}^2\text{-sec-°F/in}}$	519.0
						$\frac{\text{Btu}}{\text{ft}^2\text{-hr-°F/ft}}$	1.730
Specific heat per unit mass	S	$L^2T^{-2}\tau^{-1}$	joules/kg-°C	$\frac{\text{g-cal}}{\text{g-°C}}$	4,185	Btu/lb-°F	4,185
Thermal diffusivity	D_t	L^2T^{-1}	m²/sec	cm²/sec	10^{-4}	ft²/hr	2.581×10^{-5}
Surface emissivity	h	$MT^{-3}\tau^{-1}$	watts/m²-°C	$\frac{\text{g-cal}}{\text{cm}^2\text{-sec-°C}}$	4.185×10^{4}	$\frac{\text{Btu}}{\text{hr-ft}^2\text{-°F}}$	3.130

Values of some fundamental constants

Velocity of light in vacuum	$c = 2.998 \times 10^{8} \approx 3 \times 10^{8}$ m/sec
Permittivity of space	$\epsilon_0 = 8.854 \times 10^{-12} \approx \frac{1}{36\pi} \times 10^{-9}$ farad/m
Permeability of space	$\mu_0 = 4\pi \times 10^{-7}$ henry/m
Intrinsic impedance of space	$\eta_0 = 376.7 \approx 120\pi$ ohms
Charge of an electron	$e = 1.602 \times 10^{-19}$ coulomb
Rest mass of an electron	$m = 9.108 \times 10^{-31}$ kg

C. Typical Acoustic Properties of Fluids

Fluid	Temperature, °C	Density, kg/m^3	Velocity of sound, m/sec
Air (0.76 m Hg)†	0	1.29	331
Neon†	0	0.88	435
Hydrogen†	0	0.089	1270
Steam (0.76 m Hg)†	100	0.60	405
Water†	15	1000	1450
Alcohol†		801	1234
Pentane (C_5H_{12})‡	18	632	1052
Glycerin‡	22	1260	1986

† C. D. Hodgman, "Handbook of Chemistry and Physics," pp. 1140, 1148, 1317, 1359, Chemical Rubber Publishing Company, Cleveland, 1935.

‡ D. E. Gray (ed.), "American Institute of Physics Handbook," pp. 3-71—3-72, McGraw-Hill Book Company, Inc., New York, 1957.

Index

Units Conversion Table

Electricity and magnetism

Quantity	Symbol	Dimensions	To obtain the value in rationalized mks units:	Multiply the value in cgs electrostatic units:	By:	Multiply the value in cgs electromagnetic units:	By:
Charge	Q	Q	coulombs	statcoulombs	$10^{-9}/3$	abcoulombs	10
Electric field	$\mathbf{E}$	$MLT^{-2}Q^{-1}$	volts/m or newtons/coulomb	statvolts/cm	3×10^4	abvolts/cm	10^{-6}
Electric displacement	$\mathbf{D}$	QL^{-2}	coulombs/m²		$10^{-5}/12\pi$		$10^5/4\pi$
Electric potential	V, v	$ML^2T^{-2}Q^{-1}$	volts	statvolts	300	abvolts	10^{-8}
Current	I, i	QT^{-1}	amp	statamp	$10^{-9}/3$	abamp	10
Current density	$\mathbf{J}, \mathbf{j}$	$QL^{-2}T^{-1}$	amp/m²	statamp/cm²	$10^{-5}/3$	abamp/cm²	10^5
Conductivity	σ	$M^{-1}L^{-3}TQ^2$	mhos/m	statmhos/cm	$10^{-9}/9$	abmhos/cm	10^{11}
Conductance	G	$M^{-1}L^{-2}TQ^2$	mhos	statmhos	$10^{-11}/9$	abmhos	10^9
Resistance	R	$ML^2T^{-1}Q^{-2}$	ohms	statohms	9×10^{11}	abohms	10^{-9}
Permittivity	ϵ	$M^{-1}L^{-3}T^2Q^2$	farads/m		$10^{-9}/36\pi$		$10^{11}/4\pi$
Capacitance	C	$M^{-1}L^{-2}T^2Q^2$	farads	statfarads	$10^{-11}/9$	abfarads	10^9
Magnetic field	$\mathbf{H}$	$L^{-1}T^{-1}Q$	amp/m		$10^{-7}/12\pi$	oersteds	$10^3/4\pi$
Magnetic flux density	$\mathbf{B}$	$MT^{-1}Q^{-1}$	webers/m²		3×10^6	gausses	10^{-4}
Permeability	μ	MLQ^{-2}	henrys/m		$36\pi \times 10^{13}$		$4\pi \times 10^{-7}$
Inductance	L	ML^2Q^{-2}	henrys	stathenrys	9×10^{11}	abhenrys	10^{-9}

Diffusion

Quantity	Symbol	Dimensions	To obtain the value in mks units:	Multiply the value in metric units:	By:	Multiply the value in English units:	By:
Mass concentration	c, C	ML^{-3}	kg/m³	g/cm³	10^3	lb/ft³	17.00
Mass flux	$\mathbf{m}_v$	$ML^{-2}T^{-1}$	kg/m²-sec	g/cm²-sec	10	lb/ft²-sec	4.880
Diffusivity	D	L^2T^{-1}	m²/sec	cm²/sec	10^{-4}	ft²/sec	0.0929
				cm²/day	1.157×10^{-9}	ft²/day	1.075×10^{-6}
Mobility of charged particle	μ	$ML^4T^{-3}Q^{-1}$	m²/volt-sec	cm²/volt-sec	10^{-4}		

Values of some fundamental constants

Velocity of light in vacuum	$c = 2.998 \times 10^8 \approx 3 \times 10^8$ m/sec
Permittivity of space	$\epsilon_0 = 8.854 \times 10^{-12} \approx \frac{1}{36\pi} \times 10^{-9}$ farad/m
Permeability of space	$\mu_0 = 4\pi \times 10^{-7}$ henry/m
Intrinsic impedance of space	$\eta_0 = 376.7 \approx 120\pi$ ohms
Charge of an electron	$e = 1.602 \times 10^{-19}$ coulomb
Rest mass of an electron	$m = 9.108 \times 10^{-31}$ kg